Human Genetics

The Benjamin/Cummings Series in the Life Sciences

F. J. Ayala: *Population and Evolutionary Genetics: A Primer* (1982)

F. J. Ayala and J. A. Kiger, Jr.: *Modern Genetics* (1980)

F. J. Ayala and J. W. Valentine: *Evolving: The Theory and Processes of Organic Evolution* (1979)

M. G. Barbour, J. H. Burk, and W. D. Pitts: *Terrestrial Plant Ecology* (1980)

L. L. Cavalli-Sforza: *Elements of Human Genetics,* Second Edition (1977)

R. E. Dickerson and I. Geis: *Hemoglobin* (1982)

L. E. Hood, I. L. Weissman, and W. B. Wood: *Immunology* (1978)

L. E. Hood, J. H. Wilson, and W. B. Wood: *Molecular Biology of Eucaryotic Cells* (1975)

J. B. Jenkins: *Human Genetics* (1983)

A. L. Lehninger: *Bioenergetics: The Molecular Basis of Biological Energy Transformations,* Second Edition (1971)

S. E. Luria, S. J. Gould, and S. Singer: *A View of Life* (1981)

E. N. Marieb: *Human Anatomy and Physiology Laboratory Manual: Cat and Pig Fetal editions* (1981), *Human Anatomy and Physiology Lab Manual: Brief Edition* (1983)

E. B. Mason: *Human Physiology* (1983)

A. P. Spence: *Basic Human Anatomy* (1982)

A. P. Spence and E. B. Mason: *Human Anatomy and Physiology,* Second Edition (1983)

G. J. Tortora, B. R. Funke, and C. L. Case: *Microbiology: An Introduction* (1982)

J. D. Watson: *Molecular Biology of the Gene,* Third Edition (1976)

I. L. Weissman, L. E. Hood, and W. B. Wood: *Essential Concepts in Immunology* (1978)

N. K. Wessells: *Tissue Interactions and Development* (1977)

W. B. Wood, J. H. Wilson, R. M. Benbow, and L. E. Hood: *Biochemistry: A Problems Approach,* Second Edition (1981)

Human Genetics

John B. Jenkins

Swarthmore College

The Benjamin/Cummings Publishing Company, Inc.
Menlo Park, California · Reading, Massachusetts
London · Amsterdam · Don Mills, Ontario · Sydney

Sponsoring Editors: James W. Behnke, Jane R. Gillen
Production Editors: Susan Harrington, Patricia Burner
Developmental Editors: Robin Fox, Amy Satran
Copy Editor: Liese Hofmann
Book and Cover Designer: Marjorie Spiegelman

Library of Congress Cataloging in Publication Data

Jenkins, John B.
   Human genetics.

   Bibliography: p.
   Includes index.
   1. Human genetics. I. Title.
QH431.J34   1983   573.2'1   82-24338
ISBN 0-8053-5010-1

45,597

The Benjamin/Cummings Publishing Company, Inc.
2727 Sand Hill Road
Menlo Park, California 94025

ISBN 0-8053-5010-1
BCDEFGHIJK HA 8987654

*To John Bruner Jenkins, III
and Soraya Parvin Jenkins*

Brief Contents

Detailed Contents

Preface

Few sciences hold our interest as intensely as human genetics. All of us have questions about our biological heritage and how that heritage affects our lives today. This book will answer many questions, but it will also raise questions for which we have no answers. It is an excursion into the genetics of the human species, an area of study that is growing more rapidly with each passing day. As a matter of fact, few areas of science are expanding more rapidly.

For decades human genetics languished in relative obscurity. It was overshadowed by the spectacular advances being made in the molecular genetics of bacteria and viruses and in the more easily manipulated genetic system of the fruitfly, *Drosophila*. But human genetics also had the burden of World War II to bear. In Nazi Germany, genetic principles were perverted in their application to human beings, prompting many to turn away from the field. Within the last ten years, though, the situation has changed rather dramatically. New techniques for studying chromosomes, for growing human cells in culture, and for studying the function of human DNA have resulted in some spectacular progress. We will examine some of the progress in this book, but bear in mind that even as you read this book, new and exciting advances will probably have added to our knowledge. These ongoing discoveries help make human genetics so much fun and so exciting.

This book is intended for a one-term introductory course. I do not assume that the readers of this book have any background in college-level biology or chemistry. Nor do I assume that this book will be followed by more advanced studies in biology, though it may.

However, some students may indeed have some prior biology and chemistry courses and may actually be planning to be biology majors. Those in this latter category may choose to skip sections or skim them for quick review.

Organization

The organization of *Human Genetics* reflects the approach I use in classrooms. It is logical and works well, but this is not to say that it is the only organization possible. The first six chapters deal with various aspects of what we might call "classical genetics." That is, they discuss the basic principles of inheritance as formulated by Mendel and applied to humans, the chromosome theory of inheritance, and some of the extensions of those basic principles. The next five chapters focus on the nature and function of the genetic material. Chapters 12 and 13 explore some of the more complicated patterns of inheritance, including human behavior, and the techniques we employ to understand them. The final chapter discusses the genetics of human populations and the biological history of the human species.

Special Features

A variety of features contribute to the usefulness of this book.

- *Coverage of recent advances in genetics* is thorough, and includes accessible treatment of gene structure and function on the molecular level, somatic cell genetics, the relationships of viruses and chromosome abnormalities to cancer, and modern genetic techniques and procedures. I devote an entire chapter to metabolic disorders and hemoglobin variation, and another to the genetics of the immune system.

- *Special topic boxes* are found in almost every chapter. These highlight a variety of subjects, ranging from the technical to the controversial, and provide an element of choice in selecting material for study.

- Many *illustrations* appear throughout. They include over 300 drawings and 100 photographs, many published for the first time.

- *Learning aids in each chapter* include *key terms* in boldface within the text, *chapter summaries* in outline form, *lists of key terms and concepts* at the ends of chapters, *review questions and problems* (with answers provided), and *references* to further reading. At the back of the book is a comprehensive glossary.

Acknowledgments

This book is a very special effort involving many very special people. The outstanding people at Benjamin/Cummings helped to make the writing of this book a unique pleasure. I especially want to thank Jim Behnke, Jane Gillen, Margaret Moore, Patricia Burner, Sue Harrington, Jo Andrews, and Amy Satran for their monumental efforts on behalf of this project. A very special debt of gratitude is owed to Robin Fox for developmental editing of the manuscript, and to Duane E. Jeffery and Joyce Maxwell, who read the manuscript from cover to cover and offered numerous valuable suggestions for improvement. The quality and accuracy of my writing was greatly improved by their thoughtful comments. Duane Jeffery wrote a number of the boxes. In addition, C. K. James Shen, of the University of California, Davis, carefully proofread the entire page proof for the book, on a tight schedule. Dorothy Sivitz, one of my students at Swarthmore, was an invaluable aid in the final phases of the writing. And my colleagues at the Children's Hospital of Philadelphia, especially Beverly Emanuel, were helpful, stimulating, and encouraging throughout this project. To all of these people and numerous others, I want to say thank you.

John B. Jenkins
Swarthmore, Pennsylvania

Reviewers

Dee Baer, University of California, Berkeley

Isiah Amiel Benathen, Kingsborough Community College, City University of New York

Carl Huether, University of Cincinnati

Duane Jeffery, Brigham Young University

Annette Lipshitz, Nassau Community College

Joyce Maxwell, Cal State University, Northridge

Jeffrey Morse, University of Puget Sound

John Mullins, Michigan State University

Howard Rosen, Cal State University, Los Angeles

C. K. James Shen, University of California, Davis

John Stubbs, San Francisco State University

Hayden Williams, Golden West College

Human Genetics

1 Introduction

A Boy Named Michael

Michael entered this world a healthy, cheerful baby born of young parents who could not have been happier or more optimistic about the future. But Michael's future was to be bleak, for he was born with cystic fibrosis (CF). The symptoms were not immediately apparent, and when they did begin to appear, they were so general that no one suspected their cause. At six months of age, Michael was operated on for an intestinal obstruction, but he continued to be malnourished despite a healthy appetite. The obstruction and the persistent malnutrition led physicians to suspect CF, and further testing confirmed their fears.

As Michael grew, he was by all standards a terrific kid. Intelligent, sensitive, and humorous, he was every parent's dream. But his medical problems multiplied. He was still malnourished despite a voracious appetite. He suffered from deficiencies of vitamins A, D, K, and E. His bowel movements were exceptionally large and smelly. At age three, Michael began to experience recurring symptoms of bronchitis. Thick mucous secretions collected in his lungs, creating a painful emphysemalike condition. Over the next few years, his lung problems became more severe and placed a tremendous strain on his heart.

At age seven, Michael died of congestive heart failure. Death did not come quickly or easily to this little boy, who with his parents fought a brave battle against this unrelenting, genetically rooted disease. But come it did, and in a way that is typical for almost all who suffer from CF.

CF: A Case of Simple Inheritance

What caused Michael's tragedy? We are just now beginning to understand the underlying biochemical defect that leads to cystic fibrosis. Almost all the CF symptoms can be traced to malfunctions in the body's mucus-secreting glands. Pancreatic ducts, which carry enzymes to the intestine, become clogged with mucus, restricting the flow of digestive substances. Thus, food is not properly broken down and utilized, and malnutrition occurs. Excessive mucus in the lungs plugs up the smaller air passages, impairing the exchange of oxygen and carbon dioxide and creating severe breathing difficulties. Most people with CF die before their nineteenth birthday either of lung disease or of heart failure caused by the enormous strain of pumping blood to improperly functioning lungs.

The increased production of mucus that causes the CF symptoms may be associated with an abnormal form of an enzyme called NADH dehydrogenase, though this is not yet proven. **Enzymes** are proteins that catalyze, or assist, biochemical reactions; NADH dehydrogenase is but one of the tens of thousands of enzymes in the human body. No organism can live without enzymes, for without their assistance, the chemical processes on which life is based would occur exceedingly slowly or not at all. As Michael's case demonstrates, the lack of a single enzyme or its occurrence in an abnormal form can severely disrupt the body's normal functioning. For reasons we do not fully understand, the abnormal form of NADH dehydrogenase may cause a calcium buildup in some of the gland cells, and this in turn may lead to abnormal mucus production.

Abnormal NADH dehydrogenase is the product of a specific abnormal gene. **Genes** are the units of hereditary material that carry the encoded instructions for an organism's development and biologic functioning. Each of our genes is a copy of a gene in one of our parents. Why, then, did neither of Michael's parents display any symptoms of CF? Like nearly all gene products, NADH dehydrogenase is coded for by *two* alternative forms of the gene, called **alleles**, in every individual, one derived

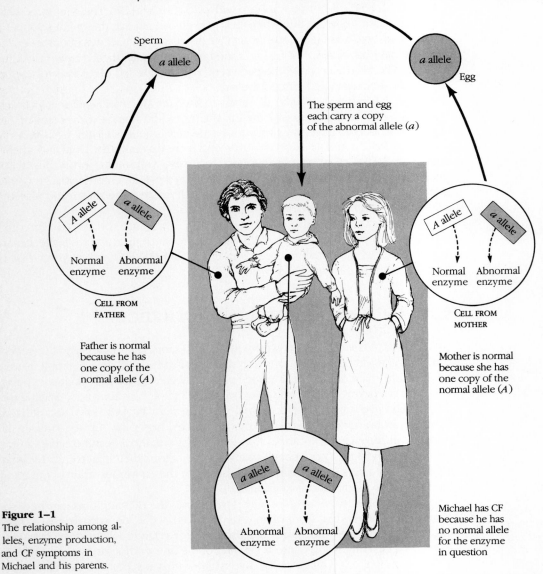

Sperm

a allele

a allele

Egg

The sperm and egg each carry a copy of the abnormal allele (*a*)

A allele   *a* allele

Normal   Abnormal
enzyme   enzyme

CELL FROM
FATHER

A allele   *a* allele

Normal   Abnormal
enzyme   enzyme

CELL FROM
MOTHER

Father is normal
because he has
one copy of the
normal allele (*A*)

Mother is normal
because she has
one copy of the
normal allele (*A*)

a allele   *a* allele

Abnormal   Abnormal
enzyme   enzyme

Michael has CF
because he has
no normal allele
for the enzyme
in question

Figure 1–1
The relationship among alleles, enzyme production,
and CF symptoms in
Michael and his parents.

from the mother and one derived from the father. A person who has even one normal allele will produce enough of the normal enzyme to ensure proper functioning of the mucous glands and will not experience any CF symptoms. In other words, the effects of the abnormal allele will be overcome, or masked, by the effects of the normal one. When two alleles code for different forms of the same trait, such as normal and abnormal enzyme production, and the effects of one allele are masked by the effects of the other, we call the masking allele **dominant** and the masked allele **recessive**. (A dominant allele is usually indicated by an uppercase italic letter and the recessive allele by a lowercase italic letter.) Each of Michael's parents carried one dominant (*A*) and one recessive allele (*a*) for the enzyme, but they were quite unaware of this until each of them passed a copy of the recessive allele on to their child. Without a copy of the dominant normal allele, Michael's body could not produce the normal form of the enzyme—and the symptoms of CF appeared. Figure 1–1 shows the relationship among alleles, enzyme production, and CF symptoms in Michael and in his parents.

The inheritance of CF follows a simple pattern called **Mendelian**, after Gregor Mendel, who first described dominant and recessive traits in peas. Such patterns were well understood long before geneticists had any idea why a trait might be dominant or recessive, or indeed, what a gene might be. Thanks to these patterns, we can make certain statistical predictions. A genetic counselor, knowing that Michael's parents must each carry one recessive allele for CF, can tell them that any future children of theirs will have one chance in four of inheriting two such alleles and thus of having the disease. Each child will have two chances in four of inheriting one dominant and one recessive allele and one chance in four of inheriting two dominant alleles (Figure 1–2). Put the other way around, every child will have three

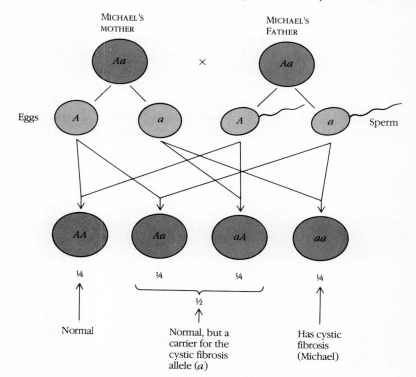

Figure 1–2

Michael's mother and father are carriers of an abnormal allele, *a*. We can make statistical predictions about the genetic makeup of their offspring.

chances in four of being free of CF symptoms, although some of these normal children will still be able to pass the defective allele on to their own offspring. Such counseling can provide Michael's parents with valuable understanding, but in the end only they can make the painful decision of whether or not to risk having another doomed child.

Genetic Aberrations and Our Understanding of Genes

Michael's case illustrates two aspects of human genetics: (1) the hereditary patterns that have been known for many decades, and (2) the biochemical expression of genes that is the subject of current research. As we examine both aspects in greater detail, we will often use enzymes and other proteins as examples, and quite a few of the traits we study will be genetic diseases. You may wonder, perhaps, why we focus so much attention on these relatively rare phenomena. The answer is partly that unusual conditions are often of medical importance, and their investigation holds the promise of practical benefit to many people. But equally important to students of genetics, though perhaps not to the Michaels of the world and their parents, is the fact that genetic abnormalities teach us a great deal about normal inheritance.

Genes, of course, determine normal as well as abnormal traits, similarities as well as differences. The genes we receive from our parents ensure, in fact, that we are fundamentally like them—that we are human beings rather than elephants or pine trees. Nevertheless, it is our differences that we are most interested in when we try to discern the patterns and mechanisms of inheritance. It is useless to try to observe a pattern of inheritance for hair color if everyone in the observed population has black hair. But if albinism (a total or partial absence of normal pigmentation) occurs in the same population, we can note the occurrence of albinos in successive generations of a given family. The family tree, or **pedigree**, will tell us something about the mechanism of inheritance in general. Obviously, though, not all variations are abnormalities. We are all aware of common, normal differences among people—differences in height, skin color, hair texture, blood type, and so on. Why, then, are we so interested in obscure traits and strangely named enzymes and the genetic diseases that cause so much misery? To understand the usefulness of such examples, you must first have some idea of what genes do.

Most genes code for **polypeptides**, the molecular chains that compose proteins. Each polypeptide chain is made up of smaller molecular units called **amino acids**, of which about 20 kinds are critical to living organisms. The structure of the gene ensures that a particular polypeptide chain has the right number and sequence of amino acids; we will describe how it does this in Chapter 7. A single polypeptide chain may function as a protein by itself, or it may combine with other polypeptides to form more complex proteins. Proteins carry out a variety of functions in the body, but most of them are either **structural proteins**, which make up the various parts of our cells, or enzymes, which catalyze our biochemical activities. Other types of proteins include the globin part of hemoglobin, antibodies (both of which we will discuss later), and some hormones. Our development from a single cell (the fertilized egg) to a complex organism with diverse tissues and individual characteristics

Figure 1–3
Abraham Lincoln's characteristic long limbs and fingers may have been caused by a genetic disorder known as the Marfan syndrome. (The Bettmann Archives, Inc.)

is largely the result of the physical and chemical interaction of our proteins with one another and with other substances in their environment.

To study simple inheritance and its molecular mechanisms, we must focus our attention on individual proteins and the individual genes that give rise to them. Most of our visible traits—for example, hair color, head shape, and body size—are not determined by a single gene or pair of genes, but are the end result of the interactions of several or many gene products. We are often unable to find a simple pattern of inheritance for such traits, because it turns out that we are really trying to follow several genes at once. On the other hand, widespread effects can sometimes be traced to a single gene, as in the case of cystic fibrosis. Metabolic problems caused by a single variant gene are of great interest to geneticists, not only because they follow an easily traceable hereditary pattern, but also because they allow us to investigate how genes actually work.

The effects of a variant gene may be striking or unnoticeable, harmful or harmless, local or widespread. But an organism is a delicately balanced piece of machinery, and when anything in its functioning is altered, chances are that the consequences will be serious. In humans, variant genes cause abnormalities of the eye, ear, and nose; blood and skin diseases; abnormal bone structure; muscle and nerve cell degeneration; and a host of other problems. Some genetically caused conditions are interesting for historical as well as biological reasons. The erratic behavior of George III, king of England, during the American Revolution was perhaps due to porphyria, a disease caused by an aberrant gene that prevents the normal breakdown of hemoglobin. The symptoms of porphyria are persistent nausea and spells of violent, destructive behavior. Abraham Lincoln may have had a genetic disorder called the Marfan syndrome, characterized by unusually long limbs and fingers ("spider fingers"), heart defects, and eye abnormalities (Figure 1–3). The underlying cause of the Marfan syndrome is not yet known, but it appears to involve abnormal synthesis of collagen or some other connective-tissue protein. By studying the biochemical effects of the aberrant genes responsible for such conditions, we can deduce the function of their normal counterparts.

There is yet another reason why medical problems crop up so often in the study of human genetics. In many ways, humans are not at all suitable as objects of genetic analysis. Most of our knowledge of genetics is derived from the breeding of organisms that have a short generation time and produce many offspring, allowing us to collect a great deal of statistically significant information in a brief time. The fruit fly, whose genetic makeup is better known than that of nearly any other species, has a generation time of only 10 days and produces hundreds of offspring per pair. Human beings, of course, cannot be mated experimentally, and as they have long generations and few children, little information can be gathered in the lifetime of an observer. A primary tool for the study of human genetics is the family pedigree, which depends on recordkeeping over several generations or on people's memories of their ancestors' traits. Usually only medical disorders or very exceptional characteristics receive such attention. Fortunately, the basic mechanisms of gene activity are the same for peas, mice, fruit flies, and people, so that much of what we learn about other species applies to ourselves as well. But our knowledge of specifically human traits is based largely on medical histories.

Difficult though the gathering of data may be, we persist in trying to unravel the mysteries of human genes. One reason for doing this is that we are intensely curious about our own biology; another is that the implications of human genetics are so enormous. We have mentioned the medical implications already, and we will discuss some of the social ones later. Because so much of our future depends on self-knowledge, it is imperative that we understand the principles governing our inheritance.

Inheritance: Some Ancient Ideas

The systematic study of genetics began with Mendel in the middle of the nineteenth century, but people certainly speculated about inheritance long before then. After all, everyone knows that people usually resemble their relatives and that the closer the relationship is, the closer the resemblance tends to be. But it is equally apparent that offspring differ from parents in many ways, and sometimes to a rather extreme degree—a couple with brown eyes and brown hair, for example, can produce a child with blue eyes and red hair. The problem, clearly, is how to explain both the similarities and the differences.

Interesting theories have been offered over the centuries to explain how characteristics are inherited. Aristotle proposed that the environment controls genetic changes, and that physical, intellectual, and personality characteristics acquired by persons during their lifetime are passed on to their offspring. According to this belief, if a man devotes a great deal of time to composing music, this effort will somehow affect his genetic material, which will transmit a musical tendency to any children he might have. This theory of acquired characteristics influenced scientific thought for over 2000 years. It certainly explains how similarities persist in a family and, suitably qualified, can also account for differences between one child and another—perhaps the father was not devoting himself to his music shortly before the conception of his nonmusical son. Still, there are some obvious objections to the idea, such as the common observation that children of cripples usually have normal limbs. An age-old response to such objections is that an organism has an "inner need" to assume a certain form, and this need is also transmitted genetically. Thus, the "inner need" to have strong, well-formed limbs will usually overcome any acquired tendency to be crippled. Though the idea that acquired characteristics are inherited is wrong, it remains a handy explanation to persons uneducated in genetics.

People have always been interested in how sex is inherited, and they have come up with a variety of theories about it. Some used to believe that the sex of the child derived from the father, with sperm from the right testicle being male-determining and sperm from the left being female-determining. Others believed that it was all controlled by the mother, the eggs from the right and left ovaries producing male and female children, respectively. Still others held that the direction of the wind or the phase of the moon at the time of conception was responsible for the baby's sex. There has also been considerable disagreement about the relative genetic contributions of the mother and the father. Throughout much of Western history, scholars maintained that the mother made no contribution at all, but served only as an incubator for the embryo, which was formed in one of the father's testicles (testes) (Figure 1–4). On the other hand, some Melanesians believed that the father made no genetic contribution, his only function being to open a passageway, by means of sexual intercourse, for the spirits responsible for conception.

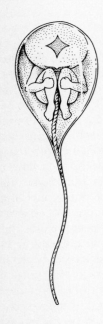

Figure 1–4
Throughout much of Western history it was believed that embryos are already formed in the father's sperm.

These ideas seem absurd to us, but they were proposed by people who had no means of observing the microscopic cells involved in conception, let alone the genetic makeup of those cells. Some ancient people actually made some quite accurate observations about inheritance, based solely on what we would call family pedigrees. In the Talmud (a compilation of Jewish religious law and folklore), we find a remarkable set of rules, dating from about A.D. 200, concerning the circumcision of male infants in families in which other infants had bled to death after the operation. The excessive bleeding, or hemorrhaging, was caused by the failure of blood to clot, a genetically determined trait called hemophilia. As we will see when we talk about sex-linked inheritance, a female can carry a gene for hemophilia and pass it on to her son, but her own blood will clot normally unless she carries two such genes. A male, on the other hand, is always a "bleeder" if he carries a hemophilia gene. The ancient Jews had no knowledge of genes, of course, but they obviously detected the sex-linked pattern of inheritance, with the disease passing through the mother to the son. The Talmud rules that if a woman has lost two sons to the "bleeder's disease," her later-born sons are exempt from the circumcision rite. Furthermore, the sons of any sisters of such a woman are also exempt, though the sons of any nonbleeder brothers are not. These rules show a clear recognition that nonbleeder females, unlike nonbleeder males, can be carriers of hemophilia. Seventeen centuries were to pass before scientists could explain this phenomenon.

Human Genetics and the Interpretation of Data

Modern genetic research began in earnest in 1900, when Mendel's laws were rediscovered after several decades of neglect. Biologists immediately began to look for human traits that followed a Mendelian pattern, and W. C. Farabee identified brachydactyly (abnormally short fingers) as such a trait in 1905. Archibald Garrod had suggested in 1902 that alkaptonuria (a metabolic defect that causes the urine and cartilage to blacken on exposure to oxygen) was inherited as a Mendelian trait, but this was not confirmed until 1908 after more families had been analyzed.

The Garrod study raises an important point: It is not always easy to collect enough information about human genetics to warrant a reliable conclusion. For example, it was once thought that the ability to roll up the sides of the tongue was a simple inherited trait. About 70% of people of European ancestry have this ability; the remaining 30% do not. But in 1952 it was shown that tongue-rolling can be learned or that it can at least be controlled by environmental factors. Only when sufficient data became available could we relegate tongue-rolling to its proper place; it is a curiosity, essentially useless as a trait for genetic analysis.

Over the past 80 years the number of human traits that we know to be genetically determined has grown enormously. As you may suspect from the earlier discussion, many of these traits involve individual proteins such as hemoglobin and various enzymes, and a number of them are relatively rare anomalies. Many other traits, while clearly inherited, are exceedingly difficult to analyze. Certain behavioral patterns, for example, are known to be inherited, at least in part, but we do not know how many genes are involved in determining such patterns, nor can we adequately assess the influence of the environment. If so simple a "genetic" trait as tongue-rolling turns out to be highly subject to environmental influence, we must surely exercise the greatest caution in attempting to determine the genetic components of such complex characteristics as personality traits and intelligence. Unfortunately, the history of human genetics has been marred by many investigations that purport to show that specific traits are genetically determined when actually they are not. These reports have often reflected the authors' own prejudices and helped create an unfavorable climate for further investigation.

Francis Galton, a cousin of Charles Darwin's, wrote several books during the latter 1800s that made a lasting contribution to the field of human genetics. Among his positive achievements were the invention of the correlation coefficient, a statistical tool used in genetics and other fields today, and the modern system of fingerprint classification. But Galton is best remembered as the founder of **eugenics**, the study of methods for improving the genetic makeup of the human species. Many of the premises of eugenics were attractive but wrong. For example, Galton studied exceptional achievements in certain families, such as musical composition in the Bach family, and he concluded that such achievements were the result of biologic inheritance. He largely ignored the influence of the environment in these cases, even though he himself coined the expression "nature versus nurture" to reflect the role the environment may play in the determination of some traits.

We can soften our criticism of Galton somewhat by pointing out that he wrote at a time when the basic principles of inheritance were known only to Mendel. But later, others made equally outrageous assertions in the face of a general understanding of fundamental genetic laws. During the first few decades of the twentieth century, the geneticist and eugenicist C. B. Davenport made what he thought was a strong case for the biologic basis of social achievement. He contrasted such prominent families as the Lees of Virginia and the Tuttle-Edwards of Connecticut, which had produced several presidents, senators, and generals, with the dismally antisocial persons who continued to appear in the Jukes and Kallikak families, subjects of early genetic studies. Davenport did not consider the family environment in his analysis; he attributed these differences solely to genes. In the early literature of eugenics we find arguments for the biologic determination of prostitution, criminality, drunkenness, and slothful living, and advocacy of legal restrictions on the reproduction of the genetically "unfit" persons exhibiting these traits. Many supporters of eugenics, including some geneticists, also viewed certain races or classes (their own, of course) as biologically superior to others and considered mating across racial or class boundaries as detrimental to the genetic makeup of the superior group.

These scientifically unfounded ideas have often provided the ideological framework for nationalism, racism, and class privilege. Although serious geneticists began to withdraw their support for eugenics by 1915, the eugenics movement had a powerful influence on legislation in the United States from the 1880s to the 1930s. Some states passed laws requiring sterilization of criminals and the insane, while others legislated against racial intermarriage, to protect the "purity" of the white race. Congress passed laws severely restricting immigration from everywhere but northern Europe; the last of these laws was not repealed until 1965. In Europe, racist pseudobiology climaxed in the horror of Nazism, which proclaimed the genetic superiority and right of political ascendency of the "Aryan" peoples (those who, according to the Nazis, were Caucasian Gentiles—preferably Nordic). Under the leadership of Adolf Hitler, the Nazis tried to exterminate the entire Jewish and Gypsy populations of Europe, and they also killed millions of others who were not to their genetic or political liking.

The question of nature versus nurture—of the relative contributions of genes and the environment—is far from resolved, and it is still being discussed in connection with alleged racial differences. In the United States today, the answers we give to questions about the relation of race to ability and intelligence may well affect social policy in such areas as education, housing, and hiring. Because the implications of this discussion are so enormous, it is vital that we recognize the limitations of our data and be aware of unjustified interpretations. This book will deal with some of the questions raised here, but you will see that many of them still lack answers. Whenever we study people, it becomes particularly difficult to separate fact from prejudice, science from pseudoscience. Human genetics, perhaps more than any other science, is subject to misinterpretation and misuse. When people cite genetics in support of social ideas, we must make quite sure that they know what they are talking about before accepting their evidence.

Today our understanding of human genetics offers those who carry genetic defects the choice of refraining voluntarily from reproduction, a choice based on a clear assessment of their own situation. Although such a decision does little to "improve the genetic stock" of humanity—we will see later how difficult it would be to try to eliminate defective genes through selective breeding—it may contribute to the well-being of those concerned. But human genetics offers far more for tomorrow. We are already able to compensate for the biochemical defects at the root of a few genetic diseases, and there is every reason to hope that we will eventually be able to correct or compensate for many others. We do not need to improve our species through artificial selection, even if it were possible to do so, for if we can treat afflicted people, their defective genes will probably no longer be a major issue. In the future, the birth of a Michael may simply be an interesting event from the geneticist's point of view rather than a cause for despair.

Key Terms and Concepts

Enzyme	Recessive	Polypeptide
Gene	Mendelian inheritance pattern	Amino acid
Allele		Structural protein
Dominant	Pedigree	Eugenics

References

Dunn, L. C. 1962. Cross currents in the history of human genetics. *Am. J. Hum. Genet.* 14:1–13.

McKusick, V. A. 1975. The growth and development of human genetics as a clinical discipline. *Am. J. Hum. Genet.* 27:261–273.

Sturtevant, A. H. 1965. *A History of Genetics.* Harper and Row, New York.

2 Mendelism: Inheritance as Probability

In 1865, an obscure Austrian monk named Gregor Mendel (Figure 2–1) presented a report on his breeding experiments with garden peas, and in it he proposed the principles on which the entire modern science of genetics is founded. Yet the era of modern genetics did not begin until 1900, when three researchers, working independently, made observations similar to Mendel's and called attention to his long-neglected study. During the next few years, the principles of heredity derived from the humble pea were found to apply to all sexually reproducing organisms, including humans.

Why was there a 35-year lag between Mendel's discoveries and their acceptance? The delay probably occurred because Mendel's research and conclusions were scientifically premature. His work was not appreciated by those few who read it, for they could not link it to any existing ideas about cells or inheritance. Mendel postulated the existence of hereditary "factors"—what we now call genes—that are transmitted intact from one generation to the next. But his model of the factors was a mathematical, not a physical, one. In 1865, chromosomes, the carriers of the factors, still awaited discovery. By 1900, the study of cell structure had advanced dramatically, and the movements of chromosomes during cell division were known. When Mendel's principles were rediscovered, they fit into an existing scheme of knowledge.

A still more fundamental change—a revolution in biological thinking—also helped pave the way for the ultimate appreciation of Mendelism. This breakthrough was the recognition of evolution through natural selection, an idea first proposed in 1859. Although theories of heredity were advanced long before this time, biologists became much more interested in inheritance once they recognized the fact of evolution. Natural selection explained how variation within a species leads to the emergence of new species, but it did not explain how the variation itself arises and is transmitted from one generation to the next. As the storm of controversy over evolution subsided, scientists, armed with their new knowledge of chromosomes, turned their attention to filling this gap in their understanding of biologic change. The time was ripe for the rediscovery of Mendel's work.

The idea that genes are particles, or factors, distinguishes Mendelism from most earlier theories of heredity. Before Mendel's ideas were accepted, it was generally believed that genetic material from parents was somehow blended in their off-

Figure 2–1
Gregor Johann Mendel.

spring. On the basis of casual observation, this belief seemed reasonable enough: People do seem to be blends of their ancestors. However, if the hereditary units themselves were blended like paint, we would find it almost impossible to analyze heredity. By systematically focusing on one pair of traits at a time, Mendel showed that the hereditary units do not mix, but retain their identities from generation to generation. It was this discovery that opened the way for all future developments in genetics.

Mendel's Time: Biology in the Nineteenth Century

Although Mendel's work has had a profound impact on modern biology, it did not emerge from a void. Others before him made observations and performed experimental matings that, properly analyzed, could have led to the formulation of the principles of inheritance. The fact that they did not suggests that the early observers were handcuffed by a set of preconceived ideas. In the next few pages we will discuss some of the influences that shaped biologic thinking in the nineteenth century and show how those influences worked against an earlier understanding of inheritance. We will also consider why Mendel, alone among the plant breeders and biologists of his time, was able to see a significant and predictable pattern in his observations.

Naturphilosophie

An important influence on nineteenth-century biology was the philosophy known as **Naturphilosophie**. Led by such prominent Germans as G. W. F. Hegel and Johann von Goethe, this school of thought held that nature is unity, that an organism is the expression of its interacting parts, and that the study of parts separate from the whole is hardly worthwhile. Plant breeders who adhered to this view of nature usually did not study a single part or trait of an organism, because they believed its attributes in isolation would be an inaccurate reflection of its attributes as part of the whole. But genetic analysis requires focusing on specific traits. If we started with the assumption that the color of the eyes is related to the length of the limbs, the color of the hair, the blood type, and so on, we would never be able to understand how eye color is inherited. Plant hybridizers before Mendel collected data consistent with the modern laws of inheritance, but their background in *Naturphilosophie* seems to have prevented them from analyzing their data piece by piece.

Two nineteenth-century developments spelled doom for the ascendancy of *Naturphilosophie*. In 1808, John Dalton formulated the atomic theory, which stated that all matter is composed of basic units called atoms. And during the 1830s and '40s, Matthias Schleiden and Theodor Schwann demonstrated that organisms are composed of basic units called cells. As scientists began to focus their attention on these minute components of matter, the influence of *Naturphilosophie* diminished. The advent of the cell theory was critical to the development of genetics, and ultimately the study of the cell and its parts was integrated with the principles of Mendelism.

Special creation

Of the views dominating biology during Mendel's intellectually formative period, none were more important than those of the great Swedish botanist Carolus Linnaeus, who in the eighteenth century founded the modern system of biologic classification and nomenclature. Linnaeus believed in **special creation**, the doctrine that every species was specially created by God in its present form. This was not a new idea, of course—it was the teaching of both Catholic and Protestant Christianity—but Linnaeus' scientific prestige helped preserve it as a powerful influence on biology.

In his early years, Linnaeus held that variation within a species represented only minor deviations from an ideal, or standard, type and could therefore never lead to a change in the type itself from one generation to the next. But the results of plant breeding seemed to contradict the notion of a fixed type, for breeders could produce new varieties by selecting plants with the most desirable combinations of traits for reproduction. An obvious question raised by such artificial selection is whether a similar process can occur naturally. Linnaeus eventually concluded, on the basis of extensive observation, that species do sometimes change naturally, though never so much as to evolve into new types. For example, dogs cannot change into cats, but only into different dog species. Still, an uneasiness about variation persisted into the nineteenth century. So most breeders tried to interpret the patterns of variation within the context of special creation and ideal types. No one before Mendel thought to seek new principles governing inheritance by doing experiments involving inheritance patterns.

Josef Kölreuter, for example, was one of the most important of the pre-Mendelian plant hybridizers who studied variation. He was the first to follow crosses systematically through more than one generation and to record the inheritance patterns of specific traits. In 1760 he published a remarkable little book summarizing his studies of various plants, including tobacco and carnations. In one of his experiments, Kölreuter crossed white-flowered carnations with red-flowered ones and noted that the offspring all had pink flowers. When he interbred the pink-flowered plants, they produced red-, pink-, and white-flowered offspring in a ratio of 1:2:1. This ratio was a clear demonstration of Mendel's principles of gene segregation, as we will see, but Kölreuter had blinders on. He said that the hybrid was an unnatural creation, artificially constructed by human hands and therefore unstable. Special creation demanded that such an organism revert to its natural form. After Kölreuter, several other breeders published experimental results similar to Mendel's—some even worked with peas—but they did not understand their significance.

Mendel was familiar with the work of Kölreuter and other plant breeders, and he undoubtedly formulated his own experiments with their results in mind. Yet his conclusions were entirely different from theirs. It is likely that he believed in special creation, at least at first; but unlike his predecessors, he did not use creationist theory as a starting point for interpreting the data. Perhaps the difference between his approach and theirs can be attributed to his training in physics at the University of Vienna. The biologists of Mendel's time were steeped in the tradition of deductive reasoning, a form of logic that explains particular observations in terms of general principles that are assumed to be true—like special creation. Physicists, on the other hand, had long used the inductive approach characteristic of modern experimental science, deriving general principles from particular observations. In designing experiments to test a hypothesis based on his observations, Mendel approached variation in the style of a physicist. While biologists and breeders glossed over variation, Mendel probed it.

Evolution

One of the earliest challenges to special creationism was the evolutionary theory of Jean Baptiste Lamarck. This eminent French naturalist proposed in 1801 that species change from one generation to the next so as to become better suited to their environments. To explain how such changes can occur, Lamarck offered a modification of the Aristotelian idea of acquired characteristics (discussed in Chapter 1). He believed that use or disuse affects the size and shape of body parts and that the organism passes on the resulting changes to its offspring. Though Lamarck presented impressive evidence for evolution, his explanation of its mechanism was unsatisfactory, and biologists were unconvinced. Furthermore, Lamarck was mercilessly hounded by special creationists.

About half a century later, Charles Darwin and Alfred Russel Wallace offered a different explanation of how species become adapted to their environments, and Darwin stated the detailed case for their theory in *The Origin of Species* (1859). Variation, Darwin explained, exists naturally in every species, and those individuals best adapted to their environment survive longer and produce more offspring than those that are less well adapted, so that their relative numbers increase in each generation. The gradual accumulation of the most favorable characteristics over thousands of generations often leads to the emergence of new species. So convincing was the evidence for **natural selection**, as Darwin called this process, that virtually all biologists had accepted the reality of evolution well before the end of the century. For several decades after the publication of Darwin's work, biologic research and debate focused on evolution. Biologists sought evidence in comparative anatomy, embryology, and paleontology for changes in species over millions of years, but they had little interest in horticultural experiments designed to show how variation is transmitted over a generation or two. It is ironic that the preoccupation with evolution contributed at first to the neglect of genetics, for, as the biologists eventually realized, an understanding of inheritance mechanisms is essential to a complete understanding of the evolutionary process.

Darwin himself did attempt to deal with the problem of how variations arise and are transmitted from one generation to the next, but in doing so he went back to the old idea that acquired characteristics are inherited. Like Lamarck, Darwin accepted the ancient Greek theory of **pangenesis**, which held that hereditary material originating in each body part of the parents gave rise to a similar part in their offspring. He proposed that each part of the body produces hereditary units called gemmules, which are transported via the circulatory system to the gonads (testes or ovaries) and packaged into **gametes**, or sex cells (sperm or eggs). The gemmules respond to environmental stress and "inner need" in specific and directed ways and thus are the source of variation.

Francis Galton, Darwin's cousin, rejected the ideas of both pangenesis and the inheritance of acquired characteristics. He suggested that gemmules are permanent residents of the sex cells and that variation arises from random chemical changes in the gemmules. This modification of the gemmule theory comes remarkably close to the modern picture of genes. Why, then, did Galton's scheme not provide a foundation for modern genetics? One answer lies in its complexity: Galton's ideas were in this respect even more premature than Mendel's. Galton was interested in heredity as it affects evolution, so he concentrated on the problem of continuous variation in a population. That is, instead of examining an either/or situation like brown mice versus albino mice (**discontinuous variation**), he tried to explain why rabbits, for example, may be black, white, or various shades of gray. **Continuous variation**, as we now know, is the result of the interaction of several pairs of genes. Galton's model involved the interaction of many gemmules and was essentially correct, but it could not be verified by the experimental methods of the time.

Like the creationists, the early evolutionists tried to fit variation into their larger theory. Mendel, on the other hand, was apparently unconcerned with the place of variation in the grand scheme of things. He simply tried to find a pattern in the transmission of variation from one generation to the next, and he studied the most straightforward possible situation—alternative forms of traits determined by single pairs of alleles (alternative forms of a gene). Only after this simplest kind of inheritance was understood could others go on to study more complex genetic patterns.

The Experiments

Mendel's experiments were brilliant in their simplicity and profound in their meaning. Beginning in 1856, three years before the publication of Darwin's *The Origin of Species*, Mendel experimented with garden peas over a period of eight years. He carried his breeding experiments through four generations of plants, recording his findings with painstaking care and analyzing each experiment statistically to assess the significance of the patterns he observed. In this way he was able to show that certain pairs of contrasting traits were inherited in a mathematically precise and predictable fashion.

Mendel's choice of the garden pea as his experimental organism was most fortunate—or perhaps we should say, carefully thought out. First of all, this plant has several very well-defined characteristics, or traits, each determined by a single pair of alleles. Mendel chose to study seven pairs of contrasting traits (Figure 2–2): (1) round versus wrinkled seeds; (2) yellow versus green seed interiors (cotyledons); (3) gray versus white seed coats; (4) full versus constricted seed pods; (5) green versus yellow seed pods; (6) axial versus terminal flowers; and (7) tall versus dwarf plants. Because these traits always appear in one form or the other and do not blend to produce intermediate types, Mendel could trace them through successive generations without ambiguity or confusion. Moreover, each pair of traits is inherited essentially independently of the others. The structure of the experiments is thus remarkably clear.

Second, the garden pea is normally **self-fertilizing**. That is, each flower contains both stamens, or male parts, and a pistil, or female part (Figure 2–3), and pollen from the stamens normally falls on the pistil of the same flower because the petals tightly enclose these reproductive parts, excluding pollen from other flowers. Cells within the pollen grains give rise to sperm, which fertilize the eggs, or ova, within the pistil. Cross-fertilization (fertilization of one plant by the pollen of another) is unusual in the garden pea except when an experimenter intervenes. The importance of self-fertilization in genetics is that after several generations it results in **pure lines**, or **pure strains**. In other words, the plants become genetically homo-

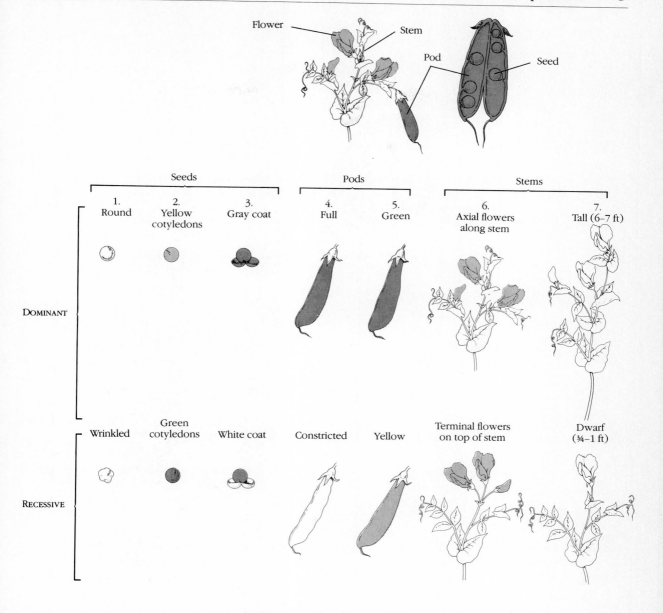

Figure 2-2
The seven contrasting traits of the garden pea that was used by Mendel.

geneous and **breed true**: Each plant characteristically produces offspring like itself. Since Mendel was assured of having pure-line parental plants and could selectively cross whichever plants he wished, he could be certain of the genetic background of the offspring.

As a result of his work with peas, Mendel proposed two simple principles that have remained the cornerstones of modern genetics. One, the **principle of segregation**, describes the behavior of a single gene or allelic pair during the formation of gametes. The other, called the **principle of independent assortment**, deals with the simultaneous inheritance of more than one pair of alleles. In addition, Mendel introduced the concept of dominant and recessive alleles.

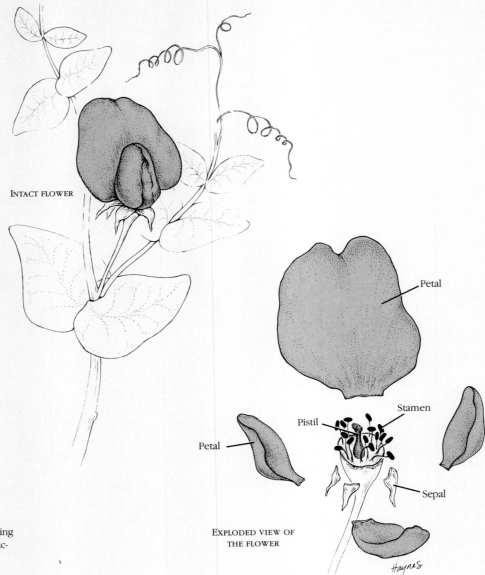

INTACT FLOWER

Petal

Pistil

Stamen

Petal

Sepal

EXPLODED VIEW OF
THE FLOWER

Haynes

Figure 2–3
Garden pea flower, showing
male and female reproduc-
tive parts.

The principle of segregation

Mendel studied each of his seven pairs of traits separately by means of a **monohybrid cross**. This is a cross in which the experimenter is interested in a single difference between the two parents (a cross between a round-seeded pea plant and a wrinkle-seeded one, for example), and the offspring of such a cross are called **monohybrids**. A cross involving two differences is a **dihybrid cross**, and one involving three differences is a **trihybrid cross**. We will use Mendel's cross of tall and dwarf pea plants to illustrate a monohybrid cross and the principle of segregation.

The monohybrid cross: Results and hypothesis   According to most early views of inheritance, crossing a tall plant with a short one should produce a plant of intermediate height. But when Mendel crossed a true-breeding tall pea plant with a true-breeding dwarf one, the first-generation offspring were all tall, whether the male parent was tall and the female dwarf or vice versa. This cross is shown in Figure 2–4a. Geneticists label the parental generation P and subsequent generations F_1, F_2, and so on, F standing for filial (from the Latin *filius*, "son"). When the first filial generation (F_1) tall hybrid plants were self-fertilized (a procedure genetically the same as crossing two identical hybrid F_1 plants), the offspring, called the second filial generation (F_2), consisted of 787 tall and 277 dwarf plants, very close to a ratio of 3 to 1 (Figure 2–4b). Monohybrid crosses involving each of the other characteristics produced similar ratios of one trait to the other.

Mendel proposed that a plant's height is determined by "factors" (now called genes), and that each factor may occur in either of two forms, which specify the alternate traits. We have already been referring to these alternative forms as alleles of a gene; and, as noted in Chapter 1, one of the alleles, called dominant, conceals the other, called recessive. If a plant has one or two dominant alleles, it will display the dominant trait; if it has two recessive alleles it will display the recessive trait. Thus, if A and a represent the dominant and recessive genes, respectively, the allelic combinations AA and Aa produce the dominant trait, and aa produces the recessive trait.

At the heart of Mendel's theory was the thesis that the parents carry a pair of alleles and that the two members of an allelic pair always remain separate and distinct. They segregate (separate) from each other when the gametes are formed, each allele going to one gamete. This is the principle of segregation. If we say that a plant has the genetic makeup Aa, this is really a shorthand way of saying that each of its cells except the gametes contains the allele pair Aa. When its gametes are formed, half of them will receive the dominant allele A and the other half the recessive allele a. A plant with only one kind of allele can, of course, produce only one kind of gamete. When an egg is fertilized by a sperm, the resulting **zygote**, or fertilized egg, will have a complete allelic pair for the characteristic of height, one allele contributed by each parent.

The parental plants (P) in Mendel's cross were true-breeding and he therefore correctly assumed that each had only one kind of allele for the characteristic in question. That is, the tall plants had the genetic makeup DD and the dwarf plants, dd. All the hybrid offspring of such plants must have the genetic makeup Dd and thus be tall (Figure 2–5a). However, self-fertilizing the hybrids (or crossing one hybrid with another) produces three different allelic combinations in the following ratio: 1DD:2Dd:1dd. The reason for this ratio is easy to see if you use the device called a **Punnett square**, which is shown in Figure 2–5b. Since the combinations DD and Dd both result in tall plants, the F_2 generation has three tall plants to one dwarf.

Some genetic terminology   At this point it will be useful to review some modern genetic terms (in addition to the word *gene* itself). As we have discussed, alternate forms of a gene are called alleles. In peas, there are two alleles determining height, a dominant one (D) specifying tallness and a recessive one (d) specifying dwarfness. It is customary to represent alleles with the initial letter of the name of the abnormal or atypical trait. Thus, tallness and dwarfness in peas are usually represented as D and d, because *dwarf* is the name of the atypical condition. The same convention applies whether the abnormal trait is dominant or recessive. Note, however, that the

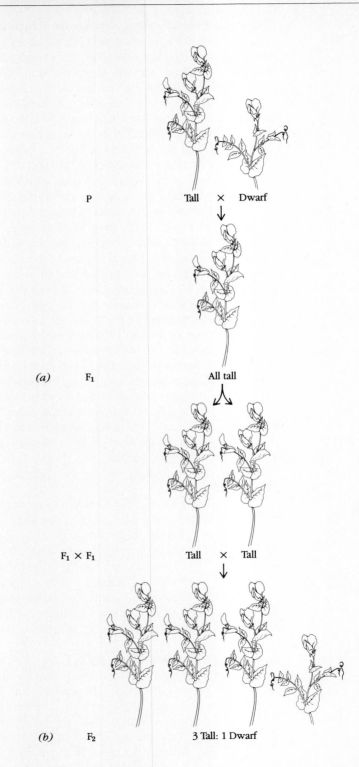

P   Tall   ×   Dwarf

(a)   F₁   All tall

F₁ × F₁   Tall   ×   Tall

Figure 2–4
A monohybrid cross showing the pattern of inheritance of pea plant height: tallness versus dwarfness, where tall is dominant over dwarf.

(b)   F₂   3 Tall: 1 Dwarf

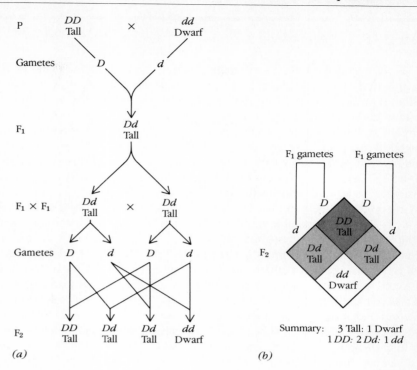

Figure 2–5
(*a*) Monohybrid cross between a pure-line tall pea (*DD*) and a pure-line dwarf pea (*dd*), showing three generations (P, F$_1$, and F$_2$). (*b*) A Punnett square. This is a convenient method for finding the different ways the alleles of the parents (outside the boxes) can be combined in the offspring (inside the boxes). This Punnett square gives the results of a cross between two *Dd* plants.

dominant allele is represented by the uppercase letter and the recessive allele by the lowercase letter.

When two gametes containing identical alleles unite, we refer to the resulting zygote, and to the organism that develops from the zygote, as a **homozygote**. More commonly we say that the zygote or organism is **homozygous** for the allelic pair or trait in question. Likewise, when two gametes containing dissimilar alleles unite, we call the product of their union a **heterozygote**, or **heterozygous** individual. A tall pea plant may be homozygous (*DD*) or heterozygous (*Dd*), but a dwarf plant is always homozygous (*dd*).

Homozygous and heterozygous are terms describing the **genotype**, or genetic makeup, of an organism. The genotype, in combination with environmental influences, determines the **phenotype**, or appearance, of an organism. The F$_2$ plants in Mendel's experiments had three genotypes (*DD*, *Dd*, and *dd*) but only two phenotypes (tall and dwarf). The first two genotypes both determined the same phenotype. When we talk about a 3:1 ratio in the F$_2$, we are talking about the ratio of the phenotypes.

The F$_3$: testing the hypothesis To prove that the 3:1 phenotypic ratio was indeed the result of allele segregation followed by random fertilization, with the allele for tallness dominant over the allele for dwarfness, Mendel self-fertilized all plants of the F$_2$ generation. Since the F$_2$ dwarf plants should be homozygous recessive (*dd*), he predicted that they would produce all dwarf F$_3$ progeny. That is, in fact, what happened. Mendel further predicted that one-third of the F$_2$ tall plants would be homozygous dominant (*DD*) and would thus produce only tall progeny. The remaining two-thirds of the F$_2$ tall plants, he said, would be heterozygous (*Dd*) and these, when

Box 2–1   *Mendel's Data—Too Good to Be True?*

For 36 years after the rediscovery in 1900 of the data reported by Gregor Mendel, geneticists examined and reexamined them. Mendel's paper is a classic of deductive logic, and many a student has marveled at the tightly condensed arguments, the straightforward discussion, and the beautiful way the numerical results drove one to the author's simple, but elegant conclusions. But eventually there were those who wondered if perhaps Mendel's reported numbers did not fit his proposals *too* well—had the honorable monk cheated? His principles were correct enough; it was just the excessive closeness of his reported data to the ideal ratios that seemed bothersome. For instance, if a friend told you he flipped a coin 10 times and got 5 heads and 5 tails you would not be overly surprised. But if he said he flipped the coin 1000 times and scored exactly 500 heads and 500 tails, you might wonder if he were cheating a bit. And if he claimed his second set of 1000 flips also came out 500:500, you would be very suspicious indeed; you would have expected a 50:50 ratio, of course, but a ratio of 500:500 actual numbers is just too close to believe.

Some of Mendel's data clearly show this excess precision. Indeed, there are problems even more strange. Most critical is his reporting of a beautiful 2:1 ratio, the results predicted by his principles from a particular series of experiments. But the experimental design he employed should have altered the observed ratio to 1.69:1. How, then, did he find his claimed 2:1?

Let us illustrate. The original analysis of this problem was published in 1936 by Sir Ronald A. Fisher, the illustrious British statistician. His analysis uses data combined from a series of actual crosses; for simplification, we will represent it here as coming all from one cross, the tall × dwarf plants. The principle is still valid.

The F_2 of the tall × dwarf cross was comprised of tall plants and dwarf plants in a 3:1 ratio. The genotypic ratio, however, should be 1 *AA*:2 *Aa*:1 *aa*. But the *AA* and *Aa* plants were phenotypically indistinguishable. Mendel needed to demonstrate that among the tall F_2 the *AA* genotype was indeed only half as common as *Aa*; only

by so doing could he demonstrate that his principles for genotypes were valid. The most direct data would be obtained from performing a testcross; all the *AA* plants when mated to *aa* would give all tall offspring, whereas *Aa* plants mated to *aa* would give a 1:1 ratio of tall:dwarf among their offspring. So when any tall F_2 plant was testcrossed and yielded any dwarf offspring, it would be seen to be *Aa*; if it produced only tall offspring it would be classed as *AA*.

But testcrosses demand much work; one has to laboriously control the fertilizations of each flower. Mendel perceived that it was much easier to let each plant fertilize itself as peas normally do; an *AA* plant would thus yield an *AA* × *AA* cross and give all tall progeny, but an *Aa* plant, yielding an *Aa* × *Aa* cross, would produce a 3:1 tall:dwarf ratio. But as before, the presence of any dwarf offspring at all would reveal a given plant to be *Aa*.

But how many offspring should one raise from each self-fertilized tall F_2 plant to be sure of its genotype? One must consider both the labor demanded and the garden space. Mendel settled on taking 10 peas from each tall F_2 plant. That is where he should have encountered problems. Consider the *Aa* × *Aa* cross. What is the chance that the first pea chosen from this plant will give rise to a tall offspring? $\frac{3}{4}$, obviously. And the chance that the first *two* peas chosen will *both* give rise to tall plants? $\frac{9}{16}$, obviously. So what, then, is the chance that all 10 peas from an *Aa* × *Aa* plant will give rise to tall plants? The answer is $(\frac{3}{4})^{10} = \frac{59049}{1048576}$, or 0.056. In other words, 5.6% of the *Aa* F_2 plants would be misclassified as *AA*; the *Aa* group would thus be diminished and the *AA* group enlarged. The resulting ratio should be 1.69:1, not the "ideal" ratio of 2:1.

Fisher analyzed Mendel's reports for 600 plants; 400, by our description, should have been *Aa* and 200 *AA*. But when alteration by the experimental design is taken into account, the expected numbers become 378:222, the 1.69:1 ratio. Yet Mendel reported 399:201, an almost perfect 2:1 ratio! Is this serious? Yes indeed; one can demonstrate that the experiment thus designed would only rarely give such a ratio as Mendel reported.

The same problem exists in other experiments in Mendel's paper as well. *Did* Mendel cheat? Did he contrive his data?

Apparently not. One gains confidence that Mendel is being truthful by other reported results; when ratios are bad, he still reports them. He reports faithfully two crosses that should each have given 3:1 ratios, but gave instead 43:2 and 32:1. He reports experiments with another type of plant, *Phaseolus,* and indicates that the results do not always fit his expectations. The evidence is strong that Mendel was honest. So what produced the strange data?

An almost amusing array of "answers" has been suggested over the years, ranging from that of a well-meaning but dishonest assistant who manufactured the results he knew his master wanted (there is no evidence for such), to biologic idiosyncrasies of peas as a research tool. A currently popular suggestion has to do with details of pollen production in peas and the way that the pollen will brush off on a bee; this would theoretically upset strictly random fertilization patterns and possibly yield the aberrant ratios reported.

This suggestion seems flawed, since Mendel, foreseeing this problem, had designed control experiments to detect any such perturbations and found they were insignificant. The more likely explanation comes from the tedious work involved in the study; Mendel had very large numbers of plants to score, and apparently he just quit counting when he had sufficient numbers in the appropriate ratios to give a good demonstration of his principles. Had he completed the analysis of the entire sample, the ratio would likely have been closer to the altered ratio predicted by the design of the experiments.

There thus is no justification in any of this to throw doubt on Mendel's integrity. The problem will be appreciated by anyone who has done this type of research. But its analysis has provided a classic case of "design bias," plus a warning to subsequent researchers to be carefully aware of how experimental design may, innocently but surely, alter the results they should expect. Mendel thus supplied to his genetics followers considerably more insight than he probably realized.

---

self-fertilized, should produce a generation composed of three tall plants to one dwarf, just as their heterozygous parents did. Mendel's predictions were borne out in the F_3.

The testcross Mendel also performed a type of cross now called a **testcross**, in which a plant of unknown genotype is crossed with a homozygous recessive plant. For example, a tall pea plant (*DD* or *Dd*) is crossed with a dwarf pea plant (*dd*). If the tall parent is homozygous dominant (*DD*), the offspring will all be heterozygous (*Dd*) and tall. But if the tall parent is heterozygous, half the offspring will be heterozygous, and half will be homozygous recessive (*dd*)—that is, half will be tall, and half will be dwarf (Figure 2–6). Thus, in a testcross, each F_1 phenotype represents only a single genotype.

Incomplete dominance You will recall that when Kölreuter crossed red-flowered and white-flowered carnations, he got an F_1 of pink-flowered plants, and that when he crossed these hybrids he got an F_2 consisting of one-fourth red, one-half pink, and one-fourth white. As you can see from Figure 2–7, where *r'* stands for red and *r* for white, these results are easily explained by the principle of segregation. The only difference between this situation and the one we have already discussed is that color in carnations exhibits **incomplete dominance**, so that the heterozygous flower (*r'r*) is intermediate in color between the parents. The genotypic ratios in the F_1 and the F_2 are exactly the same as in Mendel's experiments, but in this case each of the three genotypes produces a distinct phenotype. The phenotypic expression of

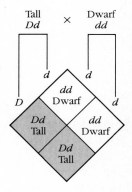

Summary: 1 Tall: 1 Dwarf

Figure 2–6

Testcross between a heterozygous F (tall plant) and a homozygous recessive P (dwarf plant), showing the 1:1 phenotypic and genotypic ratios.

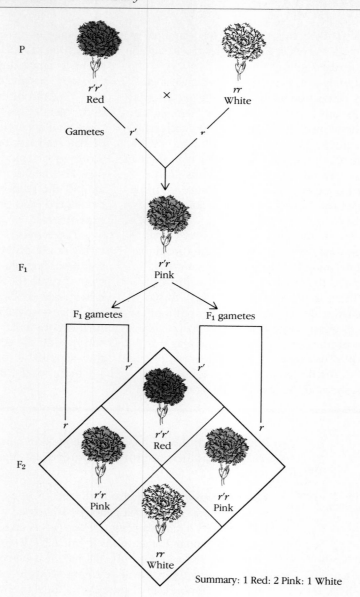

Figure 2–7
Monohybrid cross exhibiting incomplete dominance in carnations.

Summary: 1 Red: 2 Pink: 1 White

the alleles is "mixed" in the heterozygote, but the alleles themselves remain distinct and segregate cleanly, as is shown by the F_2 ratios. Note that for alleles showing incomplete dominance, we do not use capital letters, but use lowercase letters with subscripts or superscripts, such as r'.

Although phenotypic ratios may differ from Mendel's because of different dominance and recessive relationships, these ratios do not contradict Mendelian principles. In fact, they reinforce them.

Random fertilization and probability   Nowhere in the discussion of segregation have we said what a gene is, or how it determines a given trait, or why one allele

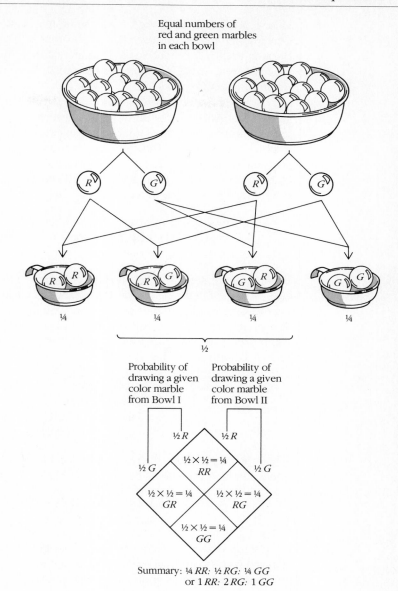

Figure 2–8

Probability. The probability of a given pairing of red and green marbles is the same as the probability of a given pairing of *D* and *d* gametes (see Figure 2–5). The checkerboard shows how to find the probability that two independent events will both occur.

Equal numbers of red and green marbles in each bowl

Probability of drawing a given color marble from Bowl I

Probability of drawing a given color marble from Bowl II

Summary: ¼ *RR*: ½ *RG*: ¼ *GG*
or 1 *RR*: 2 *RG*: 1 *GG*

should be dominant over another. Mendel did not attempt to deal with these questions, nor could he possibly have done so given the state of biologic knowledge in the nineteenth century. His approach to the hereditary "factors" was a probabilistic one: if he assumed that the factors occur in pairs, that the members of a pair segregate into different gametes, that the gametes combine randomly, and that one type of factor is dominant over the other, he could then predict phenotypic ratios on the basis of the laws of chance.

We can use a nonbiologic example to demonstrate the simple probability involved in the random combination of gametes. Imagine two bowls, labeled I and II, each filled with equal numbers of red and green marbles (Figure 2–8). With eyes closed, you reach into bowl I with your left hand and bowl II with your right, drawing one marble from each, and you repeat this procedure over and over, depositing each pair of marbles in a separate cup. In this way you will come up with four different kinds of pairs: red(I)–red(II); red(I)–green(II); green(I)–red(II); green(I)–green(II).

Now, each time you reach into bowl I, the chance, or probability, that you will draw a red marble is one in two ($\frac{1}{2}$), and so is the probability that you will draw a green marble. Exactly the same probabilities apply to your drawing a red or a green marble from bowl II. The two drawings are **independent events**; that is, the outcome of one has no effect on the outcome of the other. But what is the probability of drawing a bowl I red and a bowl II red? Such a probability is the *product* of the probabilities of the independent events:

probability of getting red(I) + red(II)
= probability of getting red(I) × probability of getting red(II)
= $\frac{1}{2} \times \frac{1}{2} = \frac{1}{4}$, or one in four

By the same reasoning, the chance of getting any of the other pairings listed above is also $\frac{1}{2} \times \frac{1}{2} = \frac{1}{4}$. But the second and third pairings are just two different ways of achieving the same result, red plus green; when you inspect the cups, you will count them as a single combination, and you will find you have twice as many cups with this combination as with either of the other two. So the probabilities of getting the various combinations are

red–red:	red(I)–red(II)	$= \frac{1}{2} \times \frac{1}{2} = \frac{1}{4}$
red–green:	$\begin{cases} \text{red(I)–green(II)} \\ \text{green(I)–red(II)} \end{cases}$	$\begin{aligned} &= \frac{1}{2} \times \frac{1}{2} = \frac{1}{4} \\ &= \frac{1}{2} \times \frac{1}{2} = \frac{1}{4} \end{aligned} \Big\} \;\; \frac{1}{2}$
green–green:	green(I)–green(II)	$= \frac{1}{2} \times \frac{1}{2} = \frac{1}{4}$

When the probabilities of the different combinations are added together, they must equal 1, which represents the total number of possibilities:

$$\frac{1}{4} + \frac{1}{2} + \frac{1}{4} = 1 \text{ (or, } 0.25 + 0.50 + 0.25 = 1,$$
$$\text{or, } 25\% + 50\% + 25\% = 100\%)$$

All this does not mean, of course, that out of every four draws you will get one red–red pair, two red–green pairs, and one green–green pair. The laws of chance are generalizations that apply to large numbers of random events. In any one series of four draws, you might get four red pairs or three green pairs and one red–green pair. But the greater the number of draws, the more likely you are to achieve a ratio close to the predicted one.

The analogy of the marbles describes exactly what Mendel said was happening when he crossed two heterozygous pea plants. Each parent produces equal numbers of D and d gametes, just as each bowl contains equal numbers of red and green marbles (see Figure 2–5). Since the male and female gametes pair randomly, you can see that the probability of any two kinds of gametes combining at a given fertilization is the same as the probability of any two kinds of marbles combining in a given draw. Thus, a large number of fertilizations results in a genotypic ratio of approximately $\frac{1}{4}$ $DD:\frac{1}{2}$ $Dd:\frac{1}{4}dd$, or 1:2:1. The dominance of D over d results in the phenotypic ratio of three tall plants to one dwarf plant.

The principle of independent assortment

In addition to his other experiments, Mendel performed dihybrid and trihybrid crosses. By thus mating plants that differed in two or three pairs of alleles, he established that all seven of the characteristics he was studying (see Figure 2–2) were inherited independently of one another.

Interpreting a dihybrid cross A dihybrid cross between a pure-line pea plant with round-yellow seeds and a pure-line pea plant with wrinkled-green seeds produced F_1 progeny that had all round-yellow seeds. This cross established round as dominant over wrinkled, and yellow as dominant over green. We will use the following symbols:

round: *W*
wrinkled: *w*
yellow: *G*
green: *g*

Self-fertilizing the F_1 hybrids produced an F_2 with the phenotypic ratio of

9 round and yellow
3 round and green
3 wrinkled and yellow
1 wrinkled and green

Mendel recognized that the results could be interpreted as the combination of two separate monohybrid crosses, as shown in Figure 2–9. (Note that the checkerboard in the center of the figure is *not* a Punnett square. What we are combining here is not gametes from two parents, but two sets of genotypes from different crosses.) You can see that the hypothetical combination of two monohybrid crosses yields the same ratios that Mendel obtained experimentally in the dihybrid cross.

Mendel concluded from such comparisons that the inheritance of one pair of traits in no way affects the inheritance of the others: each pair of alleles *assorts*, or is distributed to different gametes, independently of every other pair. This is the principle of independent assortment. There is no special tendency for the allele for round-seededness to be inherited together with the allele for yellow-seededness, as a casual observer might guess from the dihybrid phenotypes. The members of the different gene pairs are mixed randomly in the gametes, the gametes combine randomly to form zygotes, and the phenotypic ratios are determined by the dominance of one member of each allelic pair over the other.

Gametes, genotypes, and phenotypes Figure 2–10 shows the results of a dihybrid cross. The parents, being homozygous for both traits, produce only one kind of gamete each, *WG* and *wg*. The F_1 dihybrid is heterozygous for both traits—it has the genotype *WwGg*—and therefore produces four kinds of gametes: *WG*, *wG*, *Wg*, and *wg*. According to the principle of independent assortment, these four types should be formed in equal numbers, so each of them has one chance in four ($\frac{1}{4}$) of being involved in a given fertilization.

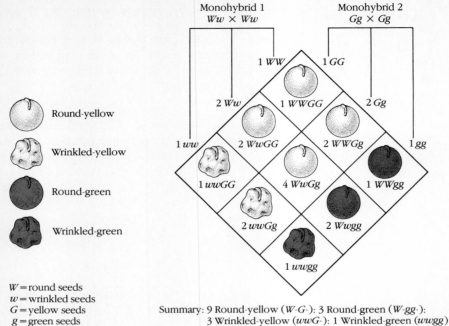

Figure 2–9
Hypothetical combination of two monohybrid crosses. The checkerboard shows the combinations that result from superimposing one set of genotypes on the other. An actual dihybrid cross ($WwGg \times WwGg$) yields the same ratios of genotypes and phenotypes.

Round-yellow

Wrinkled-yellow

Round-green

Wrinkled-green

W = round seeds
w = wrinkled seeds
G = yellow seeds
g = green seeds

Monohybrid 1
$Ww \times Ww$

Monohybrid 2
$Gg \times Gg$

Summary: 9 Round-yellow ($W\text{-}G\text{-}$): 3 Round-green ($W\text{-}gg\text{-}$):
3 Wrinkled-yellow ($wwG\text{-}$): 1 Wrinkled-green ($wwgg$)

If an F_1 plant is self-fertilized (or crossed with an identical dihybrid), the chances of a specific kind of egg and a specific kind of sperm meeting are $\frac{1}{4} \times \frac{1}{4} = \frac{1}{16}$. We can represent this cross by means of a Punnett square (Figure 2–10) in which each subdivision represents one of the 16 possible combinations. You can see, however, that some combinations are the same as others, so there are only nine different genotypes. If we arrange these genotypes according to the phenotypes they exhibit, we come up with four phenotypic classes, in a ratio of 9:3:3:1.

Suppose we are dealing with more than two pairs of traits? There are very simple formulas for predicting the number of kinds of gametes, the number of genotypes, and the number of phenotypes. We can demonstrate these by means of a trihybrid cross. In the cross shown in Figure 2–11, we are considering seed shape (W for round, w for wrinkled), seed color (G for yellow, g for green), and plant height (D for tall, d for dwarf). The F_1 hybrid ($WwGgDd$) produces eight different kinds of gametes, since three pairs of alleles can be mixed in eight ways. If we self-fertilize the F_1, the chances of any two kinds of gametes being involved in a given fertilization are $\frac{1}{8} \times \frac{1}{8} = \frac{1}{64}$, so there are 64 possible combinations of maternal and paternal genes. But these combinations include a great deal of redundancy, so the total number of different genotypes is only 27. If we count up the phenotypes, we find that this cross produces an F_2 phenotypic ratio of 27:9:9:9:3:3:3:1. Notice that there are eight kinds, or classes, of phenotypes.

Comparing the dihybrid and the trihybrid crosses, you can see that in each case the number of different kinds of gametes produced by the F_1 heterozygote is 2^n, where n is the number of independently assorting allelic pairs. The number of genotypes in the F_2 is 3^n, and the number of phenotypes is 2^n, provided that one allele of every pair is dominant over the other. In a case of incomplete dominance,

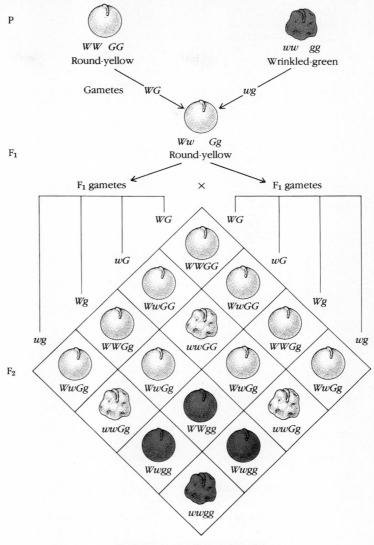

Figure 2–10
Dihybrid cross between a
pure-line pea plant with
round-yellow seeds and a
plant with wrinkled-green
seeds. A summary of the
phenotypic and genotypic
ratios is included.

Summary: 9 Round-yellow: 3 Round-green:
3 Wrinkled-yellow: 1 Wrinkled-green

where the heterozygous condition produces a distinct phenotype (e.g., pink carnations), there will be 3^n phenotypes. This information is summarized in Table 2–1.

The importance of independent assortment The independent assortment of allelic pairs greatly increases the potential for genetic variation in a given population. If, in Mendel's studies, all seven of the dominant traits were inherited together, and all seven of the recessive traits likewise, there would be only two kinds of gametes, three genotypes, and two phenotypes. Every dwarf plant would have wrinkled seeds, gray seed coats, green seed pods, and so on. But if the seven pairs of traits can be mixed in any fashion, 128 phenotypes are possible. In later chapters we will

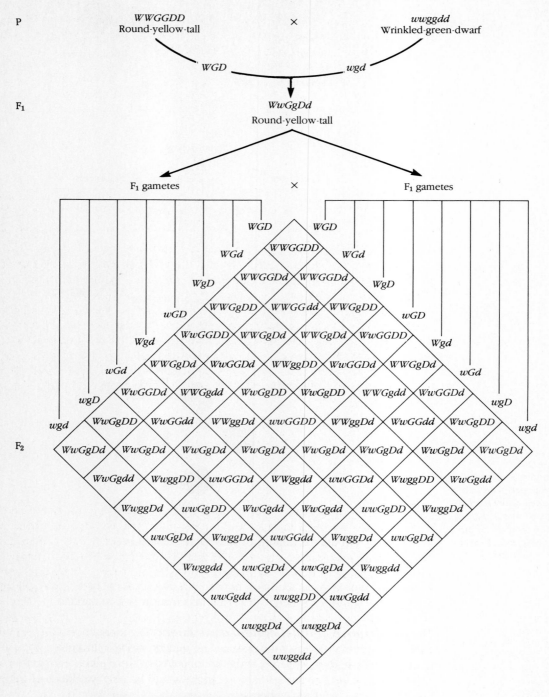

P   WWGGDD   ×   wwggdd
     Round-yellow-tall   Wrinkled-green-dwarf

              WGD   wgd

F₁   WwGgDd
                      Round-yellow-tall

              F₁ gametes   ×   F₁ gametes

Summary: 27 Round-yellow-tall: 9 Round-yellow-dwarf:
9 Round-green-tall: 9 Wrinkled-yellow-tall: 3 Round-green-dwarf:
3 Wrinkled-yellow-dwarf: 3 Wrinkled-green-tall: 1 Wrinkled-green-dwarf

Table 2–1   The number of different gametes produced by the F_1, and of different genotypes and phenotypes in the F_2, of a cross between parents differing in a given number of genes

Gene Pairs	Kinds of Gametes	Kinds of Genotypes	Kinds of Phenotypes[a]
1	2	3	2
2	4	9	4
3	8	27	8
4	16	81	16
n	2^n	3^n	2^n

[a]For dominance; if there is no dominance, there are as many kinds of phenotypes as of genotypes.

discuss the many exceptions to the principle of independent assortment—in some cases members of different allelic pairs do stick together during gamete formation. But human genes often assort independently, and this is one basis of the enormous potential for genetic variation in the offspring of a single couple.

Mendelian genetics is concerned with the prediction of genotypes and phenotypes. Even today, when a great deal of genetic research is devoted to understanding the chemistry of the gene, probability calculations like the ones we have just discussed continue to play an extremely important role. Before the end of this chapter, you will see how such calculations are used in genetic counseling. First, however, you must become acquainted with a few more of the concepts and symbols used in human genetics.

Pedigree Analysis: Mendelism in Humans

Mendel's methodical crosses were made possible by the ease with which parental types can be controlled in peas. The net results of his studies, the principles of segregation and independent assortment and the concept of dominance and recessiveness, have been repeatedly shown to apply to a multitude of plants and animals, including humans. But since humans do not lend themselves to the types of experimental procedures that peas do, special methods must be used in the analysis of human inheritance patterns.

The analysis of pedigrees is a classic method of genetic analysis in humans. Using carefully constructed pedigrees, we can determine whether a given allele is behaving as a dominant or a recessive and whether it is exhibiting independent assortment relative to other genes. We can also determine whether the expression of a trait is related to a person's sex, a complication that did not arise in Mendel's experiments.

The pedigree consists of a set of symbols that conveys information about the incidence of a trait in successive generations. Table 2–2 summarizes those symbols most commonly used in a pedigree chart. As we discuss dominant and recessive

Figure 2–11
(Facing page.)
Trihybrid cross between a pure-line pea plant homozygous for round seeds (*WW*), yellow seeds (*GG*), and tallness (*DD*) and a pure-line pea plant homozygous for wrinkled seeds (*ww*), green seeds (*gg*), and dwarfness (*dd*).

Table 2–2	Symbols commonly used in pedigree analysis

Symbol	Function
◯ or ♀	Normal female
▢ or ♂	Normal male
◯—▢	Single bar indicates mating
I ◯—▢ II ◯ ▢ ◯ 1 2 3	Normal parents and normal offspring, two girls and a boy in birth order indicated by the numbers; I and II indicate generations
▢ ◯ ◯	Single parent as presented means partner is normal or of no significance to the analysis
◯═▢ ◯ ▢	Double bar indicates a consanguineous mating (mating between close relatives) Fraternal twins (not identical)
◯ ◯	Identical twins
②and ⑥	Number of children for each sex
⬤ and ◼ ↑ ↑	Darkened square or circle means affected individual; arrow (when present) indicates the affected individual is the proband or propositus, the index case, the beginning of the analysis
◐ and ◫ ◑ and ◨	Autosomal[a] heterozygous recessive
⊙	X-linked carrier[b]
⊘ and ▨	Dead
↓•	Aborted or stillborn

[a]Having a chromosome other than a sex chromosome (discussed in Chapter 3).
[b]A sex-linked carrier (discussed in Chapter 3).

inheritance patterns, you may find it helpful to construct Punnett squares or pedigrees that illustrate specific points.

Dominant inheritance patterns

The easiest inheritance pattern to analyze is that of a trait determined by a single dominant allele. Such a trait usually appears in successive generations of a given family, whether the trait is rare or common in the population. This is so because its appearance does not depend on the coming together of two alleles. That is, a child needs to inherit only one dominant allele, from only one parent, to display a dominant trait.

For relatively common traits, like the inherited ability to taste PTC (phenylthiocarbamide), it is easy to find cases in which both parents express the trait. To determine that the trait is dominant, we examine pedigrees that include such cases. If both parents have the trait and they have a child who does not, this is evidence that the trait is dominant; if it were recessive, both parents would have to be homozygous to express the trait, and all their children would express it too. In the former case, we therefore suggest that the parents are heterozygous and that the child lacking the trait is homozygous recessive.

A family member who does not display a dominant trait is normally homozygous recessive and cannot pass the trait on. Therefore, a person with a dominant trait normally has at least one parent who also expresses it. Exceptions may occur if the dominant allele fails to express itself in the parent, or if the trait in the child is the result of a new mutation, or change in a gene. We will discuss both of these phenomena later in the book.

An example of a clearly dominant trait is dentinogenesis imperfecta, a rare tooth defect (Figure 2–12a). The teeth of affected persons have a brown or blue opalescent appearance caused by abnormal dentine formation, and the crowns of the teeth wear down easily. A pedigree of a family with a long history of the condition is shown in Figure 2–12b. Notice that the trait appears in successive generations. Notice too that about half of each affected person's children have inherited the trait, pointing out the heterozygous nature of the affected persons. People who do not express the trait do not pass it on to any of their offspring. These are all criteria for the inheritance of traits determined by dominant alleles.

The pedigree in Figure 2–12b clearly demonstrates the dominant mode of inheritance. But what if the trait in question is caused by a recessive allele? Try to interpret the same pedigree assuming that this is the case. You will have to assume that all of the spouses (I1, II5, III1, and III4) were carriers of the mutant allele and that all affected persons were homozygous:

This interpretation would, in fact, explain the roughly 1:1 ratio of affected to nonaffected people in the pedigree. But it is not justified to assume that four unrelated people are carriers of a rare recessive allele that has an incidence in the population of 1/8000. It is far more reasonable to suggest that the trait is the result of a dominant allele:

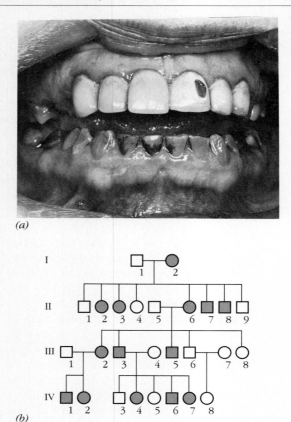

Figure 2–12
(*a*) *Dentinogenesis imper-fecta* in the family shown in (*b*). (*b*) Pedigree of *dentinogenesis imperfecta*, an autosomal dominant disorder of dentine formation.

Recessive inheritance patterns

The pattern of inheritance for a recessive trait is quite different from that of a dominant trait. By definition, a trait is recessive if both members of an allelic pair are required for the trait to be expressed. This means that *both* parents of a child with the trait must have *at least* one allele for the trait. A recessive allele may persist undetected in a family for many generations; only when it combines by chance with another recessive like itself will it produce a distinct phenotype. Thus, the pedigree pattern for recessive traits is strongly affected by two factors: (1) the frequency of the allele in the population, and (2) the extent of **inbreeding**, or mating between relatives, in a family carrying the recessive allele.

The parents of a child who expresses a recessive trait may be homozygous or heterozygous; that is, they may or may not express the trait themselves. For recessive alleles that are rare in the population, we usually find that affected children are born to parents who lack the trait themselves (except in families where there is inbreeding), because the chance of two such alleles coming together in successive generations is exceedingly small. In these cases, the recessive pattern is easy to detect: the trait typically skips one or more generations. An example of a rare recessive trait is

(a)

Figure 2–13
(a) Two albino brothers and their normally pigmented sister. All three siblings are members of the Caribe Cuna Indian population of Panama. The Cuna call them "moon children," and believe they are caused if either the mother or father looks at the moon too much during pregnancy.
(b) A pedigree for albinism.

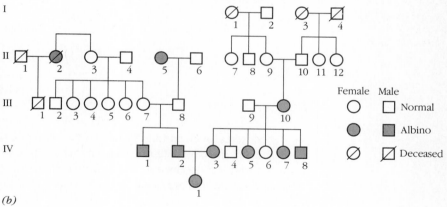

(b)

albinism, a condition in which a person cannot form the pigment melanin (Figure 2–13a). In the United States, the frequency of the trait in Caucasians is about 1/20,000. A pedigree for recessive albinism is shown in Figure 2–13b. The parents of some of the affected persons were both nonalbino, though they were both carriers of the allele that causes this condition.

　　What if a recessive trait is relatively common, like red hair in Scotland or blood type O? Then one or both parents of an affected person may well express the same trait, and we cannot tell that the trait is recessive without further information. We must look through the family pedigree for two kinds of matings:

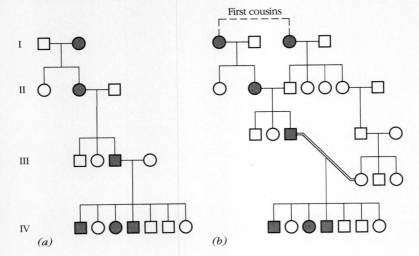

Figure 2–14
Pedigrees for alkaptonuria.
(*a*) An incomplete pedigree, which does not show that individuals II3 and III4 are related. (*b*) The complete pedigree, showing their relatedness.

1 Both parents have the trait. As the following pedigree shows, if the trait is indeed recessive, the parents must be homozygous and *all* their children will express the trait (left). If the pedigree reveals a case in which both parents have the trait and one of their children does not, we know that the trait is dominant (right). If there is no such case, the evidence is good that the trait is recessive.

<table>
<tr><td>Trait is recessive</td><td>Trait is dominant</td></tr>
<tr><td>$aa \times aa$</td><td>$Aa \times Aa$</td></tr>
<tr><td>aa</td><td>$\frac{1}{4} AA, \frac{1}{2} Aa, \frac{1}{4} aa$</td></tr>
<tr><td>(all affected)</td><td>($\frac{1}{4}$ unaffected)</td></tr>
</table>

2 Both parents lack the trait. As the following pedigree shows, if *any* of the children of such parents have the trait, the parents must be heterozygous and the trait must be recessive (left). If there is no case in which the parents lack the trait but produce a child who has it, the evidence is good that parents lacking the trait are homozygous recessive and the trait is dominant (right).

<table>
<tr><td>Trait is recessive</td><td>Trait is dominant</td></tr>
<tr><td>$Aa \times Aa$</td><td>$aa \times aa$</td></tr>
<tr><td>$\frac{1}{4} AA, \frac{1}{2} Aa, \frac{1}{4} aa$</td><td>$aa$</td></tr>
<tr><td>($\frac{1}{4}$ affected)</td><td>(all unaffected)</td></tr>
</table>

Inbreeding has a conspicuous effect on the pedigree of rare recessive traits. This is evident from Figure 2–14, a pedigree for alkaptonuria (a disease characterized by the patient's urine turning black when exposed to air). From the information presented in the incomplete pedigree of the trait (Figure 2–14a), we could reasonably assume that alkaptonuria is a dominantly inherited trait, because for such a rare trait we would infer that individuals II3 and III4 were unrelated and homozygous for the normal allele. But on closer analysis of this family, we find that there were two marriages between cousins (Figure 2–14b). Now the recessive pattern is clear: both spouses (II3 and III4, Figure 2–14a) were related and carriers of a rare recessive allele. This example shows that to make a correct genetic analysis, we need a pedigree that is as complete as possible.

In every pedigree we have discussed, the trait in question has appeared in approximately equal numbers of males and females. One of the characteristics of the traits we have discussed is that they are unrelated to an individual's sex. There are, however, recessive traits that occur chiefly in males, and we will analyze the pedigrees of these traits when we discuss sex-linkage in Chapter 4.

Probability Assessments: A Key to Genetic Counseling

Probability assessment is one of the chief tools of the genetic counselor. When a couple seeks advice about a genetic problem, the counselor must construct a pedigree for each of the pair, to try to establish their genotypes. He or she then calculates the probability of the trait in question appearing in any of the couple's future children.

Some basic probability statements emerged in our earlier discussion of Mendelism. We noted that to find the probability of two independent, random events occurring, we multiply the probabilities of either of them occurring alone. Whether we are talking about drawing a red and a green marble from a 50–50 mix or about tossing two heads in a row (or first heads, then tails) or about giving birth to a girl and then a boy (or to two girls in a row), the odds are $\frac{1}{2} \times \frac{1}{2} = \frac{1}{4}$. Of course, there are many situations in which the probability of a single event occurring is something other than $\frac{1}{2}$, and the operation in these cases is exactly the same. But for the moment let us stick to cases in which each of two alternatives has an equal ($\frac{1}{2}$) chance of happening.

Suppose you plan to have three children and want to know the odds that all will be boys. This is a simple calculation:

$$\text{probability of 3 boys} = \tfrac{1}{2} \times \tfrac{1}{2} \times \tfrac{1}{2} = \tfrac{1}{8}$$

The same calculation, of course, applies to getting three girls. In addition, the probability of getting two boys and one girl (or two girls and one boy) *in a specified order,* is also $\frac{1}{8}$:

Probability of boy + girl + boy =

$\frac{1}{2}$	$\times$	$\frac{1}{2}$	$\times$	$\frac{1}{2}$	$=$	$\frac{1}{8}$
Chance of a boy		Chance of a girl		Chance of a boy		Chance of these 3 events occurring in *this* order

Table 2–3 is a list of all possible sequences of three children. You can see that the chance of getting two boys and one girl in *any* order is $\frac{3}{8}$—three times the chance

Table 2–3   Determining the probabilities of obtaining various combinations of boys and girls in a three-child family

Sequence	Probability of Sequence	Combination	Probability of Combination
Boy + boy + boy	$\frac{1}{2} \times \frac{1}{2} \times \frac{1}{2} = \frac{1}{8}$	3 boys	$\frac{1}{8}$
L Boy + boy + girl	$\frac{1}{2} \times \frac{1}{2} \times \frac{1}{2} = \frac{1}{8}$		
M Boy + girl + boy	$\frac{1}{2} \times \frac{1}{2} \times \frac{1}{2} = \frac{1}{8}$	2 boys + 1 girl	$\frac{3}{8}$
N Girl + boy + boy	$\frac{1}{2} \times \frac{1}{2} \times \frac{1}{2} = \frac{1}{8}$		
Boy + girl + girl	$\frac{1}{2} \times \frac{1}{2} \times \frac{1}{2} = \frac{1}{8}$		
Girl + boy + girl	$\frac{1}{2} \times \frac{1}{2} \times \frac{1}{2} = \frac{1}{8}$	2 girls + 1 boy	$\frac{3}{8}$
Girl + girl + boy	$\frac{1}{2} \times \frac{1}{2} \times \frac{1}{2} = \frac{1}{8}$		
Girl + girl + girl	$\frac{1}{2} \times \frac{1}{2} \times \frac{1}{2} = \frac{1}{8}$	3 girls	$\frac{1}{8}$

of getting all boys or all girls—because there are three ways to get this combination (labeled L, M, and N).

Notice what we have done here: to find the probability that L or M or N will occur, we have added the probabilities of any one of them occurring. When we deal with independent events and want to know the chance of *all* of them occurring, we *multiply* the chances of each of them occurring alone (chance of L × chance of M × chance of N = chance of L *and* M *and* N). However, if we want to know the chance that *any one* of them will occur, we *add* the chances of each of them occurring (chance of L + chance of M + chance of N = chance of L *or* M *or* N). In other words, the less fussy we are about what we want, the better our chances of getting it. If you want three children, all girls, your chances are one in eight, but if you want three children of the same sex regardless of whether they are girls or boys, your chances are: chance of three girls + chance of three boys = $\frac{1}{8} + \frac{1}{8} = \frac{1}{4}$. On the other hand, if you want to have first three boys and then three girls, or if you hope that you will have three girls and your neighbor will have three boys, your chances of getting your wish are only $\frac{1}{8} \times \frac{1}{8} = \frac{1}{64}$.

Now, suppose you plan to have seven children and want to know the probability of having two boys and five girls. You could make a table like Table 2–3, listing the possible sequences, but that would be a lot of work. Fortunately, there is a formula—the **binomial equation**—that will give you all the possible sequences of boys and girls in any number of births (or of heads and tails in a series of coin tosses, or of any pair of alternative, random events in a sequence):

$$(p + q)^n = 1$$

where

p = probability of one event occurring at a given trial (e.g., a girl, heads)
q = probability of the alternative event occurring at a given trial (e.g., a boy, tails)

Box 2–2　　*How to Determine Coefficients in the Binomial Expansion*

Determining the coefficients in equations such as the one on page 38 can be quite a chore. But the coefficients are important inasmuch as they tell us how many ways we can obtain a particular combination of events. In other words, $21p^2q^5$ means that there are 21 ways to get a group of children composed of 2 boys and 5 girls. To facilitate the generation of coefficients, it may be more useful for you to employ Pascal's triangle which gives the coefficients for all the terms in the binomial expansion of $(p + q)^n$:

$n = 0$ 1
$n = 1$ 1   1
$n = 2$ 1   2   1
$n = 3$ 1   3   3   1
$n = 4$ 1   4   6   4   1
$n = 5$ 1   5   10   10   5   1
$n = 6$ 1   6   15   20   15   6   1
$n = 7$ 1   7   21   35   35   21   7   1

Notice that each number is obtained by adding the coefficients nearest to it in the line above; notice too that each line has one more coefficient than the *n* number (i.e., the number of coefficients is $n + 1$). This means that if a couple plans to have seven children, they will produce one of eight different combinations of male and female offspring, from all seven boys to all seven girls, and six combinations in between. The .probabilities, for example, for each combination of seven children are:

Boys	Girls		
7	0	p^7	= 1/128
6	1	$7p^6q$	= 7/128
5	2	$21p^5q^2$	= 21/128
4	3	$35p^4q^3$	= 35/128
3	4	$35p^3q^4$	= 35/128
2	5	$21p^2q^5$	= 21/128
1	6	$7pq^6$	= 7/128
0	7	q^7	= 1/128

If you wish to memorize a simple formula, you can determine the probability of any combination of events. The formula is the general term for the binomial expansion:

$$\frac{n!}{x! \, (n - x)!} \, p^x - q^{n-x}$$

where

n = number of trials or events
$n!$ = "n factorial" = $n \times (n - 1) \times (n - 2) \times \ldots \times 1$
     (Note that $0! = 1$)
x = number of times that p occurs
p = probability of an event occurring
q = probability of the alternative event occurring

Returning to our earlier problem of assessing the probability of five boys and two girls in a seven-child family, we find that

$$\text{probability} = \frac{7!}{2! \, 5!} \, (\tfrac{1}{2})^2(\tfrac{1}{2})^5$$

$$= \frac{7 \times 6 \times 5 \times 4 \times 3 \times 2 \times 1}{2 \times 1 \times 5 \times 4 \times 3 \times 2 \times 1} \, (\tfrac{1}{2})^2(\tfrac{1}{2})^5$$

Canceling the common factors in the numerator and denominator, we get

$$\frac{7 \times 6}{2 \times 1} \, (\tfrac{1}{2})^2(\tfrac{1}{2})^5 = \tfrac{21}{128}$$

n = number of trials or events
1 = total possible sequences of p and q

　　　To find the chances of having two boys and five girls, we write the equation as $(p + q)^7 = 1$, where p is the chance of a boy at each birth ($\tfrac{1}{2}$), q is the chance of a girl at each birth ($\tfrac{1}{2}$), and 7 is the number of births. We expand the binomial as explained in Box 2–2:

$$(p + q)^7 = p^7 + 7p^6q + 21p^5q^2 + 35p^4q^3$$
$$+ 35p^3q^4 + 21p^2q^5 + 7pq^6 + q^7 = 1$$

Each term of the expanded binomial represents a group of possible sequences. The first term means there is one sequence that produces seven boys, the second term means that there are seven sequences that produce six boys and one girl, and so on. The term we are after is $21p^2q^5$, which tells us there are 21 ways to get a series of two boys and five girls. Substituting numbers for p and q, we get

$$21 \left(\tfrac{1}{2}\right)^2\left(\tfrac{1}{2}\right)^5$$
$$= 21 \left(\tfrac{1}{2}\right)^7$$
$$= 21/128 \text{ chances of getting 2 boys and 5 girls in a series of 7 births}$$

If the probabilities of the two alternative events occurring are something other than $\tfrac{1}{2}$, we simply substitute other numbers in the equation. Suppose both members of a couple carry a recessive allele for cystic fibrosis and they want three children. For each birth, the probability of their having a normal child (p) is $\tfrac{3}{4}$, and the probability of their having an affected child (q) is $\tfrac{1}{4}$ ($Aa \times Aa \rightarrow \tfrac{1}{4}AA + \tfrac{1}{2}Aa + \tfrac{1}{4}aa$). For three births,

$$(p + q)^3 = p^3 + 3p^2q + 3pq^2 + q^3 = 1$$

Substituting numbers for p and q, we find the following probabilities:

$$\begin{aligned}
\text{3 normal children} \quad &= p^3 = \left(\tfrac{3}{4}\right)^3 = \tfrac{27}{64}\\
\text{2 normal, 1 affected} &= 3p^2q = 3 \times \left(\tfrac{3}{4}\right)^2 \times \tfrac{1}{4} = \tfrac{27}{64}\\
\text{1 normal, 2 affected} &= 3pq^2 = 3 \times \tfrac{3}{4} \times \left(\tfrac{1}{4}\right)^2 = \tfrac{9}{64}\\
\text{3 affected} \quad\quad\;\; &= q^3 = \left(\tfrac{1}{4}\right)^3 = \tfrac{1}{64}
\end{aligned}$$

The chance that this couple will have *at least* one affected child is the sum of the last three probabilities: $\tfrac{27}{64} + \tfrac{9}{64} + \tfrac{1}{64} = \tfrac{37}{64}$. Thus, the odds are slightly greater than even that one or more of their three children will have CF and slightly less than even ($\tfrac{27}{64}$) that all three will be normal. It must be emphasized that the couple's chance of producing an affected child remains one in four at each birth, regardless of how many affected (or normal) children they have already produced, just as the chance of a coin coming up heads remains $\tfrac{1}{2}$ at every toss regardless of how many heads have already come up.

This kind of calculation enables us to analyze human pedigrees accurately and to predict the likely outcome of matings between known genotypes. For example, consider Figure 2–14 again. The probability that couple II2–II3 will have an affected child is $\tfrac{1}{2}$ ($Aa \times aa \rightarrow \tfrac{1}{2}aa$ and $\tfrac{1}{2}Aa$). If this couple planned to have three children, the probability that all three would be normal would be $\tfrac{1}{2} \times \tfrac{1}{2} \times \tfrac{1}{2} = \tfrac{1}{8}$. In other words, the chances are $\tfrac{7}{8}$ that they will have *at least* one affected child out of three, and, as you can see, they did indeed have one. (You can use the binomial expansion to determine their chances of having a specific number of affected children.) Probability calculation is essential to the work of genetic counselors, who must base their advice on the likelihood that certain events will occur. Of course, it is the parents themselves who must eventually decide how much risk they are willing to take, but the counselor must provide them with a clear understanding of the risk.

So far we have dealt only with traits that follow the straightforward principles of Mendelian inheritance. Many traits have a more complex inheritance pattern, and we will consider some of the complexities in Chapter 4. But many of the deviations from Mendelism are easier to understand if we first know something of the mechanics of Mendelism. So before we go deeper into the details of human inheritance patterns, we will take a look, in the next chapter, at their physical basis: the movements of chromosomes.

Summary

1 The views of ***Naturphilosophie*** and **special creationism**, influential in the early nineteenth century, may have hindered the development of genetics before Mendel. Mendel's work was ignored from 1865 to 1900, a period dominated by the conflict over special creation versus evolution. With the acceptance of evolution, the need to understand heredity became pressing. The discovery of chromosomes allowed a physical interpretation of Mendel's statistical model and opened the way for the rediscovery of Mendelism.

2 Mendel's research generated the principles and concepts fundamental to modern genetics:

(a) Characteristics are determined by pairs of alleles, one derived from each parent.

(b) One allele, or alternative form of the gene, can be dominant and mask the effects of the other, which is recessive. A trait is recessive if two identical alleles are required for its expression.

(c) The **principle of segregation** states that the members of an allelic pair always remain distinct, and that they segregate (separate) from each other during the formation of **gametes** (sex cells). A gamete thus contains only one member of the allelic pair.

(d) The **principle of independent assortment** states that members of different allelic pairs assort (are distributed) independently of each other when gametes are formed. Though there are many exceptions to this rule (to be discussed later), independent assortment provides an enormous amount of genetic variation.

(e) Gametes combine randomly during fertilization. The ratio of **genotypes** (gene combinations) in the offspring can be determined from the laws of probability if we know the genotypes of the parents. The ratio of the **phenotypes** (evident traits) is determined by dominance and recessiveness.

3 Analysis of family pedigrees allows investigators to see the operation of Mendel's principles in the human species and to determine the genotypes of individuals.

(a) Dominant traits usually appear in every generation. A person with a dominant trait normally has at least one parent with the trait, but parents with the trait may produce a child without it.

(b) Recessive traits usually skip generations, but if a trait is common in a population, or if inbreeding has occurred in a family, it may appear in successive generations. If two parents have a recessive trait, all their children will have it; but a child with the trait may have parents without it.

4 The laws of probability can be used to predict the likely outcome of specific human matings. Thus, geneticists can offer advice to persons with certain traits in their family line.

Key Terms and Concepts

Naturphilosophie

Special creation

Natural selection

Pangenesis

Gamete

Discontinuous variation

Continuous variation

Self-fertilize

Pure line

True breeding

Principle of segregation

Principle of independent assortment

Monohybrid cross

Dihybrid cross

Trihybrid cross

Zygote

Punnett square

Homozygote, homozygous

Heterozygote, heterozygous

Genotype

Phenotype

Testcross

Incomplete dominance

Independent events

Inbreeding

Binomial equation

Problems

2–1 A man is heterozygous for five genes. How many different kinds of gametes can this man produce?

2–2 Suppose the man described in Problem 2–1 marries a woman who is heterozygous for only two of the aforementioned genes and homozygous for the remaining three. On the basis of just these five genes, how many different classes of offspring could they produce?

2–3 What fraction of the sperm produced by the man in Problem 2–1 will contain all dominant or all recessive alleles?

2–4 Assume that galactosemia, albinism, and total color blindness are all due to recessive alleles. A man showing no symptoms of galactosemia or albinism and with normal color vision marries a woman with the same characteristics. They produce a child with galactosemia, albinism, and total color blindness. What are the parental genotypes?

2–5 If the parents described in Problem 2–4 decided to have more children, what fraction of them would be identical to the parental phenotypes with respect to the three traits?

2–6 A person heterozygous for six independently assorting genes is testcrossed. How many phenotypic and how many genotypic classes will this person produce?

2–7 In cocker spaniels, coat color is determined by two independently assorting genes. The presence of a dominant *A* and a dominant *B* gene determines black coat color; *aa* with a dominant *B* produces a liver-colored coat; *bb* with a dominant *A* produces a red coat; and *aabb* produces a lemon-colored coat. A black cocker spaniel is mated to a lemon cocker, and they produce a lemon pup. If this same black cocker is mated to another of its own genotype, what kinds of offspring would be expected and in what proportions?

2–8 The color and the shape of summer squash are determined by two pairs of independently assorting alleles. The *Y* allele is dominant and specifies white; its recessive allele y specifies yellow. Disc-shaped squash is determined by the dominant gene S; the spheroid variety by the recessive allele s. A cross between a white disc-shaped variety and a yellow spheroid variety produces an F_1 that is all white and disc-shaped. If these F_1 plants are allowed to self-fertilize, what phenotypic ratio would you expect in the F_2?

2–9 Cystic fibrosis, you will recall, is a serious genetic disease typified by a defect in protein metabolism. Some of the consequences of the defect are pancreas degeneration, lung destruction, and chronic respiratory infection. A man and woman, neither of whom has this disease, produce three children, all of whom have it. Is CF more likely caused by a dominant allele, or a recessive allele? Defend your answer.

2–10 through 2–18 For each pedigree shown, determine if the trait is caused by a dominant allele or a recessive allele, and determine the genotype of those individuals marked by an (*).

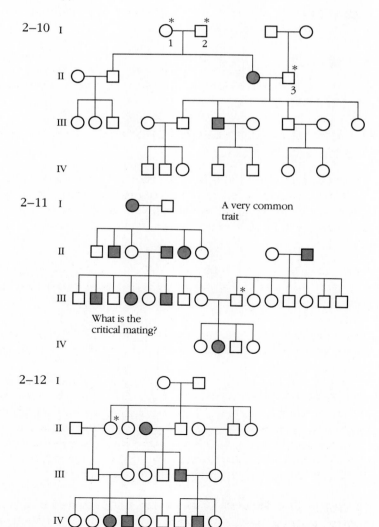

2–13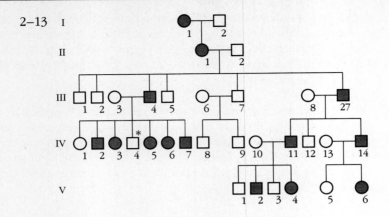

2–14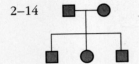

2–15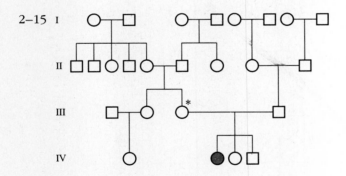

2–16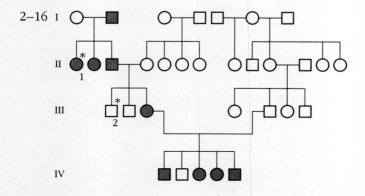

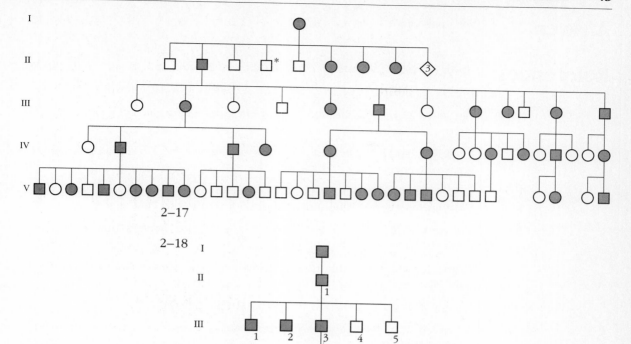

2–17

2–18

2–19   A man with the genotype *Aa* marries a woman who is also *Aa*. They plan to have five children. *A* is dominant over *a*. (a) What is the probability of this couple having three children expressing the dominant phenotype and two children expressing the recessive phenotype? (b) What is the probability of this couple having four children expressing the dominant and one child expressing the recessive?

2–20   An *Aa* man marries an *aa* woman. If they have four children, what is the probability that all four will be *aa*?

2–21   What is the probability that the couple in Problem 2–20 will have two *Aa* and two *aa* children?

2–22   A couple has five children. Determine the probability that the children will occur in the following order: girl, girl, boy, boy, girl.

2–23   Albinism is due to a recessive allele, as is the ability to taste PTC (phenylthiocarbamide). A nonalbino, nontaster woman has a father who is albino and a taster. She marries an albino taster. What types of children do you expect this couple to produce and in what proportions?

Answers

See Appendix A.

References

Boyer, S. H. 1963. *Papers on Human Genetics.* Prentice-Hall, Englewood Cliffs, NJ.

Bresler, J., ed. 1973. *Genetics and Society.* Addison-Wesley, Reading, MA.

Burdette, W. J. 1962. *Methodology in Human Genetics.* Holden-Day, San Francisco.

Crow, J. F., and J. V. Neel, eds. 1967. Proceedings of the Third International Congress of Genetics. Johns Hopkins University Press, Baltimore.

Cold Spring Harbor Symposium on Quantitative Biology, 1964. *Human Genetics.* Cold Spring Harbor Laboratories, Cold Spring Harbor, NY.

Emery, A. E. H., ed. 1970. *Modern Trends in Human Genetics.* Butterworth, London.

Harris, H., and K. Hirschhorn, eds. 1970–1981. *Advances in Human Genetics.* 11 vols. Plenum, New York.

Jenkins, J. B. 1979. *Genetics.* Houghton Mifflin, Boston.

McKusick, V. A. 1964. Approaches and methods in human genetics. *Am. J. Obst. Gyn.* 90:1014–1023.

McKusick, V. A. 1972. Heritable Disorders of Connective Tissue, 4th ed. Mosby, St. Louis.

McKusick, V. A. 1973. Human Genetics. *Annu. Rev. Genet.* 7:435–473.

McKusick, V. A. 1982. *Mendelian Inheritance in Man,* 6th ed. Johns Hopkins University Press, Baltimore.

Mendel, G. 1866. Experiments on Plant Hybrids. In Stern, C. and E. Sherwood, eds. 1966. *The Origins of Genetics.* Freeman, San Francisco.

Strickberger, M. W. 1976. *Genetics,* 2nd ed. Macmillan, New York.

Suzuki, D. T., A. J. F. Griffiths, and R. C. Lewontin. 1981. *Genetics.* Freeman, San Francisco.

Thompson, J. S., and M. W. Thompson. 1980. *Genetics in Medicine.* Saunders, Philadelphia.

3 Chromosomes: The Physical Basis of Inheritance

When Mendel formulated the laws governing inheritance, he had no idea what cellular entities caused the factors to segregate and to assort independently. Nobody did. But there had to be something inside the cell that provided a physical basis for the laws of inheritance; and soon after biologists rediscovered Mendel's laws, they showed that the "something" is the chromosomes.

The research linking inheritance to chromosomes actually began before Mendel's time. From the 1840s to the 1890s, cytological observations pointed to the contents of the cell nucleus as the hereditary material. Thus, when Mendelian genetics was reborn, it was obvious almost at once that the movements of chromosomes during gamete formation provide a physical vehicle for the segregation and independent assortment of genes.

Cell Division and the Transmission of Chromosomes

Identifying chromosomes as the carriers of heredity

Biologists' knowledge of internal cell structure was greatly enhanced by improvements in the microscope and by the development of staining techniques during the 1840s. Because various parts of the cell react with specific stains, or dyes, it became possible to see structures previously not visible. Biologists soon discovered that the nucleus (Figure 3–1) is a permanent part of the cell and that it contains a diffuse, granular-looking substance that stains darkly with certain stains (Figure 3–2). This substance they called **chromatin** (from the Greek *chromos*, color). Over the next few decades, many observers noticed that just before a cell divides, the chromatin seems to disappear, and discrete dark-staining bodies appear in the nucleus. The movements of these **chromosomes** ("color bodies"), as they were later named, were the subject of much investigation.

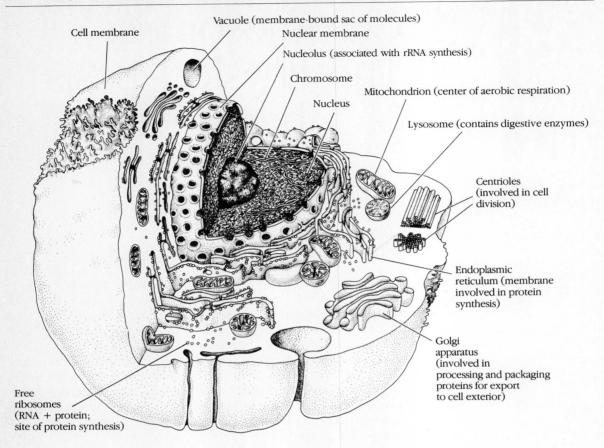

Cell membrane

Vacuole (membrane-bound sac of molecules)

Nuclear membrane

Nucleolus (associated with rRNA synthesis)

Chromosome

Nucleus

Mitochondrion (center of aerobic respiration)

Lysosome (contains digestive enzymes)

Centrioles (involved in cell division)

Endoplasmic reticulum (membrane involved in protein synthesis)

Golgi apparatus (involved in processing and packaging proteins for export to cell exterior)

Free ribosomes (RNA + protein; site of protein synthesis)

Figure 3–1
A generalized animal cell showing some of the structures commonly found.

Mitosis: Duplicating the nuclear material

By the mid-1870s cytologists realized that chromatin does not actually disappear before cell division, but condenses to form the chromosomes. They had also observed that in the early stages of cell division each chromosome is a double structure (Figure 3–3a), consisting of two half-chromosomes, or as we now call them, **chromatids**, joined at a constricted

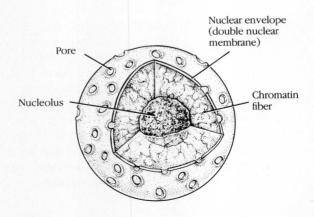

Nuclear envelope (double nuclear membrane)

Pore

Nucleolus

Chromatin fiber

Figure 3–2
A cell nucleus, showing the nucleolus, the nuclear pores in the membrane, and the diffuse chromatin material that condenses into chromosomes during cell division.

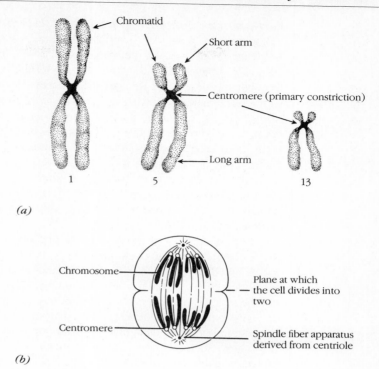

Figure 3–3
(*a*) Human chromosomes 1, 5, and 13. Each is replicated and consists of two chromatids. Chromosome 1 is a **metacentric chromosome** because the centromere is close to the middle; chromosome 5 is **submetacentric** because the centromere is set off from the middle; chromosome 13 is **acrocentric** because the centromere is at or very near the end. (*b*) During cell division, the centromere divides and the chromosomes move to opposite poles of the cell. At the time of centromere division, the chromatids are designated chromosomes.

region now called the **centromere** (also called a kinetochore). As the cell divides, the centromere of every chromosome splits, and the two chromatids separate: one goes to one of the newly forming nuclei, and the other goes to a similar nucleus at the opposite pole (Figure 3–3b). This division of the nuclear material, which occurs whenever a **somatic cell**, or nonsex cell, divides to produce other somatic cells, was later named **mitosis** (Greek *mitos*, "thread") because of the threadlike appearance of the chromosomes under low magnification.

At the end of mitosis, each **daughter cell** contains undoubled chromosomes, equal in number to the double chromosomes of the parental cell. (We consider chromatids to be chromosomes as soon as they separate from each other.) The new chromosomes then become diffuse chromatin again, but when the daughter cells are ready to divide, their chromosomes reappear as double structures. As the nineteenth-century cytologists suspected, the chromatin is replicated (duplicated) between mitotic divisions. We now know that this means not only that the amount of material is doubled, but that every gene is replicated exactly. Each of a chromosome's chromatids is identical to the other.

We will examine mitosis in more detail later in the chapter, but its essential features are simple. As a cell divides, its replicated chromosomes pull apart and half of each goes to each new cell. Daughter cells are genetically identical to each other and to the parental cell.

The discovery of mitosis strongly supported the idea that nuclear material is continuous from one cell generation to the next and that it influences the characteristics of cells. Still, the nucleus was not clearly implicated in inheritance—that is, in the contribution of parents to their offspring via gametes—until scientists observed fertilization in the late 1870s.

Fertilization: Combining two nuclei   Even before they observed fertilization, biologists had reason to suppose the nucleus was involved. Beginning with Kölreuter, experimenters had noted that in the inheritance of alternative traits, such as red or white flower color, the sex of the parents did not usually influence the outcome of a cross. **Reciprocal crosses**—for example, a white male crossed with a red female and a white female crossed with a red male—produced identical results. Yet the egg in all species is much larger than the sperm; the volume of **cytoplasm**, or material outside the nucleus, is up to 1000 times greater in the egg. If the cytoplasm were responsible for inheritance, one would expect the genetic contribution of the female to be much greater than that of the male. All this remained rather speculative, however, as long as the mechanism of fertilization was unknown. Since many nineteenth-century researchers believed that more than one sperm is required to fertilize an egg, the argument about relative cytoplasmic volumes was not entirely convincing.

Working with the gametes of plants and of sea urchins, researchers in the late 1870s observed that fertilization involves the union of a single egg with a single sperm, and that the nuclei of the two gametes fuse shortly after the cells unite. They concluded correctly that fusion of the nuclei is the critical event in fertilization and that the zygote's nucleus is composed of nuclear material from both parents. There was now impressive evidence that the chromatin is continuous not only from one cell generation to the next, but also from one generation of multicellular organism to the next. Such continuity made chromatin a promising candidate for the role of hereditary substance.

Chromosome pairs: A consequence of fertilization   A consequence of fertilization is that the zygote contains two sets of chromosomes, one from each parent. In animals, this means that every somatic cell also contains two sets of chromosomes, since every somatic cell is ultimately derived from the zygote by mitotic divisions, which reproduce the chromosomes exactly. (The situation in many plants and a few animals is more complicated.) We refer to a cell with two chromosome sets as a **diploid** cell and to its chromosome number as the diploid number, or $2n$. Each chromosome in a diploid cell, with an exception we will discuss later, has a **homologous chromosome**, or **homolog**, very much like itself in appearance and genetic information, but derived from the other parent. However, this situation is not particularly evident in the mitotic nucleus, where the chromosomes are mixed together randomly and homologs may be far apart. Though cytologists realized after observing the fusion of sex cell nuclei that zygotes contain chromatin from both parents, they did not know that chromosomes come in pairs until they observed a special kind of cell division in which homologous chromosomes pair up with each other.

Meiosis: Halving the nuclear material   The fusion of gametes presents a logistical problem. If the nuclear material of two somatic cells is combined in the zygote, one would expect the quantity of nuclear material—the number of chromosomes—to double in every generation. It is clear that this does not happen, for every species has a characteristic chromosome number. Humans, for example, have a diploid number of 46 chromosomes, meaning that they have 46 chromosomes in every somatic cell. Cytologists therefore suspected that gametes must contain only

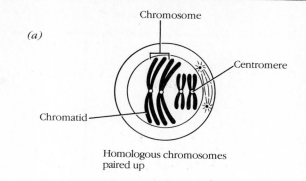

(a)

Chromosome

Centromere

Chromatid

Homologous chromosomes paired up

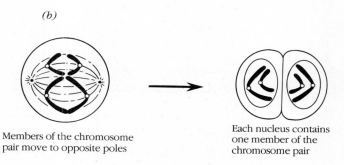

(b)

Figure 3–4
A key feature of meiotic cell division is (*a*) the pairing of homologous chromosomes and (*b*) the segregation of pair members into different nuclei.

Members of the chromosome pair move to opposite poles

Each nucleus contains one member of the chromosome pair

half the amount of nuclear material found in somatic cells, and this idea was confirmed when **meiosis** (Greek *meion*, "smaller") was first observed in 1883.

Meiosis is the type of cell division that occurs when gametes are formed. (Again, we are speaking primarily of animals, for in plants meiosis may occur at a different stage in the life cycle.) In meiosis, a diploid cell undergoes two nuclear divisions to produce four cells, each with half the usual number of chromosomes. Such cells are called **haploid**, and the haploid number of chromosomes is designated n, indicating a single set. Thus $n = 23$ in humans, and the diploid number, 46, = $2n$.

In the first meiotic division, called meiosis I, the chromosomes are arranged in pairs, each alongside its homolog (Figure 3–4a). Each chromosome is composed of two chromatids, having replicated prior to cell division. As the cell divides, one member of each chromosome pair goes to one of the newly forming nuclei and its homolog goes to the other, without any splitting of the centromeres (Figure 3–4b). Early cytologists suggested that all the chromosomes of maternal origin go to one nucleus and all those of paternal origin to the other, but, in fact, it is purely a matter of chance which member of a given chromosome pair ends up in which nucleus: the chromosome pairs assort independently. At the end of meiosis I, each new nucleus contains half the original chromosomes—one of each pair—although the chromosomes are still double-stranded. Thus, meiosis I is a **reduction division**.

In meiosis II the nuclei divide again, but this time the centromere of each chromosome divides, and the two chromatids pull away from each other, each going to a new nucleus. Meiosis II, like mitosis, is an **equational division**: the number of chromosomes in each daughter nucleus equals the number in the parental nucleus. At the end of meiosis, there are four haploid nuclei containing single-stranded chromosomes.

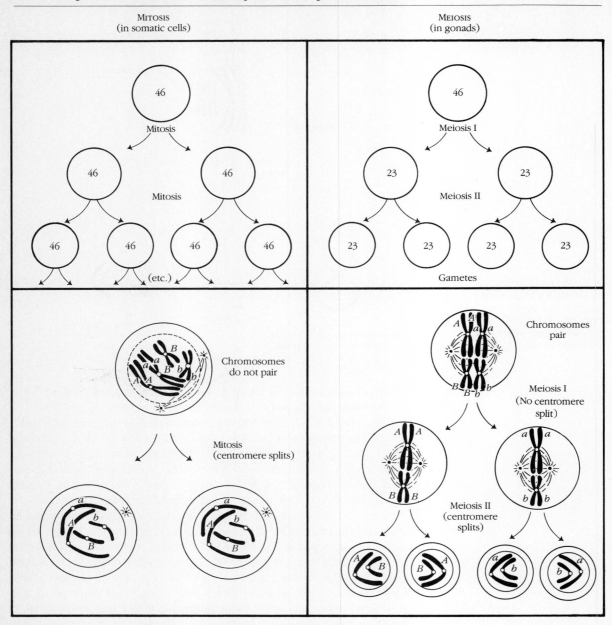

Figure 3–5
A comparison of mitosis and meiosis. The chromosomes duplicate once in each process, but meiosis entails two cell divisions, resulting in half the number of chromosomes per nucleus. Another difference is that homologous chromosomes pair in meiosis, but not in mitosis.

Whereas mitosis preserves the chromosome number of cells, meiosis reduces the number to half the original. Fertilization then restores an organism's full chromosome complement. Figure 3–5 shows the major differences between mitosis and meiosis, and Figure 3–6 illustrates the preservation of the chromosome number in a human life cycle. (Recognize that though the life cycle of plants is much more complex than that of animals, the critical event is the same: haploid gametes unite to form a diploid zygote. This means that the principles of chromosome transmission are the same in peas as in people.)

Now that we have covered the basic functions of mitosis and meiosis, let us examine these processes more closely.

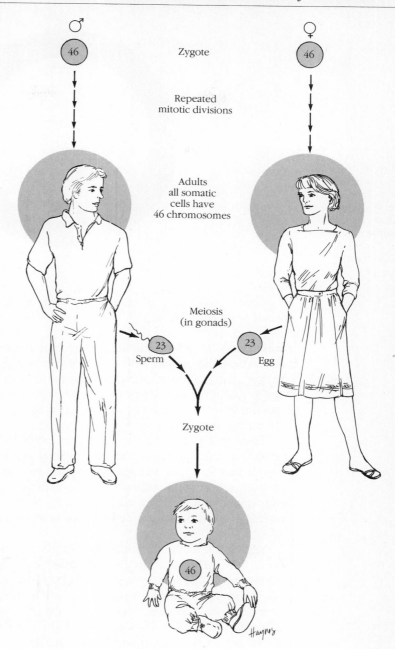

Figure 3–6
Preserving the chromosome number: the human life cycle.

Details of mitosis

The life history of somatic cells, called the **cell cycle**, consists of successive mitotic divisions (mitoses) separated by a period of nondivision called the **interphase**, during which a cell replicates its genetic material. Each interphase, then, is preceded and followed by a mitosis phase, sometimes called the M phase. Figure 3–7 shows

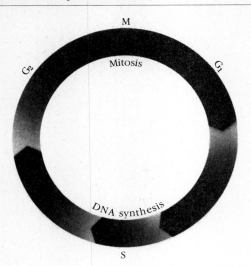

Figure 3–7
The cell cycle. The period of DNA synthesis (S) is separated from mitotic cell division by two gap periods. Growth and cell metabolism continue during the gaps, but chromosome duplication does not.

the proportion of time the cell spends in each of these stages, and you can see that mitosis is relatively short.

Interphase   The chromatin in the interphase nucleus is visible in high-magnification electron micrographs as long strands, jumbled together like strands of spaghetti in a bowl. Each chromatin strand is a very outstretched chromosome; during mitosis it will coil tightly and condense into the familiar chromosome form. The **nucleolus**, a nuclear structure that functions in protein synthesis, is clearly visible through a light microscope during interphase.

Early cytologists sometimes called interphase the "resting phase," but the term is quite inaccurate. The interphase cell is constantly engaged in its normal activities of molecular synthesis and energy production, and during this time it grows to approximately twice its original size. The chromatin, or rather, that part of the chromatin called DNA (to be discussed further on), is busy directing the synthesis of proteins and is thus directly or indirectly responsible for all the cell's growth and activity.

Replication of the chromosomes also occurs during interphase. The period during which this takes place is designated the S (for synthesis) phase, and the periods preceding and following it are called G_1 and G_2 (for gap) phases. The length of G_1 is extremely variable. It may last anywhere from a few minutes to the lifetime of the organism, depending on how often the particular kind of cell divides. The G_2 is generally quite short. During the G_2, the chromosomes are still in the form of lengthy strands, but each strand is now double.

The mitotic division   Mitosis itself is divided into four phases (Figure 3–8). These are somewhat artificial, since the process of nuclear division is actually continuous, but biologists have long divided the process into phases for convenience in description.

The first stage of division, called **prophase**, begins when the chromosomes start to coil and condense into structures visible through the light microscope. At the same time, a **spindle apparatus** begins to form in the cytoplasm. This structure

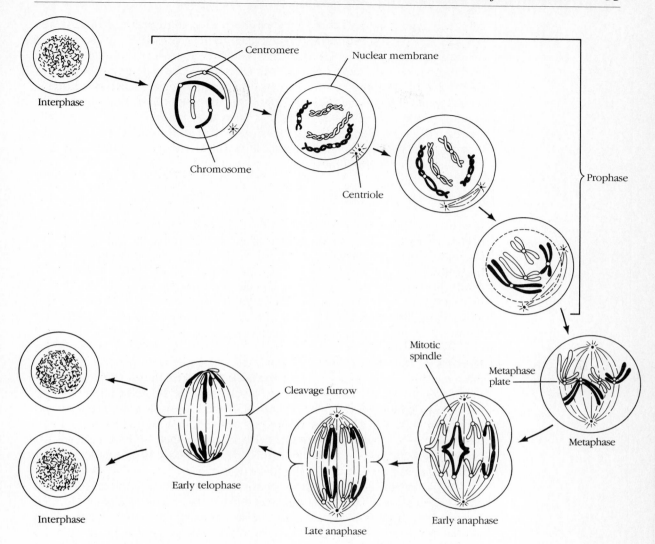

Figure 3–8
The four stages of mitosis.

consists of slender protein fibers called **spindle fibers** attached to two cytoplasmic bodies called **centrioles**; the spindle fibers will pull the chromosomes to opposite poles of the cell. By mid-prophase the chromosomes are distinct and obviously double. The material of the nucleolus disperses throughout the nuclear region during prophase and is not visible through the light microscope for the rest of the cell division.

The second stage, **metaphase**, begins when the nuclear membrane breaks down and the nucleolus is completely dispersed. The chromosomes continue to coil and thicken, and they become aligned in random sequence at the cell's equatorial region, or **metaphase plate**. Some of the spindle fibers are now seen to extend from the centrioles to the centromeres of the chromosomes, though others still extend from one centriole to the other.

The third stage, **anaphase**, begins when the centromeres divide as the **sister chromatids** of each chromosome separate and are pulled to opposite ends of the cell. After separation, sister chromatids are referred to as **daughter chromosomes**.

The final stage, **telophase**, occurs when the chromosomes have completed their movement toward the poles, and it is marked by the formation of two daughter nuclei and two cells. Nuclear membranes form around the two groups of chromosomes; the chromosomes uncoil; the spindle apparatus disappears, and **cytokinesis**, or division of the cytoplasm, occurs. Narrowly, the term *mitosis* refers only to the division of the nuclear material, but it is often used broadly to mean the entire process of cell division, including cytokinesis. The end result of mitosis (in the broader sense) is two genetically identical daughter cells, each containing the diploid number of chromosomes.

Details of meiosis

Meiosis is a lengthy process in which the chromosomes duplicate once and the nucleus divides twice, thus reducing the chromosome number by half (Figure 3–9). Like the term *mitosis*, the term *meiosis* may refer to the nuclear division alone, or to nuclear division and the accompanying cytokinesis. Meiosis in the broader sense results in the formation of four haploid cells from one diploid cell.

Each of the two nuclear divisions of meiosis has phases with the same names as those of mitosis. However, the phases of meiosis I and the corresponding phases of mitosis bear only a superficial resemblance to each other.

Meiosis I: The reduction division   The chromosomes replicate during the interphase preceding meiosis I. In **prophase I**, the first stage of meiosis I, each double-stranded chromosome comes to lie alongside its homolog and then intertwines with it in a process called **synapsis**. At the start of synapsis, each chromosome pair looks like a single structure, but as the homologs pull away from each other slightly, it becomes obvious that there are two of them and that each consists of two sister chromatids. Each intertwined pair is called a **bivalent** (for the two chromosomes) or a **tetrad** (for the four chromatids).

As the lengthy and complex prophase I continues, the homologous chromosomes seem to repel each other, but they remain attached at some points, forming X-shaped regions called **chiasmata** (singular: **chiasma**). The chiasmata apparently indicate where **crossing over**, or exchange of homologous chromosome parts, has occurred. Crossing over is extremely important genetically, but we will defer discussion of it until Chapter 5. Toward the end of prophase I, the bivalents start moving toward the metaphase plate, the nucleolus begins to disperse, and a spindle apparatus starts to form in the cytoplasm.

Meiotic metaphase I begins with the dissolving of the nuclear membrane. During metaphase I, the bivalents, their members still attached at the chiasmata, are aligned at the metaphase plate. The members of each pair point toward opposite poles of the cell, and each centromere is attached to a spindle fiber. During **anaphase I**, the homologs are pulled toward opposite poles. The centromeres have not yet divided, so each chromosome still consists of two chromatids.

When the chromosomes have reached opposite ends of the cell, **telophase I** begins. This phase is usually very short, with the chromosomes uncoiling only partially and incomplete nuclear membranes forming. In some species, including humans, cytokinesis occurs during telophase I, producing two cells. In any case, there are now two separate sets of chromosomes, each with half the number in the

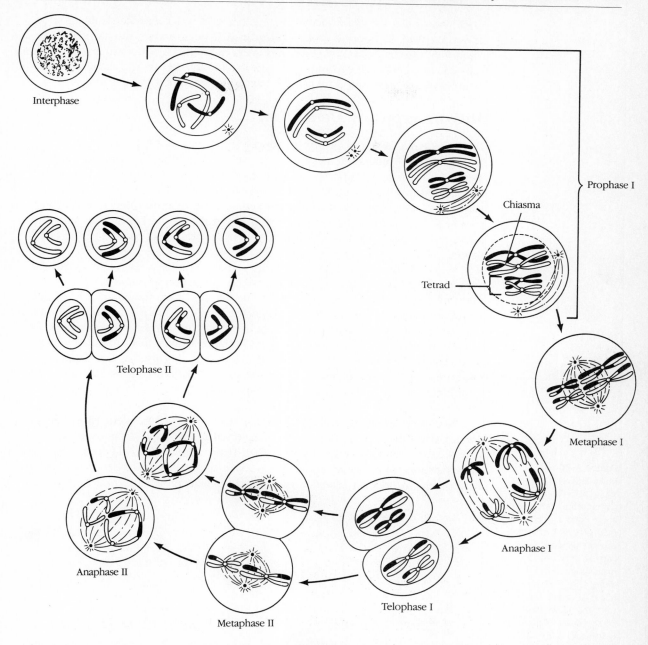

Interphase

Prophase I

Chiasma

Tetrad

Metaphase I

Anaphase I

Telophase I

Metaphase II

Anaphase II

Telophase II

Figure 3–9
The stages of meiosis.

parental cell, but each chromosome still contains twice the usual amount of genetic material. There is only a very short interphase, or none at all, after meiosis I, and there is no duplication of chromosomes.

Meiosis II: The equational division In **prophase II**, the first phase of meiosis II, the chromosomes in each group recondense and start moving toward the new equatorial regions. At **metaphase II**, the chromosomes are aligned at the metaphase plates, with spindle fibers attached to their centromeres. During **anaphase II**, sister chromatids are pulled apart (the centromeres split), and the daughter chromosomes move to opposite poles. **Telophase II** begins when the chromosomes reach the poles and cease movement. Nuclear membranes now form around each haploid complement of chromosomes, and cytokinesis occurs. The second division completed, we have four haploid cells with single-stranded chromosomes.

Meiosis explains Mendelism

In 1900, Mendelian principles were rediscovered, and in 1902 Walter Sutton and Theodor Boveri independently published papers describing the parallels between Mendel's laws and the behavior of chromosomes during meiosis. They pointed out that if one substitutes the term *chromosome* for Mendel's "factor," the terms *segregation* and *independent assortment* perfectly describe the movements of chromosomes in meiosis. In other words, one can postulate a mechanism for Mendel's principles by assuming that the factors are chromosomes, or located on chromosomes.

Consider a man heterozygous for three pairs of alleles, each pair associated with a different pair of chromosomes. If we call his genotype *AaBbCc*, then *A* is on one member of a homologous chromosome pair and *a* on the other, *B* is on one member of another pair and *b* on its homolog, and so also for *C* and *c*. If you first consider a single allelic pair (*Aa*), you can see that meiosis results in segregation of the two alleles from each other. At the start of meiosis, one chromosome of the first pair has two chromatids, each with a copy of *A*; its homolog also has two chromatids, each with a copy of *a*. As a result of meiosis, each of the four chromatids ends up in a different cell. Every daughter cell contains either *A* or *a*: the alleles have segregated.

Now consider all three allelic pairs. The basis of the independent assortment of allelic pairs is the random orientation of maternal and paternal chromosomes at the metaphase plate in meiosis I (Figure 3–10). There are eight (2^3) different ways the three pairs of chromosomes and their associated alleles can be divided between the two nuclei, and during repeated meioses all eight combinations will be formed.

Figure 3–10
Independent assortment of allelic pairs. Three pairs of chromosomes are aligned at the equatorial line in meiosis I. Each chromosome is duplicated and consists of two sister chromatids joined at the centromere. The chromosomes of maternal origin (white) carry the alleles *a*, *b*, and *c*, while those of paternal origin (black) carry alleles *A*, *B*, and *C*. The random orientation of the maternal and paternal chromosomes at metaphase I (we show two of four possible alignments here) results in the gene combinations shown. Different alignments produce different gene combinations.

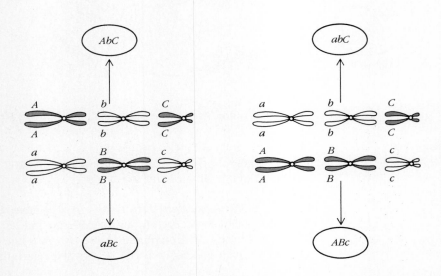

Clearly, however, allelic pairs can assort independently if they are located on different chromosome pairs. It is interesting that peas have seven pairs of chromosomes and that Mendel, who knew nothing of chromosomes, studied seven independently assorting pairs of traits. The chance that seven pairs picked at random would assort independently was extremely small. It is therefore likely that Mendel picked the traits he wished to study only after carefully screening a large number of characteristics to see which provided the simplest pattern.

The parallels between chromosomes and "factors" provided a strong argument for the **chromosome theory of inheritance**, which states that chromosomes are the carriers of the genes. This theory, now amply verified, signaled the merger of cytology and Mendelism into the unified science of genetics.

Autosomes and sex chromosomes

We mentioned earlier that there is an exception to the rule that every chromosome in a human somatic cell has a similar homolog. This exception is found in the cells of males. In 22 of a male's chromosome pairs the homologs are indeed very like each other, but the twenty-third pair consists of one fairly large chromosome called X, and one very small one called Y. The X and Y chromosomes are not homologous and are not usually paired in meiosis, except in limited regions. The corresponding pair of chromosomes in a female consists of two X chromosomes. The X and Y chromosomes are called **sex chromosomes**, and the other 44 chromosomes, which are the same in both sexes, are called **autosomes**.

It has been known for a long time that the X and Y chromosomes determine a person's sex. When gametes are formed by meiosis, the two sex chromosomes go to different cells, just like the members of every other chromosome pair (Figure 3–11). This means that each of a female's eggs will have an X chromosome, but half of a male's sperm cells will contain an X and the other half a Y. At fertilization, a zygote normally receives an X chromosome from the mother, but it may receive either an X

Figure 3–11
Segregation of the sex chromosomes at meiosis results in a single kind of egg, but two classes of sperm. Notice that it is the male that determines the sex of the offspring, not the female.

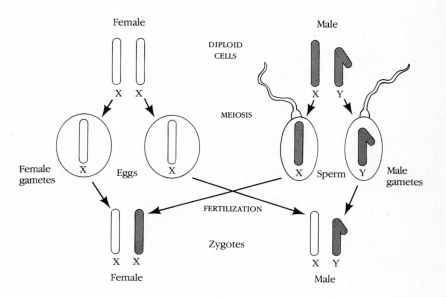

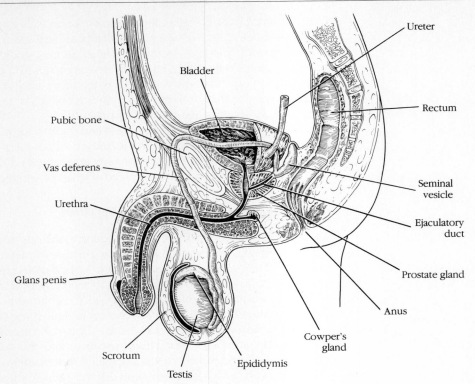

Figure 3–12
Male reproductive system. The human male reproductive organs shown in general location within the lower pelvis.

or a Y from the father and will develop accordingly as a female or a male. The formation of equal numbers of X- and Y-bearing sperm is the reason for the approximately equal number of female and male births.

Gamete formation and fertilization in humans

In humans and other mammals, meiosis occurs in the gonads—the male testes and the female ovaries—and produces sperm and ova, or eggs. The process of gamete formation is called **spermatogenesis** in the male and **oogenesis** (oh-oh-genesis) in the female. We will consider spermatogenesis first, since it is the simpler of the two processes.

Spermatogenesis In sexually mature human males, sperm cells form continuously in the seminiferous tubules of the testes (Figure 3–12). The cells that line these tubules and ultimately give rise to the sperm are called **spermatogonia** (Figure 3–13). Spermatogonia go through several mitotic divisions over a period of three to four weeks, the final division producing cells called **primary spermatocytes**. These diploid cells undergo meiosis I, which takes about 16 days and produces **secondary spermatocytes**, each containing 23 double-stranded chromosomes. The secondary spermatocytes undergo meiosis II, which also takes about 16 days, to form haploid **spermatids**, each with 23 single-stranded chromosomes (Figure 3–13b). Newly formed spermatids have an almost normal amount of cytoplasm,

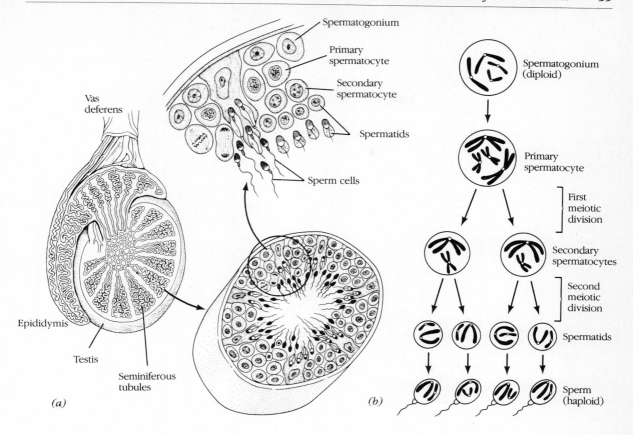

(a)

(b)

Figure 3–13
A diagram of spermatogenesis in humans. (*a*) The arrangement of seminiferous tubules in the testes, with a cross section of a seminiferous tubule and a closer view of the cells undergoing spermatogenesis. A spermatogonium is a parental cell that undergoes meiosis to produce four mature sperm cells. (*b*) Diagram of the meiosis of spermatogenesis.

but as they mature, a process that takes another 16 days, they lose nearly all their cytoplasm and develop a tail for swimming. The **mature sperm** is essentially a nucleus with a tail—a packet of genetic material capable of moving about (Figure 3–13a).

Mature sperm are stored in the epididymis (a cordlike structure at the back of the testis) until ejaculation, when they mix with secretions of the seminal vesicles and prostate gland to form semen. This mixing takes place in the latter portions of the vas deferens, or sperm duct, and the semen is ejected via the duct called the urethra. Two hundred million sperm cells may be present in a single ejaculate, about 60,000 per cubic centimeter of fluid. When these sperm are deposited in the vagina, they move quickly into the cervical canal (Figure 3–14), but only a few dozen of them manage to make the long trip to the upper part of the fallopian tube, or oviduct, where an egg may be waiting. Sperm are viable for 48 hours or so after ejaculation, so if an egg is released into the fallopian tube within roughly two days after intercourse, fertilization may still occur. Once a sperm has penetrated an egg, all other sperm are excluded.

Oogenesis and fertilization Oogenesis (Figure 3–15) in humans begins early in embryonic life. Structures called **ovarian follicles** develop in the ovaries in female embryos, each follicle containing a single cell called an **oogonium** (plural: **oogonia**). By the third month of intrauterine development in the female fetus, the

Figure 3–14
Female reproductive system. Location of the human female reproductive organs within the pelvis.

Figure 3–15
A diagram of oogenesis in humans. (*a*) A cross section of an ovary, showing the development of an ovum (egg cell) from an oogonium. An oogonium is a parent cell that undergoes meiosis to produce a single mature ovum and three polar bodies. (*b*) Diagram of the meiosis of oogenesis.

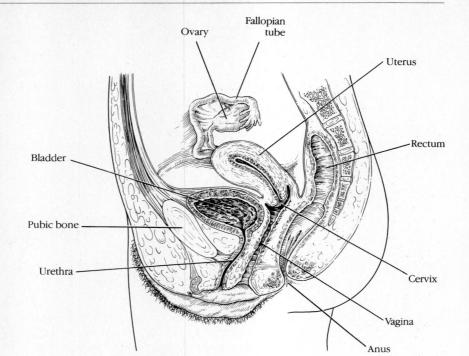

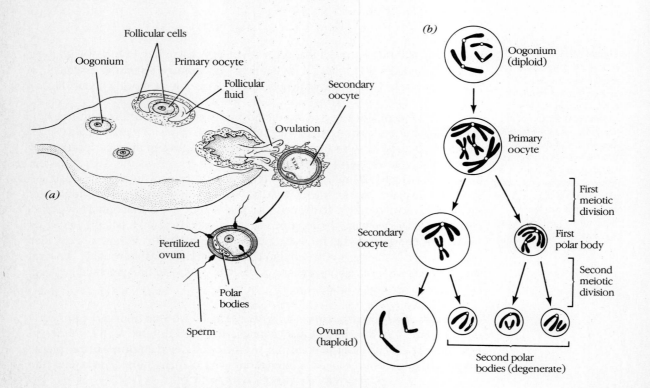

oogonia begin to differentiate into **primary oocytes**. Meiotic division of the primary oocytes thus begins while the female is still a fetus, and it reaches the late stages of prophase I by the time she is born. The primary oocytes remain in a state of suspended prophase I until the female reaches sexual maturity, and no more primary oocytes are ever produced. Thus, unlike the human male, the female is born with all her potential gametes already in a state of meiosis.

Active meiosis resumes in the female when her menstrual cycle begins. At this time she has about 70,000 ovarian follicles, each containing a primary oocyte in prophase I, and one follicle generally matures each month from puberty to menopause. In a maturing follicle, the first meiotic division is completed, but this division produces two very different cells. One, the **secondary oocyte**, contains almost all the cytoplasm of the parental cell, whereas the other, a miniscule, nonfunctional cell called a **polar body**, has virtually none. The secondary oocyte will not undergo meiosis II unless it is fertilized. The terms *egg* and *mature egg* are commonly used to mean the secondary oocyte—the product of the mature follicle—but genetically speaking, a mature ovum does not exist until after meiosis II is completed.

At ovulation, the secondary oocyte breaks out of the follicle and moves into the fallopian tube, with the polar body sticking to it. Fertilization takes place in the fallopian tube within a day or so of ovulation, and meiosis then proceeds to completion. Division of the secondary oocyte once again produces unequal cells. One is the mature ovum, whose nucleus fuses with the sperm nucleus about an hour after fertilization, forming a diploid zygote nucleus. The other is a polar body, which is extruded from the larger cell. The original polar body may also undergo meiosis II, producing two more such cells. As the zygote continues down the fallopian tube, it undergoes repeated mitotic divisions to form an embryo. Arriving in the uterus about five days after ovulation, the embryo becomes implanted in the thickened uterine lining, which supplies it with nourishment. The polar bodies simply disintegrate.

Whereas a meiotic division in the male produces four functional sperm, a completed meiosis in the female produces three nonfunctional polar bodies and one mature ovum containing all the original cytoplasm. This massive amount of cytoplasm (an egg has about 1000 times the volume of a sperm or a polar body) provides the zygote with all its metabolic needs until it reaches and attaches to the uterine wall.

If an egg is not fertilized within about 24 hours of ovulation, meiosis does not go to completion and the egg degenerates. After the egg has been released, the menstrual period occurs. This bleeding is caused by the sloughing off of the thickened uterine lining, and a new lining starts to form as another follicle begins to mature. The menstrual cycle—the reproductive events from one menstrual period to the next—is under hormonal control and takes an average of 28 days, its exact length varying from one woman to another.

The mating game: Sperm meets egg   The primary function of the mating game is to promote the union of gametes, which creates a new individual with a unique combination of genes. Many unicellular organisms reproduce by an asexual act of simple fission or mitosis, but in asexual reproduction the offspring are identical to the parents except when a **mutation**, or change in the structure of a gene, arises. It is the segregation and independent assortment of genes, followed by random combination of gametes, that gives us our biologic individuality. This scrambling of genetic material in every generation is called **genetic recombination**. The endless variation generated by recombination has provided an enormous evolutionary advantage to sexually reproducing species, since it allows for adaptation to a variety of conditions.

In humans, as in many other organisms, the union of egg and sperm has been greatly eased by the evolution of structurally complex male and female reproductive organs. Sexual intercourse, designed to bring sperm and egg together for the act of fertilization, has proved a highly successful mechanism for our species as a whole. The individual, however, may have any of a number of problems, such as sterility, that prevent intercourse from achieving its original purpose. If a couple has reproductive problems due to *male* malfunction, *artificial insemination* can be employed, a procedure in which sperm from a donor male is inserted into the vagina by means other than the penis. Fertilization can even be achieved outside the body *(in vitro fertilization)* by removing a mature egg from a female, collecting sperm from a male, and bringing these specialized cells together in a glass receptacle. Fertilization occurs in this receptacle and the zygote is then allowed to undergo a few divisions before it is implanted in the female, where it continues development. Several babies have been born by this method.

The Human Chromosome

Viewing human chromosomes

Most of the early information about chromosomes was derived from studies of nonhuman cells. Geneticists struggled for decades to develop techniques that would give them a reliable view of human chromosomes, but they encountered enormous difficulties. It was not easy to obtain human tissue to study; fixing and staining procedures produced erratic results; and it was not at all unusual to get contradictory data from seemingly identical experiments. You can appreciate the magnitude of the problems involved if you realize that until the mid-1950s geneticists did not even know the exact number of human chromosomes. Early investigators reported human chromosome complements ranging from 8 to 73, but in 1923 an authoritative study seemed to show that the normal diploid number is 48. This became the generally accepted number until 1956, when Joe-Hin Tjio and Albert Levan announced their use of new techniques to show that the correct number is 46. (A photograph similar to one in their original paper appears in Figure 3–16.) In retrospect, it seems amazing that the answer to so simple a question as "How many chromosomes does a person have?" was not known until three years after the structure of DNA was announced.

Tjio and Levan are to human cytogenetics what Mendel was to genetics. Their brilliant and painstaking improvements of existing methods for studying cells include two techniques that were particularly important to the success of their chromosome-counting work and that have since had widespread application in the study of human chromosomes. The first of these was the technique of placing dividing cells in hypotonic solutions—salt solutions of lower concentration than that of the cell contents. One reason for the earlier discrepancies in chromosome counts was the tendency of the chromosomes to overlap and stick together. Placing a cell in

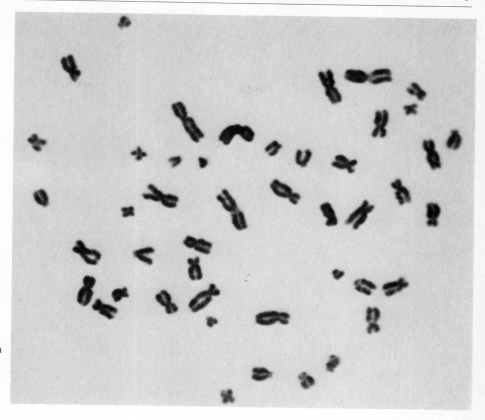

Figure 3–16
A chromosome preparation demonstrating that the diploid number in man is 46. This is a polar view of the cell at metaphase.

hypotonic solution causes water to flow inward, so that the cell swells up and the chromosomes separate from each other, becoming more easily distinguishable. The second technique that Tjio and Levan perfected was that of culturing mammalian cells. They grew lung fibroblast cells (immature connective-tissue cells) in a culture medium to obtain their material for chromosome study.

For their study of human cells, Tjio and Levan also adapted important techniques used in cytologic studies of other organisms. One such technique is the use of a stain called orcein, which stains chromosomes in a very distinct fashion, so that they are visible in more detail than with older staining procedures. Another was the arresting of dividing cells at metaphase by means of a drug called colchicine, derived from the autumn crocus. Colchicine interferes with the attachment of spindle fibers to the centromere, so that when chromosomes are lined up at the metaphase plate they cannot migrate to opposite poles of the cell. The use of colchicine allows a large number of dividing cells to accumulate for observation.

Another great advance in human cytogenetics occurred when cytologists developed a method of culturing lymphocytes (certain white blood cells) taken from the circulating peripheral blood (i.e., blood near the surface of the body). Obtaining blood samples is vastly easier than obtaining samples of other human tissues, so the problem of material for study was at least partially solved by this technique, which is now standard laboratory procedure. The most common method of harvesting human chromosomes from lymphocytes is outlined in Figure 3–17.

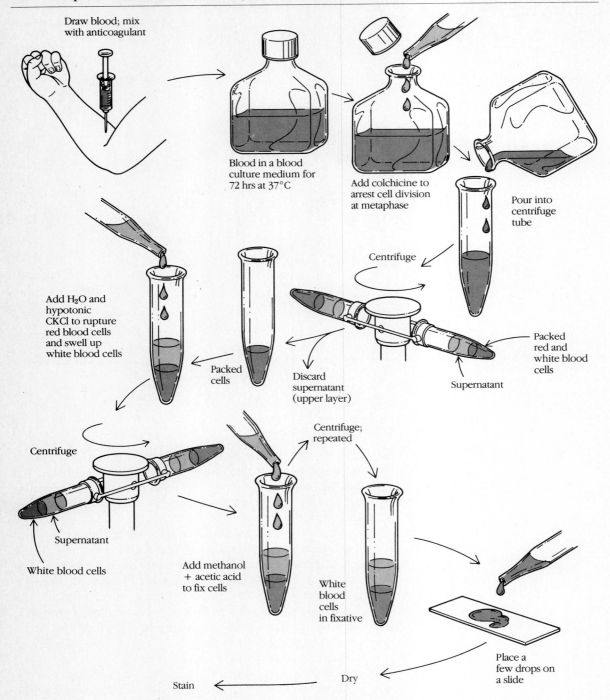

Figure 3–17
The most common method of harvesting human chromosomes from lymphocytes.

Identifying human chromosomes

Chromosome classification In recent years, human cytogeneticists have devoted major efforts to establishing uniform standards and nomenclature (names) for the classification of human chromosomes. Toward this end they have held a series of

Table 3–1 **A general classification of the human chromosomes**

Group	Chromosomes	Characteristics
A	1, 2, 3	Large, metacentric chromosomes;[a] approximately median centromere
B	4, 5	Large, submetacentric chromosomes;[a] submedian centromeres
C	6 through 12 and X (sex chromosome)	Medium-sized, submetacentric chromosomes
D	13, 14, 15	Medium-sized, acrocentric[a] chromosomes with nearly terminal centromeres. Each chromosome in this group has a small appendage, or "satellite," attached to the end of the short arm, observable in good preparations
E	16, 17, 18	Shorter than group D; metacentric or submetacentric
F	19, 20	Short; metacentric
G	21, 22, Y (sex chromsome)	Very short, acrocentric chromosomes; 21 and 22 have small satellites. Again, these satellites are observed only in better preparations

[a]See Figure 3–3a.

international conferences beginning in 1960. They have agreed that the 22 pairs of autosomes (nonsex chromosomes) should be numbered in descending order of length—the longest has the lowest number, 1—and that if two pairs are essentially the same length, the pair with the centromeres most centrally located would get the lower number. They have further decided that chromosomes should be organized into seven groups according to their size and centromere position (Table 3–1). Figure 3–18 shows a human male's **karyotype**, or chromosome makeup, arranged according to these rules of classification.

In 1960, when these classifications were being worked out, only a few chromosome pairs (1, 3, and 16) could be identified unambiguously on the basis of length and centromere position. The discovery of additional physical landmarks, such as secondary constrictions (the primary constriction is the one at the centromere), enabled cytogeneticists to identify other chromosomes. However, a true revolution in human cytogenetics occurred when the techniques called banding were discovered.

Chromosome banding During the late 1960s and early 1970s, several new staining procedures were discovered that gave the chromosomes truly distinctive appearances. These procedures caused different parts of the chromosomes to stain differently, causing the appearance of light and dark **bands**. Banding patterns are the most powerful tool now available for chromosome identification, because the pattern for every chromosome pair is unique. Furthermore, different regions of the same chromosome can be consistently identified by their banding patterns.

The first of the major banding techniques to be discovered uses quinacrine mustard, a fluorescing dye that preferentially binds to specific regions of chromosomes. When chromosomes treated with this dye are viewed under ultraviolet light,

Figure 3–18
A schematic of the normal male karyotype (chromosome makeup). Only one member of each pair of autosomes (nonsex chromosomes) is shown. The X and Y chromosomes are grouped with the autosomes according to size and centromere position.

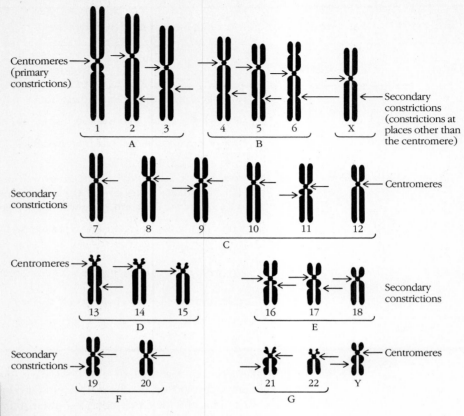

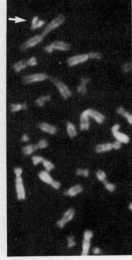

Figure 3–19
Q-banded chromosomes. Arrow points to Y chromosome, which fluoresces intensely with this stain. This karyotype also has a deletion for part of chromosome 7.

fluorescent bands of varying intensity appear, characteristic for each chromosome (Figure 3–19). These are called **Q bands** (Q for quinacrine). The most brilliantly fluorescing chromosome is the Y, which makes Q-banding an especially valuable technique for identifying the Y chromosome.

A second technique, C-banding, uses a dye mixture called Giemsa, which stains the region around the centromere of each chromosome much more deeply than the rest (Figure 3–20). The dark bands are called **C bands** (C for centromeric) and they are especially prominent in chromosomes 1, 9, and 16.

After the development of C-banding, cytologists found that by modifying the procedure, they could use Giemsa stain to show a banding pattern similar to the Q bands, but more detailed. The bands produced by the modified procedure are called Giemsa bands, or **G bands** (Figure 3–21). The G-banding technique is most successful if the chromosomes are pretreated with a protein-digesting enzyme called trypsin before the stain is applied. G-banding is preferable to Q-banding because it is less expensive and does not require a fluorescence microscope.

The last of the important banding procedures is called "reverse Giemsa," or R-banding. This technique produces bands that are in reverse contrast to the Q and G bands (Figure 3–22). That is, a dark **R band** appears where a light Q or G band would be, and vice versa. This technique is especially valuable for observing the ends of chromosomes, which come out very dark.

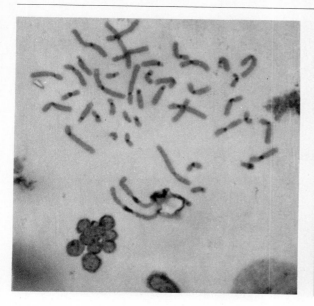
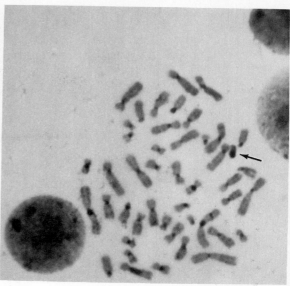

Figure 3–20
C-banded chromosomes. The Y chromosome stands out in this banding technique (arrow).

Banding revealed such intricate details of each chromosome that new standards had to be set. So at the Paris conference of 1971, cytogeneticists established a system of classification for the chromosome bands and regions. This exquisitely logical system uses numbers and letters to enable us to pinpoint any band we might be interested in. The short arm of each chromosome is called the **p arm** ("p" for French *petite*, "small"), and the long arm is the **q arm**. Regions and bands are numbered from the centromere out. To identify a band, we use a sequence of

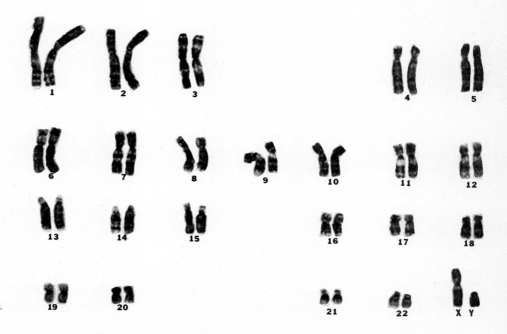

Figure 3–21
G-banded chromosomes.

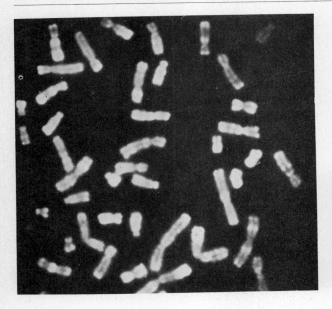

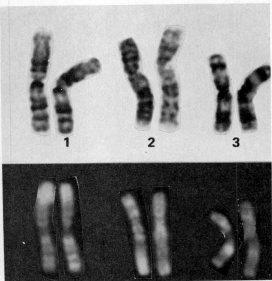

Figure 3–22
R-banded chromosomes.

four items: chromosome number, arm, region, and band number. For example, 9q34 refers to chromosome 9, the long arm, region 3, band 4. This band is indicated with an arrow in Figure 3–23. For the sex chromosomes, an X or a Y is used instead of the first number.

Chromosome structure

As soon as the chromosome was identified as the carrier of genetic information, scientists began to seek in its chemistry and structure an answer to the question "What is a gene?" Interestingly enough, they had a satisfactory answer to the question before they had a completely satisfactory picture of a chromosome, for the answer lay in the molecular structure of one constituent of chromosomes—the **DNA**.

In the early years of this century, chemical analysis showed that chromosomes contain DNA (deoxyribonucleic acid) and protein. In the 1940s, experiments in which bacteria were genetically altered by assimilation of isolated DNA demonstrated that the DNA, not the protein, was the genetically significant part of the chromatin strand. In 1953, Watson and Crick published their famous paper on the molecular structure of DNA, and their model, which has remained virtually unchanged for 30 years, quickly proved to hold the key to the genetic code. Only in the 1970s, however, did a clear picture begin to emerge of how DNA and protein are bound together in the chromosome.

Genes are DNA   Shortly after the turn of the century, some biologists proposed that the chromosome is an inert "home" structure for genes, which journey away and perform their functions in other parts of the cell. But we now know that genes are segments of chromosomal DNA, and that they direct the synthesis of proteins by means of intermediary molecules (called ribonucleic acid, or RNA) that move into

Figure 3–23
(Facing page.)
Diagrammatic representations of the 24 kinds of human chromosomes. The numbering to the left of each chromosome shows the numbering system adopted at the Paris Conference (1971) for the bands and regions observed in metaphase spreads with Q-, G-, and R-staining methods.

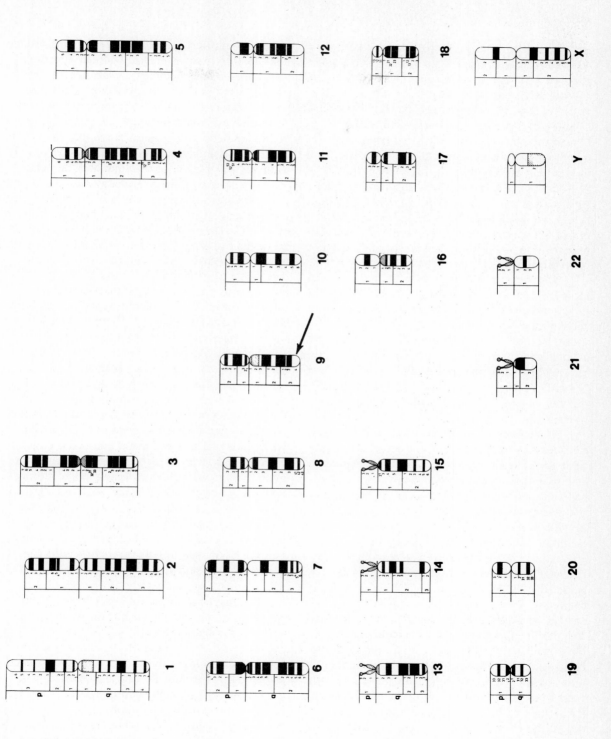

Box 3–1   Nobel Sperm Banks—A Genetic Resource?

For some decades now, we have been able to deep-freeze sperm from various animals in liquid nitrogen and store it rather indefinitely. When appropriately thawed, it functions perfectly well. Thus, the past 15 years or so have seen the establishment of "sperm banks"—repositories for human gametes. Sperm banks have been used by men about to obtain vasectomies as insurance that they could have children in the future if they chose to do so; by servicemen departing for Vietnam, so that their wives could have their children if they were killed; and as supplementary or replacement sperm by couples infertile because the husband's sperm production was insufficient for conception.

Most such repositories store any man's sperm for a fee. But a few decades ago, Nobel Prize–winning geneticist H.J. Muller proposed another, more controversial sperm bank: a repository for the sperm of men with highly esteemed qualities such as intelligence. There is, in fact, such a repository in Southern California specifically for the sperm of Nobel Prize winners in science, and a very small number of such men have actually donated their sperm to it.

Of course, there are major practical and ethical questions associated with the idea.* Are all the desirable components of personality (creativity, ambition, discipline, cooperativeness, perception, patience, and so on) really genetically based? Can one ensure that the man's best traits (rather than his worse ones) will be those passed on to his children? (As one story goes, George Bernard Shaw was once approached by a beautiful but rather empty-headed young woman who suggested that the two of them should produce a child, one who would be blessed with her looks and his intelligence. Shaw is said to have immediately declined, pointing out to her the unacceptable possibility that the child would instead inherit *his* looks and *her* brains.) Will the traits that we now consider desirable be equally prized, say, 50 years from now? Further, can one ethically select the women to be impregnated? And who will decide the environment in which the children will be raised?

Although he recognized the problems, Muller felt that a program of selected genetic breeding was worthwhile nonetheless. Realizing that high intelligence alone was an insufficient guide for selecting donors of sperm, he stated concerns about "brotherliness, loving kindness, generosity . . . joy of life aesthetic propensities, emotional vigor combined with stability, self-control, resilience, moral fortitude, independence of mind, and the balancing humility and sociality that lead one to accept fair criticism from others, and to criticize and correct oneself . . . health and vigor," and "lateness of senility." He proposed that any sperm collected be held for at least 20 years before use to give a better perspective on the contributions and personal traits of each candidate. And Muller made it clear that, to preserve genetic diversity, sperm must be obtained from highly regarded men from a broad spectrum of society. There was for Muller no one "superior" phenotype to be selected.

Even so, the proposal is enough to shake some observers. It is, in fact, selective breeding, even though it is purely private and purely voluntary. It is too reminiscent, some say, of the Nazi's *Lebensborn* movement.** The misery and mischief of that program were too great to be forgotten; we cannot risk anything that comes even close to a repeat. However, the Nobel sperm bank is not intended for production of a mass population genetically superior to its parents, while the Nazi project was a grotesque expression of "negative eugenics" perpetrated on a large population by a repressive government. Although some comment that the Nobel sperm bank program could conceivably be expanded and made mandatory by some future government, the possibility seems highly unlikely. It seems far better to judge the program on its present merits than by imagined social and political scenarios.

In favor of the program is the fact that the sperm are coming from men of high contribution as judged by their intellectual attainments. Most are older and have remained in relatively good health, which answers favorably to Muller's concern about "lateness of senility." However, there is increasing evidence that older men produce sperm carrying a higher frequency of deleterious genes and are more likely to produce children with chromosomal abnormalities. The sperm is used to impregnate women in good health and of high intelligence. Yet the program seems to have been relatively lax in even trying to determine favorable environments for the children's upbringing. Worse, the program's managers list one concern that could be alarming to some of the critics mentioned previously: they perceive the Nobel sperm bank as a token effort to reduce the disparity between the higher reproduction rates of the lower socioeconomic classes compared to the upper classes that generally produce the more influential members of society.

In short, the available evidence to most scientists indicates that a breeding program of the type envisioned by the Nobel sperm bank is too loaded with genetic difficulties, as well as ethically unacceptable and just plain impractical, to justify support. Sperm banks for couples who are otherwise unable to have children are of course valuable and justified; those designed with selective breeding in mind are not.

---

*A good introductory treatment, pro and con, of the Nobel sperm bank is found in *The Humanist* 40:61–63, July/August, 1980, and 40:27ff, September/October 1980.

**For a classic treatment, see Marc Hillel and Claire Henry, *Of Pure Blood* (New York: McGraw-Hill, 1976); also available in paperback.

---

the cytoplasm through the pores of the nuclear membrane. A later view portrayed chromosomes as consisting of genes strung together like beads on a string. In some ways, the beads-on-a-string model is not entirely inaccurate. We find that chromosomes consist of DNA and protein elaborately wound together in a complex three-dimensional structure with genes arranged linearly. This linear view of gene organization was useful, for it allowed geneticists to work out the sequence of specific genes on chromosomes long before they understood the structure or functioning of genes.

The nucleosome model   Figure 3–24 shows the current view of mammalian chromosome structure. The basic molecular components of the chromatin strand are DNA and several varieties of a class of protein called **histones**. The DNA molecule consists of two spiral strands, linked together at intervals by molecular bonds to form a double helix. The chemical structure of the DNA double helix is the basis of the genetic code, and we will describe this structure in detail in Chapter 7. A chromosome contains a continuous length of DNA, wound around individual structures

(text continues on p. 73)

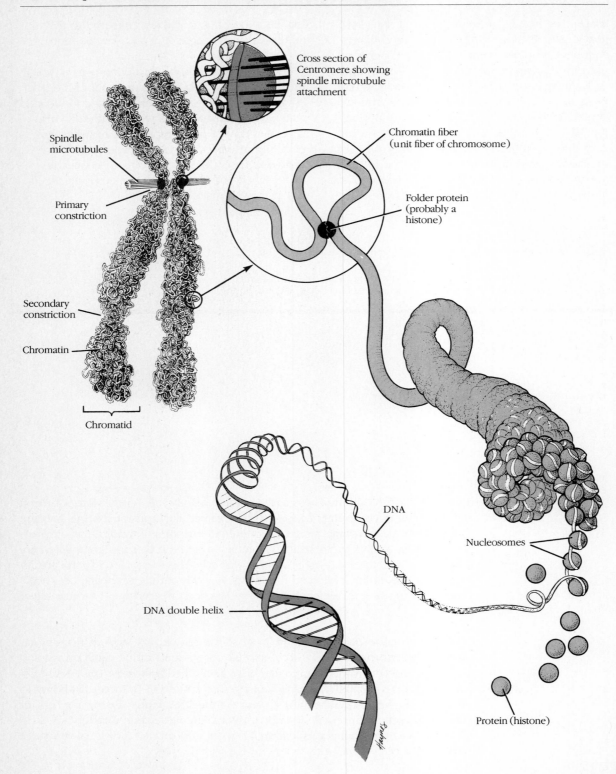

Cross section of
Centromere showing
spindle microtubule
attachment

Chromatin fiber
(unit fiber of chromosome)

Spindle
microtubules

Folder protein
(probably a
histone)

Primary
constriction

Secondary
constriction

Chromatin

Chromatid

Chromatin fiber
(unit fiber of chromosome)

DNA

Nucleosomes

DNA double helix

Protein (histone)

Figure 3–24
(Facing page.)
A diagram of the probable structure of a metaphase chromosome. Each chromatid contains a single DNA molecule. This molecule is several centimeters long and combines with histone proteins to form nucleosomes. The nucleosomes, along with some nonhistone proteins, are supercoiled to form the basic unit fiber of the chromosome (about 20 nm in diameter). Folder proteins (probably one of the histone proteins) condense the chromosome during the prophase of mitosis and meiosis. The final folding pattern produces the characteristic structural features of the chromosome, such as secondary constructions, arm length, and thickness. Two centromeres, one for each chromatid, form attachment sites for spindle microtubules. The entire metaphase chromosome, including the centromere, is replicated.

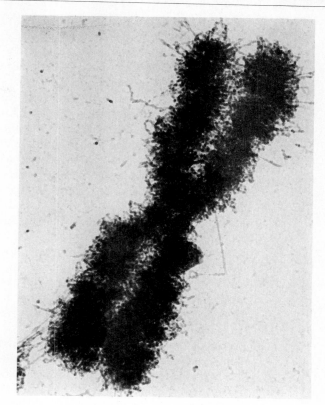

Figure 3–25
(Above right.)
A metaphase chromosome isolated from a human cell. This is a metacentric chromosome because the centromere is near the center. Note the two chromatids and the individual strands of protein-DNA (chromatin) fibers.

made up of histones, rather like a continuous thread winding around a series of spools. Each DNA–histone unit is called a **nucleosome**. The nucleosomes coil and condense into units that appear as bands or chromomeres when they are appropriately stained. Further coiling and condensation produces a chromosome similar to those we are able to view with the electron microscope (Figure 3–25).

Summary

1 Cytological studies before 1900 convinced many biologists that chromosomes carry the genetic information. When they rediscovered Mendelian principles, biologists realized that chromosomes provided a physical basis for them.

2 The **somatic cells** (nonsex cells) of humans are **diploid** ($2n$): they contain two sets of chromosomes, one from each parent. Each chromosome thus has a **homolog**, or mate. Human somatic cells have 22 pairs of **autosomes** (nonsex chromosomes), in which the homologs are nearly identical, and one pair of **sex chromosomes**. The sex chromosomes are similar (XX) in females, dissimilar (XY) in males.

3 **Mitosis** is a process of cell division in which replicated chromosomes divide equally, so that each daughter cell is genetically the same as the parental cell. Replication of the chromosomes occurs in the **interphase** between cell divisions. A replicated chromosome consists of two **sister chromatids**, attached at the **centromere;** these separate during mitosis to become daughter chromosomes. Human somatic cells are derived from the zygote by repeated mitosis.

4 **Meiosis** is a process in which a diploid cell divides twice to produce four **haploid** cells, each with a single set of chromosomes (n). In meiosis I, each replicated chromosome pairs with its homolog; the two then separate and go to different cell nuclei. In meiosis II, the sister chromatids separate and the daughter chromosomes go to different nuclei. The behavior of chromosomes during meiosis explains the segregation of alleles and the independent assortment of different allelic pairs.

5 Human gametes are generated by meiosis. **Spermatogenesis** produces four haploid **sperm** cells; **oogenesis** produces one haploid **ovum** and three haploid **polar bodies**. **Fertilization** restores the diploid number in the zygote and **recombines** parental genes in a unique way.

6 The study of human chromosomes was greatly advanced by the use of **hypotonic solutions** to separate chromosomes and by the perfection of techniques for human cell culture. These advances enabled cytologists to show that the human diploid number is 46.

7 Human chromosomes are numbered according to length and centromere position. Recent staining techniques called **banding** allow cytologists to distinguish clearly among chromosome pairs and to identify small regions of individual chromosomes.

8 The human chromosome is composed of the genetic material DNA and proteins called **histones**. A continuous DNA strand is wound around groups of histones to form units called **nucleosomes**. The string of nucleosomes coils to form the chromosomal bands.

Key Terms and Concepts

Chromatin	Cell cycle	Autosome
Chromosome	Interphase	Spermatogenesis
Metacentric chromosome	Nucleolus	Oogenesis
Acrocentric chromosome	Prophase	Spermatogonia
Teleocentric chromosome	Spindle apparatus (Spindle fibers)	Spermatocytes, primary and secondary
Chromatid	Centrioles	Spermatid
Centromere	Metaphase	Mature sperm
Somatic cell	Metaphase plate	Ovarian follicles
Mitosis	Anaphase	Oogonium
Daughter cell	Sister chromatid	Oocytes, primary and secondary
Reciprocal cross	Daughter chromosome	Polar body
	Telophase	

Cytoplasm

Diploid

Homolog, homologous
   chromosome

Meiosis

Haploid

Reduction division

Equational division

Cytokinesis

Synapsis

Tetrad

Bivalent

Chiasma

Crossing over

Chromosome theory
   of inheritance

Sex chromosomes

Mutation

Genetic recombination

Karyotype

Chromosome banding

Chromosome arm

DNA

Histone

Nucleosome

Problems

3–1 In the following diagrams of cell division, are any of the cells haploid? Explain your answer. What mitotic stages do these cells represent?

(a)

(b)

(c)

3–2 What meiotic stages are represented by the following cells?

(a)

Nuclear membrane
dissolving
(b)

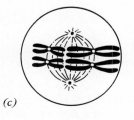
(c)

Nuclear
membrane
forming

(d)

(e)

3–3   A cell has four pairs of homologous chromosomes (*Aa*, *Bb*, *Cc*, and *Dd*). How many different kinds of gametes can this cell produce?

3–4   Below is a diagram of a cell in metaphase of the first meiotic division. When this cell completes meiosis, what meiotic products will form?

3–5   A fruit fly has eight chromosomes in its somatic cells. How many of those chromosomes does it receive from its father?

3–6   If the garden pea has 14 chromosomes, how many different kinds of gametes can it form?

3–7   What is the probability that a human sperm cell will contain only those centromeres contributed to the male by his father?

3–8   A cell has two pairs of chromosomes. (a) How would you distinguish between a mitotic prophase and a meiotic prophase I in this cell? (b) What about a meiotic metaphase I versus a mitotic metaphase? (c) A meiotic metaphase II versus a mitotic metaphase? (d) A meiotic metaphase I versus a meiotic metaphase II?

3–9   What would you conclude about the divisional stage of the cell diagrammed below?

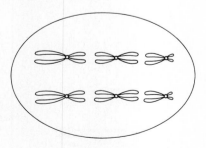

3–10   Although most cells in the human body contain 46 chromosomes, some of the liver cells contain 92 chromosomes. Propose a mechanism for this.

3–11   Would you consider either of the following two cells to be diploid? Explain.

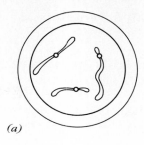

(a)

(b)

3–12 What is the main purpose of meiosis?

3–13 Why is meiosis so genetically important?

3–14 If gametes were produced by the process of mitosis, what consequences would this have for future generations?

3–15 Compare meiosis I with mitosis.

3–16 Compare meiosis II with mitosis.

3–17 If a species is haploid, such as certain fungi, can it undergo meiosis? Explain. Can an individual that is haploid undergo meiosis? Explain.

3–18 Describe the human life cycle.

3–19 Evaluate the following statement: "An egg is hundreds of times larger than a sperm, so it must be genetically more important."

3–20 How many chromosomes would you expect to find in the following human cells (remembering that the human has a diploid chromosome number of 46):
    (a) oogonium
    (b) spermatid
    (c) polar body
    (d) retinal cell
    (e) white blood cell
    (f) secondary spermatocyte
    (g) mature ovum
    (h) skin cell
    (i) brain cell
    (j) primary oocyte

3–21 How many mature sperm cells will be formed by 200 primary spermatocytes? by 200 spermatids?

3–22 How many mature ova will be formed by 200 primary oocytes? by 200 secondary oocytes?

Answers   See Appendix A.

References

Dupraw, E. J. 1970. *DNA and Chromosomes*. Holt, Rinehart, and Winston, New York.

Evans, H. J. 1978. Some facts and fancies relating to chromosome structure in man. *Adv. Hum. Genet.* 8:347–438.

Hulten, M., and J. Lindsten. 1973. Cytogenetic aspects of human male meiosis. *Adv. Hum. Genet.* 4:327–394.

Jenkins, J. B. 1979. *Genetics,* 2nd ed. Houghton Mifflin, Boston.

John, B. 1976. Myths and mechanisms of meiosis. *Chromosoma* 54:295–325.

John, B., and K. R. Lewis. 1975. *Chromosome Hierarchy*. Oxford University Press, London.

Latt, S. A. 1976. Optical studies of metaphase chromosome organization. *Annu. Rev. Biophys. Bioeng.* 5:1–37.

Mazia, D. 1974. The cell cycle. *Sci. Am.* 230:54–64 (Jan.).

Millette, C. F. 1979. Cell surface antigens during mammalian spermatogenesis. *Curr. Top. Dev. Biol.* 13:1–30.

Monroy, A., and A. A. Moscona. 1978. *Mechanisms of Fertilization*. (Curr. *Top Dev. Biol.,* vol. 12.) Academic Press, New York.

Novikoff, A. B., and E. Holtzman. 1976. *Cells and Organelles,* 2nd ed. Holt, Rinehart, and Winston, New York.

Prescott, D. M. 1976. The cell cycle and the control of cellular reproduction. *Adv. Genet.* 18:100–178.

Voeller, B. R. 1968. *The Chromosome Theory of Inheritance*. Appleton-Century-Crofts, New York.

Wang, J. C. 1982. The path of DNA in the nucleosome. *Cell* 29:724–726.

4 Autosomal and Sex-Linked Inheritance

The fragile organic links connecting one generation to the next are the chromosomes. These remarkable biological units are shuffled by meiosis, packaged into specialized cells called gametes, then combined with the shuffled chromosomes of another individual to form a new and unique complement. The transmission of the chromosomes and of the genes they comprise follows a set of principles known as Mendelism, a term that embraces the basic laws of inheritance.

Let us briefly sum up what we have learned of genes and chromosomes so far, at the same time adding a new concept. In diploid organisms, most genes occur in allelic pairs, and each pair controls a given function (usually by producing a protein). In a given individual, the members of the pair may be identical or they may code for alternative expressions of a given trait. The members of an allelic pair are found on homologous chromosomes, one chromosome derived from each parent. Furthermore, they are found at the same **locus** (plural, *loci*), or physical position on the chromosome. Until now we have defined alleles in terms of their function, but we can also define them in more physical terms: alleles are alternative forms of a gene that may be found at a single locus. Note that *single locus* refers to a position on *both* members of a homologous chromosome pair: if a characteristic is controlled by a single *pair* of alleles, it is controlled by a single locus. In the case of genes found on the sex chromosomes, alleles do not always occur in pairs, for in human males the two sex chromosomes (X and Y) are dissimilar, and thus each gene is represented by a single allele.

In this chapter we will continue our discussion of the inheritance patterns of characteristics controlled by a single locus, beginning with patterns determined by genes on the autosomes. These **autosomal inheritance** patterns are commonly of the straightforward Mendelian variety we considered in Chapter 2. But they may also exhibit deviations from Mendelian ratios, and we will consider such variations here. We will then examine the distinctive patterns of **sex-linked inheritance**—the transmission of traits determined by genes on the sex chromosomes. We will also consider the mechanisms of the dominance/recessive relationship and several other aspects of simple (single-locus) inheritance.

Genes on Autosomes

In 1872, a young boy accompanied his physician father as he made his rounds among the isolated villages of rural Long Island. As they passed through the countryside, they chanced upon a small clearing in a heavily forested area, and there came across a scene so terrifying to the young, impressionable boy that he was unable to forget it for the rest of his life. Two women, a mother and daughter, moved through the tall grass that blanketed the clearing. Both were as thin as skeletons, and both were in the throes of contortions and convulsions so violent that they looked almost inhuman. The boy was filled with pity for them.

The boy was George Huntington, and he vowed that day to uncover the cause of the bizarre and terrifying malady he had witnessed. He followed his father's footsteps into medicine and, true to his vow, discovered the genetic basis of the affliction he first saw in the forest clearing. Today the disease bears his name: Huntington's chorea, or (now preferred) **Huntington's disease** (HD). The biochemical basis of the disease is still unknown. It is characterized by a progressive deterioration of the nervous system, which causes uncontrolled movements of the limbs (in Greek, *khoreia* means dance). Eventually the patient becomes helpless and bedridden, and mental deterioration occurs; death comes about 10 years or so after the onset of symptoms. A well-known victim of HD was Woody Guthrie, the folksinger and composer of the 1930s and '40s (and father of another famed folksinger, Arlo Guthrie), who died in 1967.

Young Huntington tracked down the relatives of victims of the disease and found that there was a history of the disease in each family line. The members of an affected line were painfully aware that the disease ran in the family and that some of them would develop it, though they could not, of course, know which ones. All appeared to wait stoically for the first symptoms to appear, symptoms that they knew would spell a death sentence. From his analysis of family pedigrees, Huntington concluded that this disease was caused by a single dominant allele found on one of the autosomes. He thus related his observations to Mendel's newly established principles.

Huntington perceived that the disease is an autosomal trait because it appears with equal frequency in males and females. Furthermore, reciprocal crosses produce identical results: as far as the phenotype of the offspring is concerned, it does not matter which parent contributes which allele. These are features of traits controlled by autosomal genes. As we will see, the situation is strikingly different for traits controlled by genes on the sex chromosomes.

Autosomal dominant inheritance patterns

As you now know, if an allele is phenotypically expressed in both the homozygous and the heterozygous state, we refer to that allele and its associated trait as dominant; if an allele is phenotypically expressed only when it is homozygous, we call the allele and the trait recessive. HD is classed as a dominant trait for several reasons. About half the children of marriages between an affected person and a normal person develop HD. As Figure 4–1 shows, the heterozygous (*HD/hd*) parent with HD produces two classes of gametes in equal numbers, whereas the homozygous (*hd/hd*) normal parent produces only one. Random fertilization results in a 1:1 ratio of *HD/hd* to *hd/hd* offspring, and the *HD/hd* children will develop the disease. As is usual with dominant traits, individuals who express HD have a parent who also has the trait. Dominant traits do not characteristically skip generations, but, through chance alone, they may fail to be transmitted to offspring.

Not all rare dominant traits are as distressing as HD. An interesting but harmless trait that demonstrates dominant inheritance is woolly hair in white persons. This

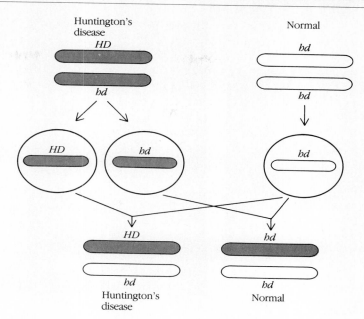

Figure 4–1
The transmission of Huntington's disease, an autosomal dominant trait.

hair has a texture superficially resembling that of the hair of blacks, but it is caused by a rare gene that occurs in certain white families (Figure 4–2). A pedigree of the woolly-hair trait reveals the features of autosomal dominance we have been discussing. Transmission is from affected parents to about half the children, and each affected child had an affected parent. The distribution of sexes among the woolly-haired people is about equal. Furthermore, the distribution of sexes among the affected children of a given couple is not determined by the sex of the parent displaying the trait. All of this suggests that we are dealing with an autosomally dominant trait.

Delayed-onset traits   The woolly-hair trait is classed as a straightforward trait exhibiting simple dominance because it follows all the rules we mentioned and makes its appearance as soon as the hair begins to grow. But HD is different. The vast majority of people with this affliction are young adults when they develop their first clinical symptoms. Many have already married, and many have had children. HD is but one of a number of traits called **delayed-onset traits**. A study of about 800 cases of HD showed that symptoms set in for the first time between the ages of 25 and 60 in over 90% of the cases, with the average age of onset in the early 40s. However, there were also unusual cases in which the symptoms first appeared before the age of 10 and after the age of 70. We do not know the reason for this great variation in the age of onset. So, to construct a correct pedigree for this trait, we must examine individuals who are past middle age, and even then we may misclassify some persons who are genetically destined to have the disease. Compounding the tragedy of a delayed-onset disease like HD is the fact that people often pass on the trait to their children before they know they have it.

Penetrance and expressivity   Another phenomenon that can complicate a dominant inheritance pattern (or a recessive one) is **incomplete penetrance**.

(a)

(b)

(c)

Figure 4–2

(a) A child with the woolly-hair trait. *(b)* In this family, the mother, who has the trait, passed it on to half of her children. *(c)* A pedigree for the woolly-hair trait. Each affected person has at least one parent who has the trait.

Penetrance is expressed as the frequency (in percent) with which a gene or genotype produces a specific phenotype. As an example of incomplete penetrance, consider the inheritance of retinoblastoma, a malignant eye tumor in children. A serious form of retinoblastoma behaves as an autosomal dominant trait. However, in studying pedigrees of this condition, we occasionally observe the skipping of a generation (Figure 4–3). The normal mother (II4) who produced the female child with retinoblastoma must have had the dominant allele for the disease, but for unknown reasons she never developed symptoms. We find this to be true in about 10% of all people carrying this gene. Since 90% of the carriers of the retinoblastoma allele do develop symptoms, we say that the penetrance of this allele is 90%.

In contrast to penetrance is the phenomenon of **variable expressivity**. In the case of a dominant gene showing variable expressivity, every individual carrying the gene may display the trait in question (i.e., penetrance may be 100%), but the *degree* of phenotypic expression varies. An example of variable expressivity is the dominantly inherited trait **neurofibromatosis** (Recklinghausen's disease), a syndrome characterized by a type of abnormal skin pigmentation called *café-au-lait* ("coffee-

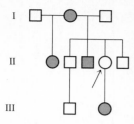

Figure 4–3
A pedigree for retinoblastoma showing incomplete penetrance. Female II4 must be heterozygous, but she does not express the disease.

Figure 4–4
Neurofibromatosis (Recklinghausen's disease). (*a*) Back of a patient with multiple tumors. (*b*) John Merrick, the "Elephant Man," on his admission to the London hospital in 1886.

with-milk") spots and by numerous tumors that develop in the central and peripheral nervous system (Figure 4–4). The film "The Elephant Man" was based on the life of a man named John Merrick who suffered severely from this disease. But though some people with neurofibromatosis have all its symptoms, others may show only one or two skin spots. Thus, neurofibromatosis is an autosomal dominant trait with 100% penetrance, but its expressivity varies. The terms *penetrance* and *expressivity* are helpful in describing certain genetic phenomena, but they do not explain them. Unfortunately, we are usually unable to describe the basis for incomplete penetrance and variable expressivity, other than variation in genetic background and environment.

Autosomal recessive inheritance patterns

As we have discussed, if an allele is not phenotypically expressed in the heterozygote, we call its pattern of inheritance recessive. *Autosomal* recessive traits are expressed only when the alleles specifying them are homozygous. This means that the affected person must receive a copy of the specific allele from each parent. If both parents are heterozygous, they can produce three classes of children:

AA: no trait; noncarrier
Aa: no trait; carrier
aa: expresses the trait

The pedigree of a recessive trait shows that the trait skips generations and that affected persons usually have nonaffected parents. Cystic fibrosis, the tragic affliction described in Chapter 1, is an example of an autosomal recessive trait; a pedigree of this trait is seen in Figure 4–5. We can conclude from this pedigree that individuals III1 and III3 were both carriers of the rare recessive allele.

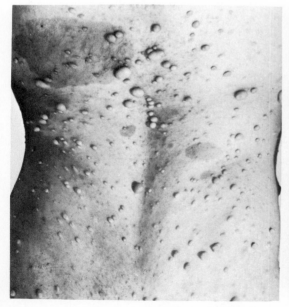

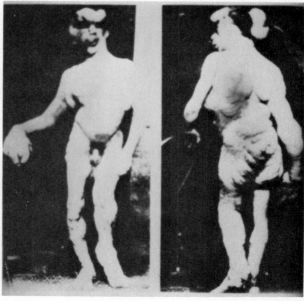

(*a*)   (*b*)

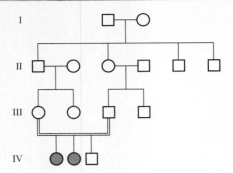

Figure 4–5
A pedigree for cystic fibrosis.

There are a number of other autosomal recessive traits that we have mentioned or that you may be familiar with. These include albinism (Figure 4–6), total color blindness (achromatopsia), and phenylketonuria (PKU), a metabolic disease associated with mental retardation.

Genetically isolated populations Although a recessive allele may be exceedingly rare in the general population, it may occur with a very high frequency in a smaller group whose members usually intermarry. An example is **Tay–Sachs disease**, a degenerative disease of the nervous system caused by an enzyme deficiency. The symptoms first appear around six months of age, and afflicted children undergo rapid deterioration, becoming blind, mentally retarded, and in other ways incapacitated. Death usually comes in early childhood. In the general population, the trait appears about once in every 400,000 births. However, in Ashkenazic Jews (Jews of Central and Eastern European origin, including most American Jews) the frequency of Tay–Sachs disease was until recently over 100 times as high—about one in every 3600 births. The incidence is now much lower owing to the detection of carrier parents and the detection and abortion of affected fetuses.

What makes the frequency of the Tay–Sachs allele so high in Ashkenazic Jews? Though we are unable to answer this question with certainty, we can suggest some

Figure 4–6
A man and young girl with albinism.

contributing factors. For unknown reasons, in the past the heterozygous carriers of the Tay–Sachs allele appear to have had more children than noncarriers. In other words, marriages involving at least one heterozygote (*Aa*) seem to be more fertile than *AA* × *AA* marriages. This discrepancy would cause the Tay–Sachs allele to multiply faster than the normal allele in a population in which its frequency was already fairly high. Interestingly, this same heterozygote advantage is also seen in the predominantly Northern European carriers of the cystic fibrosis allele.

Another factor that may have served to increase the frequency of Tay–Sachs disease in the U.S. Ashkenazic population was the great number of cousin marriages among the early immigrants. Such marriages increase the likelihood of two heterozygotes mating. But today the Tay–Sachs allele appears so often among American Ashkenazic Jews that matings between cousins are no longer a significant factor in the high frequency of children with the disease. That is, a carrier now has about the same chance of mating with another carrier when marrying a nonrelative as when marrying a relative. The Ashkenazic parents of affected children are usually not closely related; but when Tay–Sachs disease appears in the non-Jewish population, we almost always find that there has been a cousin marriage.

There are other instances of rare recessive traits that are exceptionally frequent in small, genetically isolated groups: for example, albinism in certain clans of Hopi Indians, and PKU in certain Mormon communities in Central Utah. Table 4–1 summarizes some genetic disorders commonly found in certain Caucasian ethnic groups.

Table 4–1   Some genetic disorders especially prevalent in Caucasian ethnic groups

Group/ Genetic Disorder	Phenotypic Symptoms
Caucasians in general	
Alkaptonuria	Incomplete metabolism of the amino acids tyrosine and phenylalanine (urine is dark after being alkalinated); pigmentation of cartilage
Congenital erythropoietic porphyria	Skin lesions in infants; skin rashes and pink stains on teeth; headaches and general weakness; sensitivity to barbiturates
Cystic fibrosis	Respiratory problems; lower life expectancy (see Chapter 7)
Oculocutaneous albinism	Poor vision; sensitivity to sun; increased risk of skin cancer
Rh-negative blood type	Hemolytic disease of some newborns
Tay–Sachs disease	"O-variant" form of Tay–Sachs disease affecting non-Jews
Ashkenazic Jews (from Central, Eastern, and Western Europe, North and South America, and Australia)	
Diabetes mellitus	In adults, inhibited insulin action; in juveniles, failure of beta cells to secrete sufficient insulin; elevated sugar levels
Hyperuricemia (gout)	Abnormal levels of uric acid in blood

(continued on p. 86)

Group/ Genetic Disorder	Phenotypic Symptoms
Ashkenazic Jews (continued)	
Familial dysautonomia (Riley–Day syndrome)	Deterioration of autonomic nervous system; insensitivity to some pain such as burns and puncture wounds; drooling and excessive perspiration; low intelligence; tongue without taste buds
Pemphigus vulgaris	Occurrence of blisters that heal suddenly but do not scar
Tay–Sachs disease (amaurotic idiocy)	Red spots in eye macula; after 6–9 months, rapid losses in motor skills and vision because of an accumulation of lipids; death usually between 2 and 4 years
Sephardic Jews (from areas of the Mediterranean, including parts of Turkey and the Mediterranean Sea area; also, USA)	
Cystic disease of the lungs	Occurrence of cysts in lungs
Oriental Jews (from Asia Minor, Iran, Iraq, Yemen)	
Bloom's syndrome	High risk of leukemia; red blotches on face; increased sensitivity to light; narrow head; reduced immunoglobulin
Mediterranean peoples (Armenians, Greeks, and Italians)	
Thalassemia major (Cooley's anemia)	Insufficient beta chain hemoglobin synthesis; often, premature death
Northern Europeans	
Lactose intolerance	Inability to digest lactose
Irish	
Encephalocele	Protrusion of brain tissue through cranial fissure
Amish, Icelanders	
Ellis–van Creveld syndrome	Dwarfism so that limbs and appendages are short and body normal; polydactyly; heart defects in about half of victims
Eskimos	
Kushokwin	Protein deficiency and bone deformities

Inbreeding   There are situations that may lead us to confuse a dominant inheritance pattern with a recessive one. Say an affected person mates with an unaffected, presumably unrelated person, and half their children are affected. This ratio would ordinarily indicate a dominant mode of inheritance, but on a more careful analysis of the pedigree we may find that the presumed unrelated person is in fact a relative and is thus a carrier of the recessive allele. A case in point involves a recessive trait called alkaptonuria, the rare metabolic disorder in which the urine turns black on exposure to oxygen. In 1956, a family was discovered that seemed to express alkaptonuria as a dominant trait with incomplete penetrance (Figure 4–7). This discovery flew in the face of all that was known about the disease. All other pedigrees for alkaptonuria indicated a classic recessive mode of inheritance. How could it now turn up as a dominant trait? In 1959, however, the female mate of an affected person IV2 in Figure 4–7 was found to be a relative and hence a carrier of the recessive allele.

Genes on Sex Chromosomes

Genes located on the sex chromosomes are technically called sex-linked. Such genes are not distributed equally to males and females, since females have two X chromosomes and males have one X and one Y. Because of this unequal distribution of sex chromosomes and their associated genes, sex-linked inheritance patterns differ significantly from autosomal inheritance patterns. But the vast majority of sex-linked genes that have so far been identified are X-linked, so when we say sex-linked we usually mean X-linked. Anything carried on the Y chromosome will be directly designated "Y-linked."

X-linked dominant inheritance patterns

A dominant X-linked allele presents an inheritance pattern clearly distinct from that of both autosomal genes and sex-linked recessive alleles. Suppose A is an X-linked allele for a dominant trait, and a is its recessive allele. Representing these alleles as X^A and X^a, we can show both the sexes and the genotypes of the offspring of any cross. Consider a cross between a female homozygous for a (X^aX^a) and a male **hemizygous** for A (X^AY). "Hemizygous" refers to the presence of a single allele rather than a pair; it is evident that a male must be hemizygous for all X-linked genes

Figure 4–7

A pedigree for alkaptonuria. At first it appeared to follow a dominant inheritance pattern. Then it was discovered that the female married to IV2 was a relative and probable carrier.

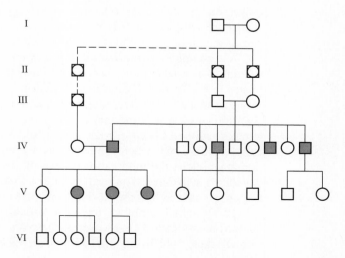

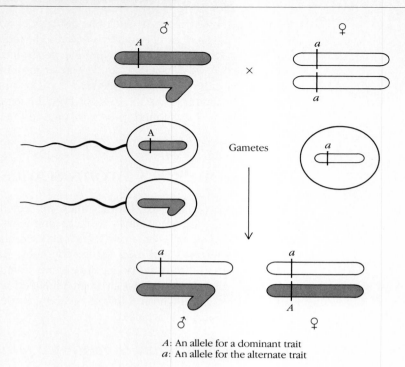

Figure 4–8
When the male is hemizy-
gous for a dominant trait
(*A*), it is transmitted from
father to daughter.

A: An allele for a dominant trait
a: An allele for the alternate trait

(Figure 4–8). The male parent in this cross is phenotypically *A*; that is, he displays
the trait in question. But all the male offspring of this cross are hemizygous for *a*
(X*ªY), so they do not express the trait. All the female offspring are heterozygous and
phenotypically *A*.

The reciprocal cross produces different results (Figure 4–9). In this cross, all
the offspring are phenotypically *A*. This pattern is quite different from one involving
autosomal alleles, where reciprocal crosses always produce identical results. The
reason for the difference, of course, is that reciprocal crosses involving sex-linked
alleles are symmetrical only with respect to phenotype; genotypically, the female
parent is homozygous and the male is hemizygous.

An example of a trait that follows an X-linked dominant mode of inheritance is
vitamin D–resistant rickets, also known as hypophosphatemia (deficiency of phosphates in the blood). From the pedigree in Figure 4–10, we see that (1) affected
males produce all affected daughters and normal sons; and (2) affected heterozygous females transmit the trait to half their children without regard to sex. If the
females had been homozygous, they would have transmitted the trait to all their
children. Note that when a female has a homozygous or heterozygous X-linked
dominant allele, the inheritance pattern in the offspring is indistinguishable from
that of an autosomal dominant. We must look at the offspring of affected males
before we can say whether a dominant trait is autosomal or X-linked.

For X-linked dominant traits, we usually find that males are more severely
affected than females. One reason for this is that the heterozygous female has one
normal allele, which may partially compensate for the debilitating effects of the

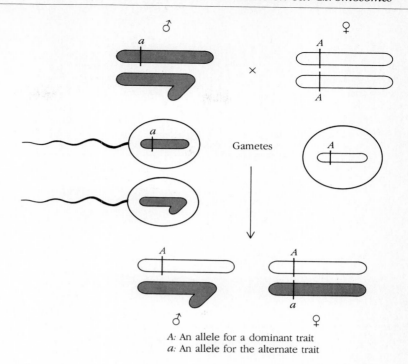

Figure 4–9
When the female is homozygous for a dominant trait (*A*), it is transmitted to both her male and her female children.

A: An allele for a dominant trait
a: An allele for the alternate trait

abnormal allele. Many X-linked dominant traits are lethal in males and nonlethal in females, as we see in the case of the severe skin disorder called incontinentia pigmenti (Figure 4–11). Only females are born with this defect; males appear to be aborted.

X-linked recessive inheritance patterns

X-linked recessive traits in both sexes produce distinctive inheritance patterns in the offspring. All males who carry the allele express it, whereas females do not, unless they are homozygous. Thus, rare X-linked recessive traits are found almost

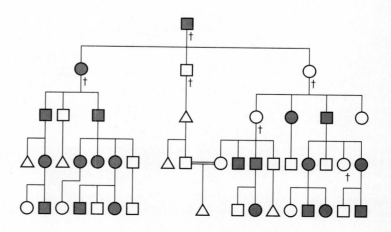

Figure 4–10
A pedigree for vitamin D–resistant rickets, an X-linked dominant trait.

Figure 4–11
The pedigree for incontinentia pigmenti, an X-linked dominant, suggests the gene is lethal in males and non-lethal in females.

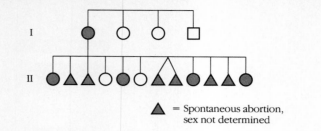

▲ = Spontaneous abortion, sex not determined

Figure 4–12
(*a*) A sex-linked recessive trait is passed from father to daughter but is not expressed in any of his children. The daughters are, however, carriers. (*b*) If the mother is homozygous for the sex-linked recessive trait, she will pass it to her sons, who will express it. Her daughters will be carriers.

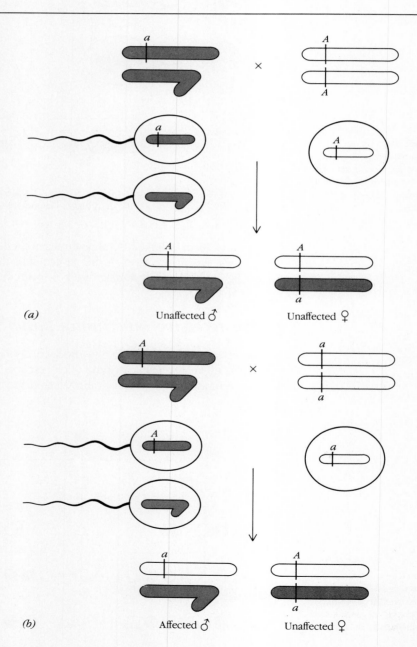

exclusively in males. Consider the crosses shown in Figure 4–12. There is only one way to produce an affected female: a female who carries at least one allele for the trait must mate with an affected male. If she is heterozygous, half the children of both sexes will be affected (Figure 4–13).

A bizarre example of an X-linked recessive trait is the one that causes the self-mutilating behavior pattern known as the **Lesch–Nyhan syndrome**. The allele responsible for this condition is mutant, so it does not produce its normal gene product, an enzyme called hypoxanthine–guanine phosphoribosyltransferase (HGPRT). Hemizygous males totally lack this enzyme. Homozygous females are not known to exist. The absence of HGPRT leads to a metabolic disorder that, among other things, causes the person to exhibit a range of compulsive, hyperexcited behaviors, including the biting off of fingertips, toes, and lips (Figure 4–14). The trait is rare and seems to affect only males. This is as expected, since affected males never reproduce. From this pedigree, we see a classic pattern of an X-linked recessive trait: (1) the trait appears exclusively in males; (2) the trait is passed from a carrier female to half of her sons. Other examples of X-linked recessive traits you may be familiar with are hemophilia and red–green color blindness.

Y-linked inheritance

Genes located on the Y chromosome follow a **holandric**, or Y-linked, inheritance pattern: they are always passed from father to son, and they never appear in females. Although nobody doubts that the Y chromosome carries some genes, at present we can point to only one gene that appears to be Y-linked. This is the H-Y antigen gene, which initiates male development; we will discuss it in detail a little later. Another possibility is a gene that causes hairy pinnae, or auricles (Figure 4–15), but recent studies of this trait show that the location of this gene on the Y chromosome is quite uncertain.

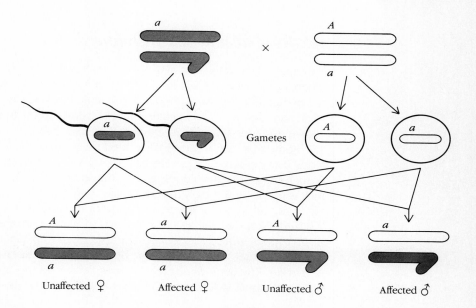

Figure 4–13
When a carrier female mates with a hemizygous male who has a sex-linked recessive trait, half of her offspring (both male and female) will have the trait.

Gametes

Unaffected ♀　　Affected ♀　　Unaffected ♂　　Affected ♂

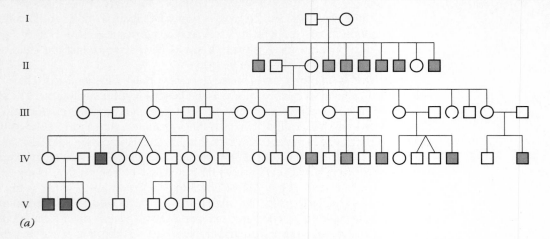

(a)

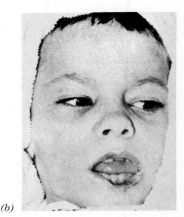

Figure 4–14
Lesch–Nyhan syndrome. (*a*) The pedigree shows the trait to be sex-linked recessive. (*b*) A child with Lesch–Nyhan syndrome. Note the damage to lower lip.

(b)

Figure 4–15
Hairy pinna. It is now in doubt that this trait is caused by a Y-linked gene, as has been thought.

Sex-limited and sex-influenced traits

There are some traits that may appear to be sex-linked but are in fact autosomal. When a gene, whether X-linked or autosomal, is expressed in only one sex, we say it is **sex-limited**. Hydrometrocolpos, a condition in which the uterus falls into the vagina and fills with fluid (Figure 4–16a), is caused by an autosomal gene and, of course, is expressed only in females (Figure 4–16b).

Some traits, called **sex-influenced** traits, are expressed differently in the two sexes. The pedigree for pattern baldness (Figure 4–17b) illustrates the inheritance of such a trait. Pattern baldness may be transmitted directly from father to son, which means that it cannot be X-linked: a male contributes no X chromosome to his son. Since not all sons are affected, it appears that the trait is autosomal rather than Y-linked. However, the trait is expressed in males who are either homozygous or heterozygous, whereas in females it is expressed only in the homozygous state. We can state this another way: the gene behaves as a dominant in males and a recessive in females. The determining factor for expression of this trait is the hormonal environment, which differs in the two sexes.

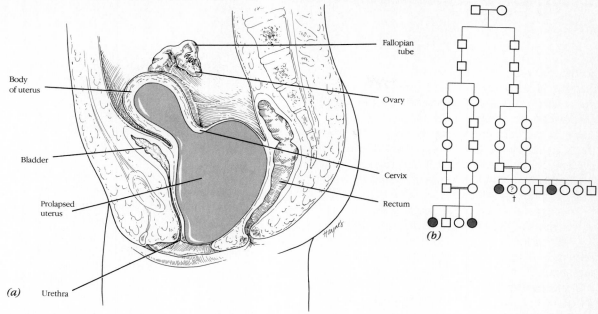

(a)

(b)

Figure 4–16
(*a*) Diagram of hydrome-trocolpos, a sex-limited trait. The uterus is prolapsed into the vaginal canal and has fluid accumulation. (*b*) A pedigree for hydrometro-colpos in an inbred family.

Sex Determination in Humans

The most obvious characteristic associated with sex chromosomes is sex itself—an individual's maleness or femaleness. But simply knowing that XX is female and XY is male does not tell us anything about how sexual development is actually determined, and a great deal of research has been devoted to clarifying this matter. Curiously, though we have learned quite a lot about the biochemical basis of sex differences, we are still not entirely certain about the relationship between sex-determining biochemistry and sex chromosomes.

(text continues on p. 96)

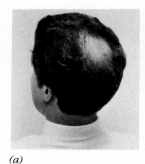

(a)

Figure 4–17
(*a*) Pattern baldness, a sex-influenced trait. The heterozygous male is bald, but the heterozygous female is not. (*b*) A pedigree of pattern baldness. Notice that only males show it in this pedigree.

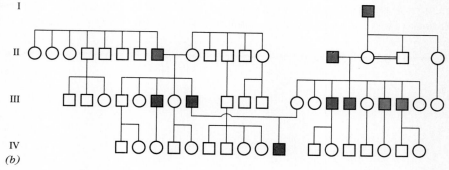

(b)

Box 4–1   Porcupine Men—A Y-Linked Gene?

So far in this chapter we have seen the different patterns of inheritance shown by genes carried on the autosomes versus those carried on the sex chromosomes. We have seen that the Y chromosome carries very few known genes at all, whereas the X chromosome is known to carry dozens. It is evident that any genetic trait that is carried on the Y chromosome would be of some interest just for that fact alone—and, indeed, several traits have been suggested. One of the most spectacular involves the so-called porcupine men of England.

The printed record begins in 1732, when the secretary of the Royal Society of London reported the appearance of a father and son at one of the society's meetings. The father was normal; the boy, about 14 years of age, was not. The father is quoted as saying that the boy's "skin was clear at his Birth as in other Children, and so continued for about seven or eight Weeks, after which, without his being sick, it began to turn yellow, as if he had had the Jaundice; from which by degrees it changed black, and in a little time afterwards thickened, and grew into that State it appeared in at present."

Edward Lambert, as the boy was named, continued to be a public sensation. By 1755 he had been dubbed a "porcupine man," and a myth began to build: ". . . he has had six children, all with the same rugged covering as himself. . . ," ". . . a race of people may be propagated by this man . . . 'tis not improbable they might be deemed a different species of mankind." Edward and his son, also named Edward and also affected, were publicly displayed. The handbills advertising them proclaimed, "Their solid Quills so numerous as not to be credited till seen. . . ," and the two

were said to be "cover'd from Head to Foot with solid Quills, except their Face, the Palms of their Hands, and Bottoms of their Feet."

A later member of the Lambert family, also displayed publicly for profit, was described like this in a handbill:

This young man is 30 years of age, covered with scales, with the exception of the face, soles of the feet, and palms of the hand, which are like those of any other man. These Scales, nearly half an inch long, are so hard and firm, that with a touch of the finger they make a sound like stones striking together; those on the stomach are short, round, and distant; those on the arms, on the contrary, approach each other like the bristles of an hedge-hog. . . . He speaks French, German, Italian and English extremely well, and will answer every question to Visitors. The great-grandfather of the singular family to which this young man belongs, was found savage in the woods of North America. The peculiarity descends only in the male line.

An even later addition to the publicity for the growing sensation took a more titillating approach: "Where the propagation of men brought into the world thus cased, is to end, —who shall tell? The biped armadilloes find no difficulty in wiving. Nay, the present Mr. Lambert says with regard to his skin, that his good lady 'rather likes it of the two.'"

Eventually, a formal pedigree of the family and the porcupine trait was published. It appeared in many early texts as an example of a Y-linked trait, one which "descends only in the male line." We reproduce it here (the question mark in the diamond designates an unknown number of normal sibs):

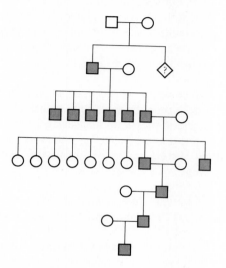

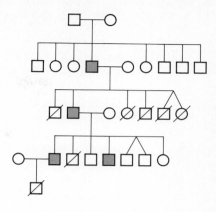

Obviously, if the pedigree is accurate, a strong case is made for linkage of the mutant gene to the Y chromosome: it appears in all males, appearing in no females, and is not transmitted by any of the females to their offspring (the families of the females in the pedigree were not shown, as all were said to be normal). But was the pedigree really accurate? The sensational nature of the case, in regard to both linkage and phenotype, prompted an intensive investigation by two distinguished geneticists, Lionel Penrose and Curt Stern. Their verdict? The original pedigree was inaccurate. The following is the most accurate (though considerably simplified) pedigree that we can construct from the relevant secular and parish records:

The oblique lines in the open symbols designate persons who were reported to have been affected with the porcupine trait, but for which solid documentation is not available. Only four "porcupine men" could definitely be established, and the trait has subsequently been lost in the family. One notices immediately, however, that some sons of affected males were *not* affected, and that some of the females in the family apparently *were* affected; both observations are incompatible with a suggestion that the gene is carried on the Y chromosome.

The condition seems likely to not be confined to the Lambert family, because other families are now known with similar but less spectacular manifestations. The known cases are clearly caused by an autosomal dominant gene; it is almost certain that the Lambert family pedigree represents only a particular allelic mutation of the gene. Spectacular it was; Y-linked it was not. And the entire matter serves as an eloquent example of the need for precise and well-documented research.

The H-Y antigen and sex differentiation

During the first month of development, the human embyro is sexually indifferent. It has primordial undifferentiated gonads, which can develop into testes or ovaries, and two potential reproductive tracts, male and female, only one of which will normally develop. The external genitals at this stage are potentially either male or female. Sex appears to be largely determined by a substance called the **H-Y antigen**. If this antigen is present, the embryonic gonads become organized into testes; if it is not, they become ovaries. Further sex differentiation—development as a male or a female—depends on the hormonal environment created by the presence or absence of testes.

The H-Y antigen, first described in mice nearly 30 years ago, is probably coded for by a gene on the Y chromosome. It is secreted by the undifferentiated gonad of XY embryos and attaches to specific receptor sites found on the gonadal cells of both sexes. The H-Y antigen stimulates the undifferentiated gonads to differentiate into testes. The developing testes then secrete male sex hormones (known as **androgens**), which include the following:

1 Testosterone stimulates the development of the wolffian duct system into the internal male reproductive tract (vas deferens, seminal vesicles, and other structures).
2 Dihydrotestosterone, which is enzymatically derived from testosterone, promotes the development of the male external genitals (penis and scrotum).
3 Müllerian-inhibiting hormone promotes the degeneration of the müllerian duct system, a system that produces the internal female reproductive system (fallopian tubes and uterus).

Figure 4–18 summarizes the development of the internal reproductive duct systems of the male and female. The development of the XX embryo as a female depends primarily on the absence of the H-Y antigen and androgens rather than on the presence of specific female hormones, for the ovary does not produce much of any hormone during embryonic life. The müllerian, or female, duct system develops into the uterus and fallopian tubes, while the external genitals become the clitoris and labia (Figure 4–19); a vagina forms between the internal and external parts of the reproductive system. In both sexes, the alternate internal duct system degenerates. In postnatal life, the ovaries produce the female hormones estradiol and progesterone, which are responsible for the development of female secondary sex characteristics during puberty and for the onset and maintenance of the menstrual cycle. Both "male" and "female" hormones are actually present in both sexes, although in very different proportions.

Figure 4–18
(Facing page.)
Development of male and female duct systems from the sexually indifferent stage. Notice that at seven weeks, the embryo has all the structures necessary to develop into both male and female.

Anomalies of sex differentiation

We have been able to piece together this picture of human sex differentiation in part because of the discovery of two groups of people who are genetic males (XY) but have the external appearance of females. In one group, XY individuals lack the enzyme that converts testosterone to dihydrotestosterone. Such people have a male internal reproductive tract because of the influence of testosterone, but they

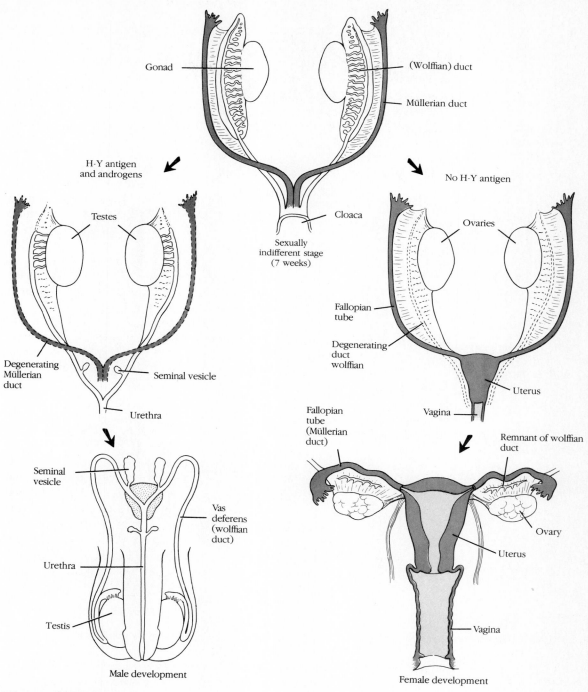

Embryonic development of male and female internal reproductive structures.

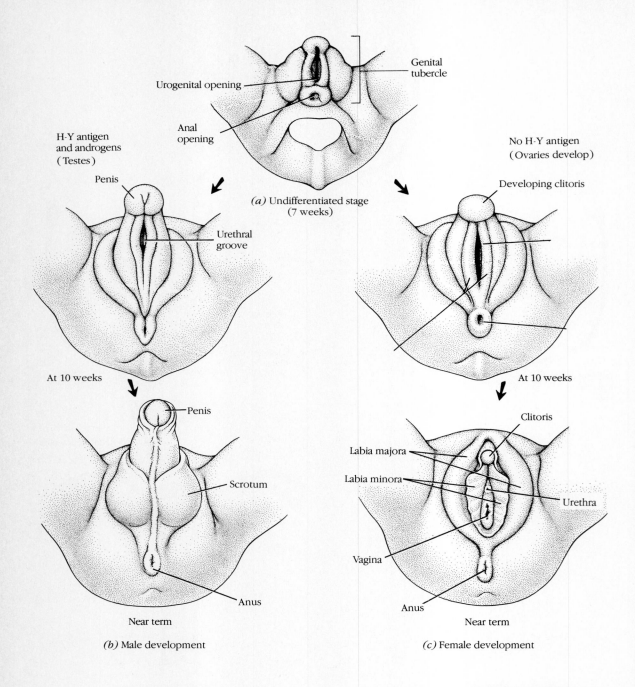

Genital
tubercle

Urogenital opening

Anal
opening

H-Y antigen
and androgens
(Testes)

Penis

Urethral
groove

At 10 weeks

(a) Undifferentiated stage
(7 weeks)

No H-Y antigen
(Ovaries develop)

Developing clitoris

At 10 weeks

Penis

Scrotum

Anus

Near term

(b) Male development

Clitoris

Labia majora

Labia minora

Urethra

Vagina

Anus

Near term

(c) Female development

Embryonic development
of male and female
external reproductive
structures.

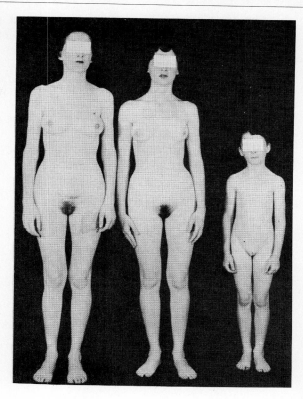

Figure 4–20
Testicular feminization syndrome. The three "sisters" in this photo, ages 25, 21, and 8, are all 46 XY genetic males—phenotypic females. All three have intraabdominal testes and blind, shallow vaginas. The syndrome is the result of an X-linked dominant gene (*Tfm*).

lack male external genitals because of the missing dihydrotestosterone, and the reproductive tract ends in a vagina. This anomalous condition appears to be inherited as an autosomal recessive.

A second group expresses the trait called **testicular feminization syndrome** (Figure 4–20). People with this trait are genetic males who develop into perfectly normal-looking females. Yet they have a blind vagina (a condition that can be surgically corrected) and testicles located in the outer labia, the inguinal canal, or the abdomen. They never menstruate and usually lack underarm and body hair and have sparse pubic hair. As adults, they generally have well-developed breasts and very feminine contours, though often with longer than average legs. Some of these persons are even employed as high-fashion models.

The cause of the testicular feminization syndrome seems to be a dominant gene on the X chromosome, *Tfm*, that alters cell membranes so that cells cannot respond to testosterone. In XY individuals who carry this gene, H-Y antigen is produced, testicles develop, and male hormone levels are normal. But the hormones must bind to specific areas on cell surfaces to promote male development, and if these areas are defective because of the *Tfm* gene, the hormones do not bind and female characteristics develop. This is an example of an X-linked sex-limited trait. A female may have the *Tfm* gene, but because her embryonic sexual development does not depend on testosterone reaching her cells, she is not affected by it.

It is evident from these cases that the presence or absence of androgens is critical to the determination of an individual's sex. Testicular development and androgen production are initiated by H-Y antigen in the XY embryo, but further male

Figure 4–19
(Facing page.)
Development of human external genitals.

Figure 4–21
A summary of human sexual differentiation.

differentiation can be blocked by genes that interfere with androgen production or effectiveness at various stages of development. Whenever such interference occurs, development continues in a female direction. Figure 4–21 summarizes human sexual development.

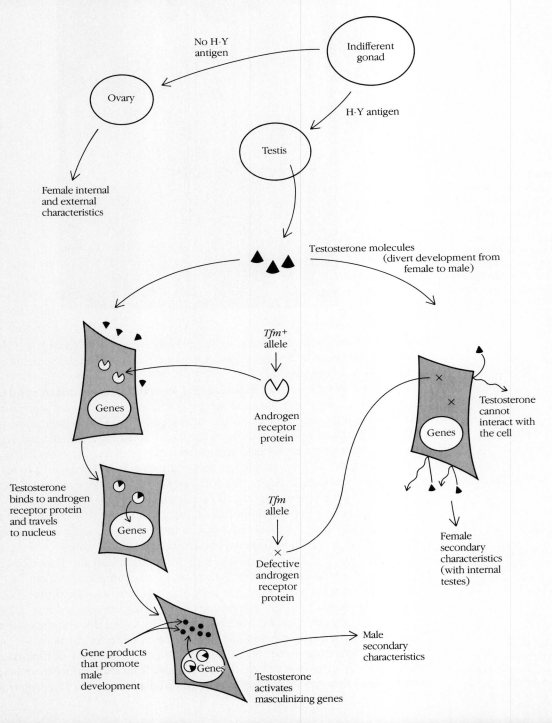

The H-Y gene: Sex-linked or sex-limited?

If the H-Y antigen initiates male development, it would seem reasonable that the gene that codes for production of this antigen should be on the Y chromosome. In fact, it now seems very likely that the H-Y gene or an H-Y regulator gene is located on the short arm of the Y chromosome. But some researchers dispute this, for we have recently discovered rare cases of XX individuals who are completely normal phenotypic males! Even though some of these persons carry a tiny piece of Y translocated onto another chromosome, this discovery has led to the suggestion that the H-Y gene is located on an autosome or the X chromosome, but that its activity is controlled by a gene on the Y chromosome. According to this idea, a gene on the Y chromosome gives rise to a product that acts as a "switch" for the H-Y gene located elsewhere. In the case of the XX genetic female/phenotypic male, the H-Y gene may have been activated by another gene product that through mutation came to resemble the switch product of the Y-linked gene. Alternatively, the H-Y gene itself may have mutated so that it was in a switched on position.

Those who favor the idea that the H-Y gene is on the Y chromosome interpret the situation differently. They suggest that an XX genetic female/phenotypic male carries a small segment of the father's Y chromosome—a segment bearing the H-Y gene—on the X chromosome or on an autosome. Such translocations of segments from one chromosome to another certainly do occur during meiosis (we will discuss them in Chapter 6), but so far it has not been possible to detect any translocated Y material in all XX phenotypic males.

If the H-Y gene is on the Y chromosome, H-Y antigen production is a Y-linked trait. But if it is on another chromosome, and is merely switched on or regulated by a gene on the Y chromosome, H-Y antigen production is a sex-limited trait. Females, in other words, would also have the H-Y gene but would not normally be affected by it.

The Meaning of Dominance

Until now we have discussed traits as though they always fit neatly into either a dominant or a recessive inheritance pattern. But the members of an allelic pair can exhibit other types of relationships to each other. In fact, dominant and recessive are relative conditions, and the distinction between them is not always entirely clear. An allele we classify as dominant because it produces a definite effect in the heterozygous state may produce a still more pronounced effect when it is homozygous. Likewise, an allele we classify as recessive because it produces a noticeable effect only when it is homozygous or hemizygous may actually have a subtle expression in the heterozygote. You can see that our classification of a trait as dominant or recessive may depend on how much of an intermediate effect we perceive in the heterozygote.

Intermediate inheritance

When the phenotype of the heterozygote is clearly intermediate between the two homozygous phenotypes, the inheritance pattern is called **intermediate inheritance** or incomplete dominance. A clear example is flower color in snapdragons, where a cross between a red-flowered plant and a white-flowered one produces a

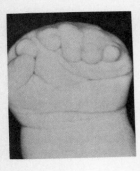

Figure 4–22
Severe hand deformity caused by homozygosity for the camptobrachydactyly allele.

pink-flowered F₁. The reason for this intermediate condition is simple: a plant with two alleles for redness produces twice as much red pigment as a plant with only one. Such cases of intermediate phenotypes are also observed in humans. If we examine the traits we call dominant and recessive carefully, we find that intermediate conditions commonly exist.

In traits classed as dominant In the case of abnormal dominant traits, we are often aware only of the heterozygous condition. This is partly because the allele for such a trait is usually rare, so the mating of heterozygotes that could produce a dominant homozygote is an extremely unusual event. In the case of woolly hair, for example, we do not know of any instance where two people displaying the trait married. In addition, a double dose of an allele that produces severe abnormalities is often lethal, so that the homozygous individuals that do occur are spontaneously aborted or die very early. For these reasons, we usually see people who may actually be intermediate in phenotype.

On rare occasions a person homozygous for a dominant trait does survive and proves to be much more severely afflicted than the heterozygote. Such a case was reported in 1972 in a family that expressed the dominant trait known as **camptobrachydactyly**. The allele that causes this affliction results in severe hand and foot malformations (Figure 4–22) as well as vaginal and urinary problems. A look at the pedigree (Figure 4–23b) will reveal that there was a first-cousin marriage

Figure 4–23
(*a*) Hand of a person with camptobrachydactyly. (*b*) A pedigree for camptobrachydactyly. Half-solid symbols are affected persons. Solid symbols are severely affected and presumed to be homozygous.

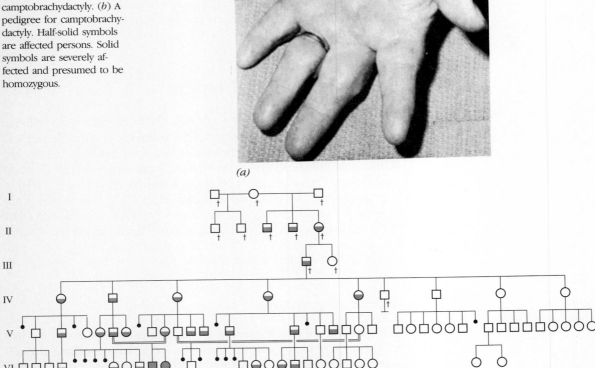

(*a*)

(*b*)

between two affected individuals (V5 and V8). This mating produced two severely affected children (VI8 and VI9), both of whom are probably homozygotes for the dominant allele. In addition to having severely malformed hands and feet, they are mentally retarded, are deaf, and suffer from a wide range of other defects.

Though we call camptobrachydactyly dominant, its inheritance pattern is just like that of color in snapdragons or carnations (see Figure 2–7). Twenty-five percent of the offspring of heterozygotes are expected to be dominant homozygotes (*DD*) and severely affected, 50% to be heterozygotes (*Dd*) and moderately affected, and 25% to be recessive homozygotes *(dd)* and normal. One of the tragedies of the case just described is that, after the birth of their first child, the parents obtained genetic counseling and were told that any future children had a 75% chance of being affected to some degree, yet they still proceeded to have another child, who was severely deformed.

Most aberrant traits classified as dominant involve proteins that are structural rather than enzymatic in their function. Suppose a normal allele, A^1, codes for a structural component of a cell—say, a cell membrane protein—and its allele, A^2, codes for an altered form of this protein (Figure 4–24a). The genotype A^1A^1 produces all normal protein and hence a normal phenotype, whereas A^2A^2 produces all abnormal protein and an abnormal phenotype. The A^1A^2 heterozygote produces both kinds of protein, but since all of such a person's cells carry some of the abnormal protein, the person will probably express the abnormal, or A^2, phenotype. We therefore call the abnormal allele dominant (*A*) and the normal one recessive (*a*). However, the abnormal condition may be somewhat neutralized in the heterozygote by the presence of the normal protein, so that the phenotype is not as extreme as in the dominant heterozygote. An explanation of this sort may apply to a condition like camptobrachydactyly.

In traits classified as recessive

Most abnormal recessive traits involve genes that code for enzymes. Here obvious intermediate forms are absent because a single dose of a normal enzyme is usually adequate to produce normal functioning. Suppose that allele A^1 codes for an enzyme that catalyzes a specific biochemical reaction and its allele, A^2, codes for a nonfunctional variant of the enzyme (Figure 4–24b). The A^1A^1 genotype produces 100 units of the normal enzyme: A^1A^2, 50 units; and A^2A^2, no units. The A^1A^2 heterozygote is clearly intermediate in enzyme production, just as the heterozygous snapdragon is intermediate in pigment production. But if 50 units of enzyme can catalyze the biochemical reaction at a rate that ensures the organism of normal, or nearly normal, functioning, the heterozygous phenotype is normal. We therefore classify the A^2 allele as recessive (*a*) and the normal A^1 allele as dominant (*A*).

An example of this situation is **galactosemia**, a serious metabolic disease that causes mental retardation, cirrhosis of the liver, eye cataracts, and other problems. It is classified as a recessive disease because the gene must be homozygous for the disease to develop. Phenotypically, *AA* is normal, *Aa* is normal, and *aa* results in galactosemia. The cause of the disease is the lack of an enzyme called galactose-1-phosphate uridyltransferase (GPUT). The *aa* genotype produces no GPUT; the *AA* genotype produces *x* units of GPUT; and the *Aa* genotype produces 0.6*x* units of GPUT. In other words, the effects of the genes are more or less additive with regard to the gene products, and the heterozygote from this point of view is intermediate in

Alleles code for structural proteins

A^1 normal protein
A^2 variant protein

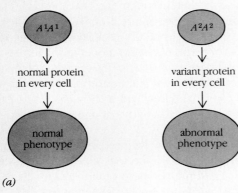

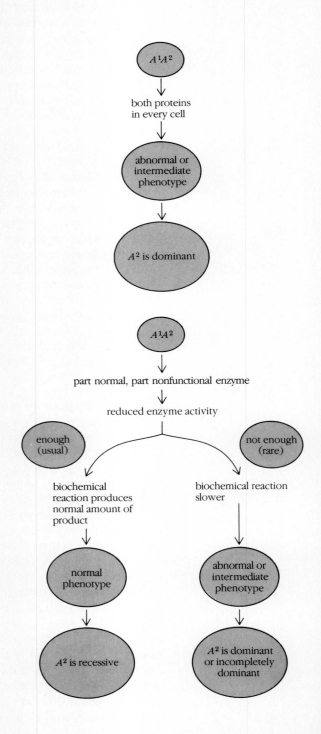

(a)

Alleles code for enzymes

A^1 normal enzyme
A^2 nonfunctional enzyme (no activity)

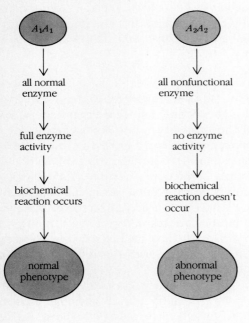

(b)

character. But since 0.6*x* units of GPUT are adequate for normal cell function, the *Aa* genotype produces an apparently normal individual.

Let us return to the hypothetical alleles A^1 and A^2. If the normal allele A^1 did *not* produce enough enzyme to ensure a normal phenotype, the A^1A^2 heterozygote would probably be intermediate in character between the normal homozygote, A^1A^1, and the abnormal homozygote, A^2A^2. Since it would then take only one dose of the abnormal allele to produce a visible result, we would classify A^2 as dominant over A^1, particularly if the abnormal homozygous condition were very rare and/or lethal. However, there are few examples of dominant abnormalities involving enzymes, such as enzymes connected with the synthesis of certain pigments.

Our classification of galactosemia and similar traits as recessive depends on using a classic definition of the term *phenotype*. The phenotype is the expression of the gene, and classically it means the *visible trait*. But the more we have learned about gene function, the less adequate this definition has become. Genes have various levels of expression, from original gene product to external manifestations, and when we speak of phenotypes we must often make clear what level of gene expression we are talking about. In the case of galactosemia, if phenotype means disease symptoms, the trait is recessive, but if it means enzyme units produced, the trait shows intermediate inheritance. The terms *dominant, recessive*, and *intermediate* turn out to have meaning only within the context of a specified level of analysis.

Codominance

In addition to dominance/recessiveness and intermediate inheritance, there exists an allelic relationship called codominance. An individual heterozygous for two codominant alleles expresses *both* traits. The most outstanding example in humans is the inheritance of blood types A and B, each of which is determined by a single allele. The I^A allele codes for the presence of a molecule, called A antigen, on the surface of the red blood cells, while the I^B allele codes for a second molecule, called B antigen. An individual homozygous for I^A or I^B has only one of the antigens, and thus has blood type A or B. However, a person who is heterozygous for these alleles has both antigens and thus has blood type AB. The heterozygous phenotype is not quantitatively intermediate but, rather, shows both parental traits.

Our classification of an allelic pair as codominant may depend, again, on the level of gene expression we are considering. An example is sickle-cell anemia, classified as either recessive or intermediate on the basis of disease symptoms, but as codominant on the basis of allele products. Sickle-cell anemia, or sickle-cell disease, is a serious blood disorder caused by an abnormal form of hemoglobin, the oxygen-transporting protein in red blood cells. Normal adult hemoglobin is called hemoglobin A (HbA), and the sickle-cell form is hemoglobin S (HbS). Two alleles for HbS are required to produce the disease, so the allele is usually classed as recessive. The HbA/HbS heterozygote is said to have "sickle-cell trait," a condition in which the individual is nearly always in normal health but may experience some disease symptoms when severely deprived of oxygen. Thus, we can also classify the gene as intermediate in its inheritance pattern if we consider its limited effects in the heterozygote. However, if we examine the hemoglobin of heterozygotes, we find it is of both types—half HbA and half HbS. If we classify the alleles according to their products, we must regard them as codominant.

Figure 4–24
(Facing page.)
The basis of some allelic relationships.

Lethal alleles

We have referred repeatedly to lethal alleles—alleles so disruptive that they cause the death of the person carrying them. The term **lethal** is sometimes restricted to alleles that cause death in the prenatal period; alleles that have their lethal effects in postnatal life are then called **sublethal**, or **semilethal**.

Identifying lethal alleles can be difficult if death occurs prenatally. A stillbirth, miscarriage, or abortion is sometimes difficult to label as a *genetically* caused death. Thus, most of the lethal alleles we are sure of cause death in various stages of postnatal life. Consider the problems when either a dominant or a recessive allele is lethal in early prenatal life:

Recessive	Dominant
AA normal	*AA* lethal (aborted)
Aa normal (carrier)	*Aa* lethal (aborted)
aa lethal (aborted)	*aa* normal

In each case, the only children born would be perfectly normal, and so we would have no immediate reason to suspect that a lethal allele was present.

There is no doubt that prenatal lethals exist, because whenever we look for them in other experimental animals, we find them. In organisms like mice and fruit flies, we are able to detect lethal alleles because we can perform specific genetic crosses. If certain predicted phenotypic classes are always missing, we suspect that one of the alleles is lethal in the homozygous or heterozygous state.

In some cases an allele is lethal when homozygous; but when it is heterozygous, it produces a visible but nonlethal phenotypic alteration. Yellow fur in mice is caused by such a gene (*Y*). Pure strains of yellow-furred mice do not and cannot exist, because the yellow homozygous condition is lethal. A cross between two heterozygotes (*Yy* × *Yy*) should produce the following results:

$$\frac{1}{4} \; YY \; \text{lethal}$$
$$\frac{1}{2} \; Yy \; \text{yellow fur}$$
$$\frac{1}{4} \; yy \; \text{normal; tan fur}$$

The ratios at birth are 2 yellow to 1 tan; the *YY* class is missing. Proof of the prenatal lethal action of the *YY* genotype is possible because we can sacrifice *Yy* female mice that have been mated with *Yy* males and examine the early embryos. About one-fourth of the embryos turn out to be badly malformed and dead, as we would predict.

Notice that we have labeled the abnormal allele as though it were dominant, and so it is, with respect to yellow fur: it takes only one such allele to produce a yellow-furred mouse. But with respect to its lethal effects, the allele is recessive—it takes two to produce death. There are many lethal alleles in humans that behave just like yellow fur in mice (although the lethal effects may not occur until postnatal life), and labeling these alleles as dominant or recessive depends on which trait we are interested in.

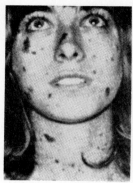

Figure 4–25

A young woman with xeroderma pigmentosa, a recessive disease caused by a defective DNA-repair system.

Xeroderma pigmentosum (XP), for example, is a disease caused by the absence of an enzyme that aids in the repair of DNA that has been damaged by radiation or chemicals. Without this repair function, cells die or become cancerous. A person homozygous for the XP allele has heavy freckling, large ulcerations in the skin, and tumorous growths in areas exposed to light (Figure 4–25). Death almost always occurs before reproductive maturity. The heterozygous person often has just the heavy freckling, without skin ulcerations or tumors. (Of course, most types of heavy freckling have other causes.) XP is considered a recessive trait, because in compiling medical histories and pedigrees we are interested chiefly in the homozygous disease state. But with respect to the simple freckling, it is dominant, just like yellow fur in mice.

On the other hand, although **achondroplastic dwarfism** exhibits the same inheritance pattern, we consider it dominant. In a person heterozygous for this condition, the long bones fail to grow properly; such a person is abnormally short, but otherwise normal (Figure 4–26). However, the dominant homozygous offspring of two achondroplastic dwarfs are invariably born dead or die soon after birth: two such alleles are lethal. In this case we are concerned chiefly with the dwarf condition, and since only one allele is required to produce dwarfism, we call the allele dominant. Classifying traits like XP and achondroplastic dwarfism as dominant or recessive is truly arbitrary. We could say, of course, that they show intermediate inheritance, like camptobrachydactyly, but the heterozygous condition, like yellow fur in mice, shows little resemblance to the lethal homozygous state—it is not visibly intermediate.

There are other lethal alleles that express their lethality when homozygous, but produce no obvious heterozygous phenotype. These alleles are classed as recessive. Tay–Sachs disease and cystic fibrosis are examples we have already discussed. Another example is the allele that causes ichthyosis congenita (Figure 4–27), a lethal

Figure 4–26

A mother with achondroplasia and her two daughters, both of whom have the trait.

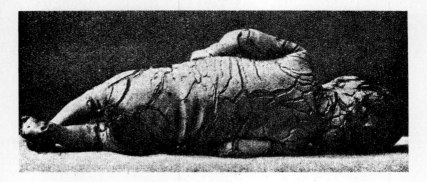

Figure 4–27
A newborn child with ichthyosis congenita, a recessive lethal condition.

condition distinguished by brittle, leathery skin with deep, bleeding fissures. Finally, there are alleles like the one for Huntington's disease, lethal in the heterozygous state. Such alleles are clearly dominant and exhibit late onset. If these alleles killed their possessors before reproductive age, they would not be passed on; instead, their existence would be maintained only through the occurrence of new mutations.

Multiple Alleles

Our discussion so far may have given you the impression that for each gene locus there are just two alternative alleles. This is an oversimplified view. Any individual, of course, can have no more than two different alleles per locus, because a person has but two copies of each locus. It is common, however, for several alternative alleles to exist in a population. One individual may be A^1A^2, another A^2A^3, a third A^4A^5, and a fourth A^6A^3. All these alleles (A^1, A^2, A^3, A^4, A^5, A^6) are variants found at the same locus. Such a collection of alternative alleles is called a **multiple allelic series**. The system of notation we have used here does not tell us anything about the dominance relationships among the alternative alleles. When we deal with particular multiple allelic series we will encounter various systems for expressing such relationships.

A classic example of a human multiple allelic series is the ABO blood group alleles. We have already mentioned the codominant alleles I^A and I^B, which code for the presence of A antigen and B antigen, respectively. But a third major allele occurs at this locus and is represented as i. The i allele does not code any antigen on the surface of the red blood cell and is therefore, as the lowercase letter implies, recessive to both I^A and I^B. In its homozygous state (ii), it produces the antigenless condition called blood type O. The existence of three alleles that determine blood type means six genotypes are possible:

ii	I^AI^A
iI^A	I^AI^B
iI^B	I^BI^B

The ABO alleles are sometimes represented by the same letters as the antigens they code for, as shown in the righthand column of the following table. Since I^A and I^B (A and B) are codominant and i (O) is recessive, the six genotypes produce four phenotypes:

Phenotype (blood type)	Genotype(s)
A	$I^A I^A$, iI^A (or AA, AO)
B	$I^B I^B$, iI^B (or BB, BO)
O	ii (or OO)
AB	$I^A I^B$ (or AB)

Although a homozygous condition is necessary to produce type O, this blood type is nevertheless very common because of the high frequency of the i allele in the population.

Pleiotropy

In the first decade or two of this century, geneticists had only a foggy notion of what a gene is and how it functions. Many of them envisioned each gene as being responsible for a particular visible trait. The living body, in this view, was a patchwork quilt of different, nonoverlapping traits, and each unit of the quilt was determined by a single gene. But these archaic ideas began to change as we came to understand that most genes function by coding for the production of specific proteins. It became clear that genes usually do not produce visible phenotypes directly. Rather, phenotypes occur as a direct or indirect consequence of the type of gene product coded for.

The gene, then, can be viewed as a blueprint for the production of a protein. If this protein acts as an enzyme, it may participate in a metabolic reaction that is part of a larger and more complex series of interconnected metabolic reactions. If a defective gene causes an enzyme to be absent, the metabolic reaction the enzyme catalyzes will not occur. Because the reactions are interconnected, blocking one may affect the entire interconnected system.

When a gene has a variety of phenotypic effects, we say that it is **pleiotropic**. Pleiotropy is simply a manifestation of the interrelatedness of cellular functions. For example, a certain recessive gene, when homozygous, results in the absence of an enzyme called phosphofructokinase (PFK). This enzyme deficiency results in a rather mild form of hemolytic anemia, a type of anemia caused by the disintegration of red blood cells. The reduced life span of the red blood cells triggers the development of other problems: there is a mild jaundice (yellowness of the skin and eyes) from bile pigments produced by the excessive breakdown of the red cells, and the spleen enlarges because of the need to produce more red cells. These secondary symptoms are labeled pleiotropic effects.

We have encountered many similar instances of pleiotropism in our study of abnormal genetic syndromes. Recall, for example, the multiple phenotypic manifestations of cystic fibrosis and of Huntington's disease. When we consider common, normal genes, we find that their pleiotropic effects are so extensive that it is difficult to separate the effects of one gene from those of another. The widespread occurrence of pleiotropy reflects the highly integrated nature of living systems.

A Problem in Genetic Counseling

Recently, a young woman came into a Philadelphia hospital seeking genetic counseling. She presented a very difficult problem: she was recently married and had just found out that her husband is her half-brother. Evidently, her father had had an affair with a woman who lived down the street, the result of which was a male child. The two children grew up together, became childhood sweethearts, and were married before it was revealed that they had the same father. The problem facing this young couple is complex and not yet solved: irrespective of laws that prohibit this degree of incestuous marriage, the couple plan to remain together and hope to have children. Their question: What is the probability that they can produce a normal child? Though their question cannot be answered until the father's genetic background has been thoroughly researched (a process he is so far reluctant to participate in), their dilemma emphasizes a problem that genetic counselors must constantly face: the genetic consequences of inbreeding.

The risks of inbreeding are apparent when we examine a brother–sister mating when their father carried a deleterious recessive allele (Figure 4–28). The probability that the father passed the deleterious allele to C is $\frac{1}{2}$, and the probability that he passed it to D is also $\frac{1}{2}$. The probability that both C and D are carriers of the recessive allele is thus $\frac{1}{2} \times \frac{1}{2} = \frac{1}{4}$. Assuming that both C and D are carriers of the defective allele, the probability that they will produce a child (E) homozygous for this allele is $\frac{1}{4}$, and the probability that E will be homozygous for the deleterious allele is

$$
\begin{array}{ccc}
\frac{1}{4} & \times & \frac{1}{4} & = & \frac{1}{16} \\
\text{probability that C} & & \text{probability that E} & & \text{probability that E} \\
\text{and D are carriers} & & \text{will be } nn \text{ given} & & \text{will be } nn \\
& & \text{that C and D are} & & \\
& & \text{carriers} & &
\end{array}
$$

Figure 4–28
(Left figure.)
Pedigree of an offspring of a
mating between a brother
and a sister (see text).

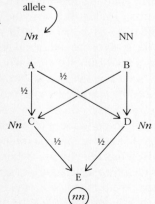

Figure 4–29
(Right figure.)
A pedigree of an offspring
of a mating between first
cousins (see text).

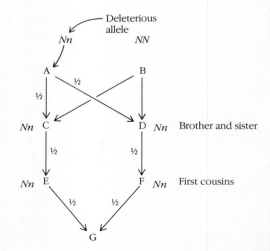

As the relationship between parents becomes more distant, the risk of having a homozygous child lessens. For example, the probability is $\frac{1}{64}$ that a mating between first cousins will produce an affected child if one of the original parents was a carrier of a particular recessive allele (Figure 4–29).

Summary

1 Genes on autosomes are transmitted equally to both sexes and are generally expressed equally in both. They are also transmitted equally *from* both sexes: **autosomal** genes do not usually produce different inheritance patterns when reciprocal crosses are performed.

2 Autosomal traits usually follow Mendelian patterns, but several complications may occur:
   (a) Some traits are **delayed-onset traits**, not expressed until long after birth.
   (b) Traits may exhibit **incomplete penetrance**. Penetrance is defined as the frequency with which a specific genotype exhibits the predicted phenotype.
   (c) Traits may exhibit **variable expressivity**; that is, the degree of expression of the genotype may vary from one individual to another.

3 Genes on the X chromosome are called **sex-linked**.
   (a) X-linked genes are passed from father to daughter and from mother to sons and daughters.
   (b) A dominant X-linked gene in a male produces a distinctive inheritance pattern in the offspring—all daughters have the trait, all sons do not—but such a gene in a female is indistinguishable from an autosomal dominant.
   (c) A recessive X-linked gene in either sex produces an inheritance pattern in the offspring distinct from that for an autosomal gene and for the reciprocal cross. An X-linked recessive trait is more frequently expressed in males, but homozygous females may also show it.
   (d) Y-linked genes are passed from father to son, a pattern of inheritance termed **holandric**.
   (e) A trait expressed in only one sex is called **sex-limited**. A trait that is differentially expressed in males and females is called **sex-influenced**. Genes for such traits may be on autosomes or on sex chromosomes.

4 Human XY embryos produce **H-Y antigen** early in their development. In the presence of this antigen, gonadal cells develop into testes; in its absence, they become ovaries. **Androgens** secreted by the testes promote the development of male characteristics; in the absence of androgens, female characteristics develop.
   (a) Such anomalies (marked deviations from normal) as XX males and XY females can be explained by the presence or absence of H-Y antigen, androgens, and specific cell receptor sites.
   (b) The gene for H-Y antigen (or its regulator) is probably on the Y chromosome.

5 Dominance and recessiveness are best understood by considering gene products.
   (a) In general, recessive traits involve enzymes, and dominant traits involve structural proteins.
   (b) The classification of a gene as dominant, recessive, or intermediate in its inheritance pattern may depend on whether we are considering gene products or visible phenotypes.
   (c) When an offspring's phenotype is intermediate between the homozygous pa-

rental phenotypes, the inheritance pattern is called **intermediate inheritance** (or incomplete dominance).

(d) Homozygous dominant traits are usually rare and very severe or lethal. The visible phenotype called dominant is often really an intermediate heterozygous phenotype.

(e) Individuals heterozygous for recessive traits may be intermediate in terms of gene products, but the visible phenotype is usually normal.

6 **Codominance** occurs when an offspring's phenotype has features of both homozygous parental phenotypes. Blood type AB is such a phenotype.

7 **Lethal alleles** cause the death of the carrier either prenatally or postnatally. They can be dominant or recessive, and some may be classified as either.

8 Several alleles of the same **locus** may exist in a population. The alleles that code for the ABO blood groups form such a **multiple allelic series**.

9 A gene can have a broad range of phenotypic effects, a phenomenon called **pleiotropy**.

Key Terms and Concepts

Locus	Holandric inheritance	Codominance
Autosomal inheritance	Sex-limited	Sickle-cell anemia
Sex-linked inheritance	Sex-influenced	Lethal (sublethal, semilethal) allele
Huntington's disease	H-Y antigen	
Delayed-onset trait	Androgen	Xeroderma pigmentosum
Incomplete penetrance	Testicular feminization	Achondroplastic dwarfism
Variable expressivity	Intermediate inheritance	Multiple alleles
Neurofibromatosis	Camptobrachydactyly	Pleiotropy
Tay–Sachs disease	Galactosemia	
Hemizygous		
Lesch–Nyhan syndrome		

Problems

4–1 A man who is a carrier of an autosomal recessive allele for cystic fibrosis produces two children, a male and a female, by two different women who are not carriers. The two children, unaware of the fact they have a common father, decide to marry and have children. What is the probability that they will have a child with CF?

4–2 Precocious puberty is an inherited trait. A pedigree for the trait is shown below. How would you interpret this inheritance pattern?

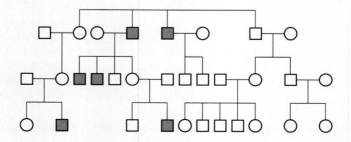

4–3 The inherited skin disease called keratosis follicularis exhibits the inheritance pattern shown below. How would you interpret this pattern?

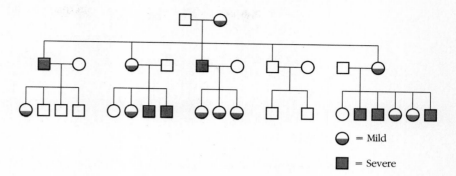

◒ = Mild

▨ = Severe

4–4 Below is a pedigree for hemophilia A in European royal houses. How would you interpret this inheritance pattern?

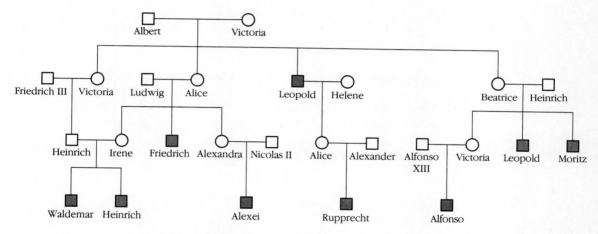

4–5 How would you interpret the following inheritance pattern for retinoblastoma?

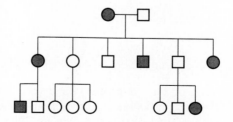

4–6 At a recent women's track meet, the winner of the 400-meter race was challenged by an opposing coach. This coach demanded a chromosome test to verify the runner's sex. The runner was 46,XX (had 46 chromosomes that included 2 Xs). Does this finding prove that the runner was female?

4–7 Interpret the following pedigree for the inheritance of a specific hair pattern.

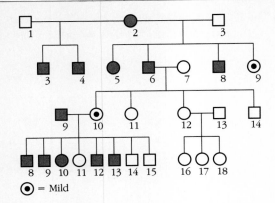

⊙ = Mild

4–8   Under what circumstances can a gene located on the Y chromosome be transmitted to females?

4–9   If an XY person was a fertile female, what kind of children would she produce?

4–10   A woman who is homozygous for a sex-linked recessive trait produces a son who is not affected. Propose an explanation for this, assuming the woman's husband is normal.

4–11   If an XX person was fertile and male, what kind of children would he produce?

4–12   Ichthyosis vulgaris is a skin disease that is transmitted in a family line as shown in the following pedigree. How would you interpret this inheritance pattern?

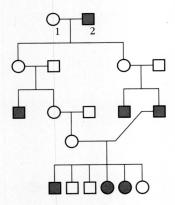

4–13   A gene for red–green color blindness is X-linked and recessive. A color-blind woman produces a non-color-blind daughter who marries a non-color-blind man. They have a son and a daughter. What is the probability that
(a) the son is color-blind?
(b) the daughter is color-blind?
(c) both children are color-blind?

4–14   How would your answers to Problem 4–13 change if the woman married a color-blind man?

4–15 A man with a genetically caused harelip marries a normal woman. They produce eight children, four sons and four daughters. One of the daughters and three of the sons have harelip. The harelip daughter marries a man without harelip and they have two boys with harelip and three daughters without it. One of the sons who has harelip produces eight children: four sons with harelip and four daughters without it; the other son with harelip produces four nonharelip daughters, two harelip sons, and two nonharelip sons. Interpret this inheritance pattern.

4–16 Draw a pedigree for Problem 4–15 and write in genotypes when they can be determined with certainty.

4–17 A couple produces a child who suffers from an X-linked disease. The couple is normal and the child is male.
(a) What are the parental genotypes?
(b) What is the probability that they will produce an affected daughter? An affected son?
(c) What is the chance that their next child will have the disease?

4–18 Pattern baldness in humans is determined by a pair of alleles (A = bald; a = nonbald). Two heterozygotes marry and produce the following offspring ratio:
  bald male: 3
  nonbald male: 1
  bald female: 1
  nonbald female: 3
(a) How would you interpret this mating?
(b) What were the parental phenotypes?

Answers

See Appendix A.

References

Dubowitz, V. 1978. *Muscle Disorders in Childhood.* Saunders, Philadelphia.

Emery, A. E. H. 1973. *Antinatal Diagnosis of Genetic Disease.* Williams and Wilkins, Baltimore.

Emery. A. E. H. 1976. *Methodology in Medical Genetics.* Churchill Livingstone, Edinburgh.

Gorlin, R. J., and W. S. Boggs. 1977. The genetic aspects of facial abnormalities. H. Harris and K. Hirschhorn, eds. *Adv. Hum. Genet.* 8:235–346.

Gorlin, R. J., J. J. Pindborg, and M. M. Cohen. 1976. *Syndromes of the Head and Neck,* 2nd ed. McGraw-Hill, New York.

Hadorn, E. 1955. *Developmental Genetics and Lethal Factors.* Wiley, New York.

Harris, H. 1975. *Prenatal Diagnosis and Selective Abortion.* Harvard University Press, Cambridge, MA.

Harris, H. 1975. *Principles of Human Biochemical Genetics,* 2nd ed. Elsevier North-Holland, New York.

Koo, G. C., et al. 1977. Mapping the locus of the H-Y gene on the human Y chromosome. *Science* 198:940–942.

McKusick, V. A. 1972. Heritable Disorders of Connective Tissue, 4th ed. Mosby, St. Louis.

McKusick, V. A. 1982. *Mendelian Inheritance in Man,* 6th ed. Johns Hopkins University Press, Baltimore.

McKusick, V. A., and G. A. Chase. 1973. Human Genetics. *Annu. Rev. Genet.* 7:435–473.

Murphy, E. A., and G. A. Chase. 1975. *Principles of Genetic Counselling.* Year Book Medical Publications, Chicago.

Nora, J. J., and F. C. Fraser. 1974. *Medical Genetics: Principles and Practice.* Lea and Febinger, Philadelphia.

Ohno, S. 1978. *Major Sex Determing Genes.* (Monographs on Endocrinology, vol. 11.) Springer Verlag, New York.

Rosenfeld, R. G., et al. 1979. Sexual and somatic determinants of the human Y chromosome studies in a 46, XYp⁻ phenotypic female. *Am. J. Hum. Genet.* 31:458–468.

Simpson, J. L. 1976. *Disorders of Sexual Differentiation: Etiology and Clinical Delineation.* Academic Press, New York.

Smith, D. W. 1976. *Recognizable Patterns of Human Malformation,* 2nd ed. Saunders, Philadelphia.

Stanbury, J. B., J. B. Wyngaarden, and D. S. Fredrickson. 1983. *The Metabolic Basis of Inherited Disease,* 5th ed. McGraw-Hill, New York.

Steinberg, A. G., ed. 1974. *Progress in Medical Genetics,* vol. 10. Grune and Stratton, New York.

Steinberg, A. G., and A. G. Bearns eds. 1961–1973. *Progress in Medical Genetics,* vol. 1–9. Grune and Stratton, New York.

Steinberg, A. G., et al., eds. 1976–1977. *Progress in Medical Genetics,* New Ser., vols. 1 and 2. Saunders, Philadelphia.

Stern, C. 1973 *Principles of Human Genetics,* 3rd ed. Freeman, San Francisco.

Vogel, F., and A. G. Motulsky. 1980. *Human Genetics: Problems and Approaches.* Springer Verlag, New York.

Wachtel, S. S. 1977. H-Y antigen and the genetics of sex determination. *Science* 198:797–799.

5 Gene Mapping

One of the problems that geneticists must solve is the location of genes on chromosomes. It is not enough simply to say that a gene is autosomal or sex-linked. If it is autosomal, on which of the autosomes is it located? And no matter which chromosome is found to house a particular genetic locus, we want to know where the locus is on the chromosome.

A woman is heterozygous for two sex-linked recessive traits, red–green color blindness and hemophilia. She marries a man who has neither trait, and they have many sons. Some of the sons express both traits, some express neither, but none expresses just one.

The cross just described is technically a dihybrid cross, because we are following the inheritance pattern for two different pairs of alleles. Although Mendel's principle of independent assortment states that the traits in a dihybrid cross are inherited independently, in this particular case the two traits always seem to stay together. The gene determining color blindness does not assort independently of the gene that determines hemophilia because both are on the same chromosome (Figure 5–1). Genes that tend to be inherited together are said to be **linked**, or to constitute a **linkage group**. Linked genes are necessarily on the same chromosome, although not all genes on the same chromosome may appear in experiments to be linked, as you will see. Genes on the same chromosome, whether linked or not, are called **syntenic**.

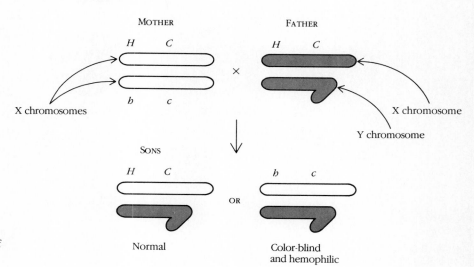

Figure 5–1
A cross showing linkage. The two pairs of alleles do not assort independently of each other.

Parental generation:

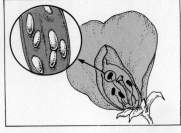 ×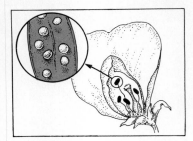

Purple flowers, long pollen grains
PPLL genotype

Red flowers, round pollen grains
ppll genotype

F₁ generation:

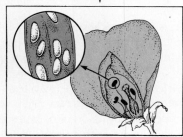

Purple flowers, long pollen grains
PpLl genotype

Self-fertilization

F₂:

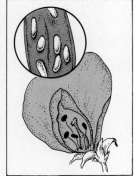  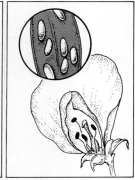 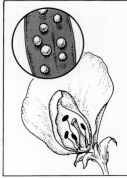

	Purple, long, *P-L-.*	Purple, round, *P-ll.*	Red, long, *ppL-.*	Red, round, *ppll.*
Observed phenotypes:	296	19	27	87
Expected from 9:3:3:1 ratio:	242	80	80	27

Since the number of genes greatly exceeds the number of chromosomes, each human chromosome must carry hundreds or thousands of gene loci. In this chapter, we will address a central problem of modern human genetics: Which genes are located on which chromosome, and how are the genes on one particular chromosome arranged relative to each other?

Gene Linkage and Recombination

Over 75 years ago, W. Bateson and R. C. Punnett discovered that two genes do not necessarily assort independently of each other. They crossed sweet pea plants that differed in flower color (purple or red) and pollen shape (long or round) and got the results shown in Figure 5–2.

Clearly, the F_2 does not exhibit the expected 9:3:3:1 ratio of phenotypes. The purple/long and red/round classes are much larger than we would expect if the genes for flower color and pollen shape assorted independently, suggesting that *P* and *L* tend to stick together during meiosis, as do *p* and *l*. Such sticking together would be easy to understand if *P* and *L* were on one member of a chromosome pair and *p* and *l* on the other. But if they are, how can the purple/round and red/long classes be accounted for? If two genes are on the same chromosome, why are they not *always* inherited together? In 1910, T. H. Morgan proposed, correctly, that alleles sometimes switch places during meiosis.

Crossing Over

Recall that during meiosis I, the members of a homologous chromosome pair synapse, or intertwine with each other, and that as they pull apart they remain connected at certain points, forming X-shaped regions called **chiasmata** (singular: **chiasma**) (Figure 5–3). Chiasmata mark the places where crossing over, or exchange of segments between nonsister chromatids, has occurred. During synapsis, a break may occur anywhere between the centromere and the end of a chromatid, and the entire segment from the break to the tip is switched with the corresponding segment of another chromatid. Morgan proposed that this crossing over of segments of nonsister chromatids is the mechanism responsible for the exchange of two members of an allelic pair.

Crossing over should not be confused with any of the abnormal transfers of chromosome material that we have mentioned from time to time and that will be

Figure 5–2
(Facing page.)
Demonstration of gene
linkage in sweet peas.

Figure 5–3
Photo of chromosomes in
late prophase$_1$, showing a
chiasma, where crossing
over of nonsister chroma-
tids has occurred.

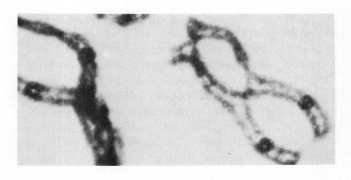

discussed in detail in Chapter 6. Translocation, for example, involves the transfer of a chromosome segment to a nonhomologous chromosome. Crossing over, however, is a normal part of meiosis and usually involves a point-for-point exchange of material between homologs. The only chromosomes that do not ordinarily engage in such exchange with each other are the X and Y chromosomes, which do not have corresponding loci. Paired X and Y chromosomes do not display any chiasmata. Two X chromosomes, however, do cross over, just like a pair of autosomes. The recombination of genes on chromosomes through crossing over multiplies enormously the already staggering possibilities for variation that arise simply through recombination of parental chromosomes in offspring.

Suppose a person is heterozygous for two pairs of alleles, *Aa* and *Bb*; *A* and *B* are on one chromosome, and *a* and *b* are on its homolog. This genotype is expressed as *AB/ab* to show which genes are linked. (Of course, it is also possible for *A* and *b* to be on one chromosome and *a* and *B* on its homolog; this genotype is *Ab/aB*.) If no crossover occurs between the *A* locus and the *B* locus, meiosis in an *AB/ab* cell will give rise to two kinds of gametes (Figure 5–4a). If, on the other hand, a crossover occurs anywhere between these two loci, meiosis will give rise to all four possible combinations: *AB*, *Ab*, *aB*, and *ab* (Figure 5–4b). In this example, gametes of the first and the last type are called **nonrecombinant**, or **parental**, **gametes**, while the other two kinds are called **recombinant gametes**.

In a given individual, some meioses will involve a crossover between the two loci in question and some will not. The frequency with which crossing over occurs depends on the distance between the two loci, for the farther apart they are, the greater the probability that a break will occur between them. If two genes are very close together on a chromosome, they may be inherited together virtually all the time, thus displaying **complete linkage**. Genes that are farther apart display **incomplete linkage**. When genes are incompletely linked, crossovers will result in the production of all four kinds of gametes, though in unequal numbers (the

Figure 5–4
The genetic results of crossing over (*a*) or no crossing over (*b*) in a person of genotype AB/ab. The chromosomes are drawn double-stranded here to show that a crossover involving nonsister chromatids produces recombinant chromatids. In the other diagrams of crossing over, each chromosome is represented by a single line, though such chromosomes would, in reality, be compound chromosomes consisting of two chromatids.

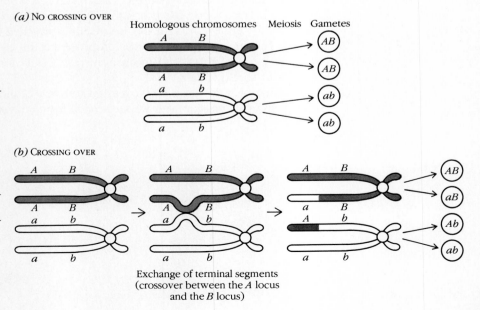

(*a*) NO CROSSING OVER

Homologous chromosomes   Meiosis   Gametes

(*b*) CROSSING OVER

Exchange of terminal segments
(crossover between the *A* locus
and the *B* locus)

parental types equally common, the recombinant types less common but in equal frequency with each other). Finally, when genes are very far apart, they undergo such frequent crossing over that they are inherited independently of each other—that is, the four types of gametes occur in equal numbers, just as if the genes were on different chromosomes. Though the genes are on the same chromosome, they do not behave as if they are linked, but they are still said to be syntenic.

Classic approaches to gene mapping

Morgan suggested that the frequency of crossing over between two loci of the same chromosome is proportional to the distance between them. One of his students, A. H. Sturtevant, transformed this seminal idea into a tool for **mapping** the positions of genes.

Gene distances and sequence   Sturtevant suggested that the frequency of crossing over between two points on a chromosome be expressed as **map units**, and he arbitrarily set 1% recombination as equivalent to one unit. If an organism has the genotype *AB/ab* and 98% of its gametes are of the parental types (*AB* and *ab*) while 2% are recombinants (*Ab* and *aB*), the frequency of recombination of the two loci is 2%, and the loci are two map units apart. In organisms like sweet peas and fruit flies, which produce large numbers of offspring, we can determine the proportion of recombinant gametes by crossing two individuals homozygous for the linked traits, then testcrossing the F_1 offspring with a homozygous recessive individual (Figure 5–5). A cross performed to determine the distance between two points, or loci, on a chromosome is called a **two-point cross**.

Suppose the data from one such cross indicate that the distance between locus A and locus B is 10 units, and the data from another cross show that the distance between locus A and locus C is 4 units (Figure 5–6). We can tell whether A is located between B and C or whether, on the contrary, C is located between A and B, if we perform yet another cross to determine the distance between B and C. In the first case, B and C will be 14 map units apart (14% recombination of the B and C loci); in the second case, they will be only 6 units apart (6% recombination). Thus, a series of two-point crosses enables us to determine the relative positions of genes on a chromosome as well as the distances between them.

Double crossovers   If a crossover between locus A and locus B is followed by a second crossover of the same two chromatids anywhere between these two loci, the effect is to return the genes to their original positions. Two crossovers thus go uncounted, because the result of a **double crossover** is indistinguishable from the result of no crossing over (Figure 5–7a). Two-point crosses therefore tend to produce an undercount of actual crossovers and a resultant underestimation of gene distances, particularly when the points are far apart and double crossovers occur in a large percentage of meioses.

To obtain a more accurate crossover count, geneticists usually perform **three-point crosses**—crosses involving three linked loci instead of two. A double crossover can often be detected in a three-point cross, because in such a cross, the middle gene locus is the only one that has switched positions (Figure 5–7b). Even with a three-point cross, the number of undetected double crossovers increases with the map distance, but if the three loci are not too far apart—that is, if they are

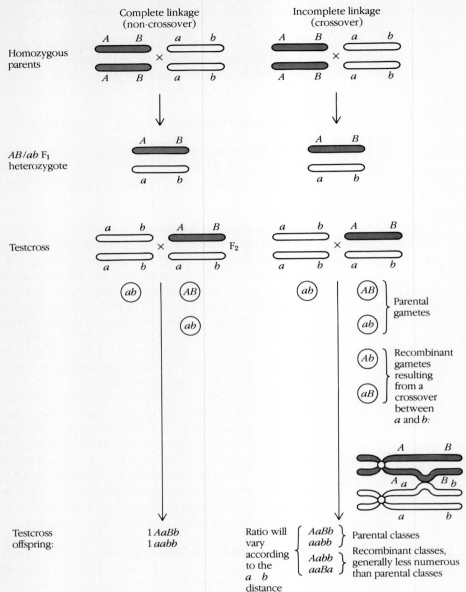

Figure 5–5
Morgan's interpretation of complete versus incomplete linkage.

on the same chromosome arm—a three-point cross can generate a relatively accurate measure of the map units separating them.

Three-point crosses involve considerably more tabulation and analysis of data than two-point crosses, but the method of operation is basically the same in both. Sturtevant used three-point crosses to determine the sequence and spacing of genes on the *Drosophila* (fruit fly) chromosome. The result was the first gene map (Figure 5–8), constructed in 1916.

Limit of mapping distances We said earlier that when syntenic genes are far enough apart, they may not display linkage. Fifty map units is the upper limit of the

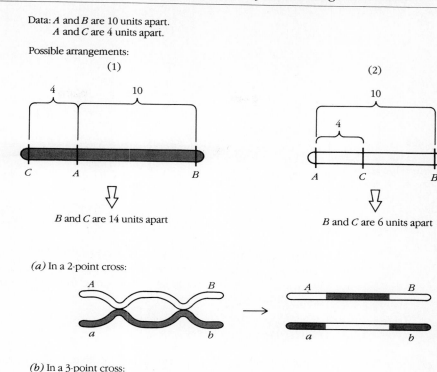

Figure 5–6
Determining gene sequence from gene distances. To choose between (1) and (2), we must determine the distance between *B* and *C*.

Figure 5–7
Double crossovers. The three-point cross detects this crossover; the two-point cross does not.

gene distance that can be determined from recombination data, because a recombination frequency of 50% or more is indistinguishable from the independent assortment of genes on different chromosomes. In other words, if loci A and B are more than 50 map units apart, an *AB/ab* individual will produce *AB*, *Ab*, *aB*, and *ab* gametes in a ratio of 1:1:1:1, the ratio produced by independent assortment. The crossing of two such individuals will produce the 9:3:3:1 phenotypic ratio we expect in any dihybrid cross—there would be no evidence that the loci are on the same chromosome. A map showing distances of 50 units or more must be constructed by piecing together shorter chromosome segments.

Detection of Gene Linkage in Humans

Deciding whether or not two human gene loci are linked is sometimes quite difficult. As you have just seen, we cannot conclude that two loci are on different chromosomes just because they display independent assortment. In humans, moreover, it is often hard even to figure out whether two genes assort independently; the small

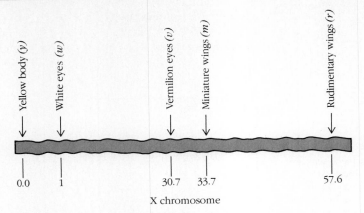

Figure 5–8
The first genetic map. Sturtevant mapped five genes on the *Drosophila* X chromosome. (Not drawn precisely to scale.)

number of offspring per pair and the frequent uncertainty about parental genotypes can make such decisions much more complicated than in sweet peas or fruit flies. But there are several tools for overcoming these problems, the classic approach being direct analysis of pedigrees.

Pedigree analysis

Linkage is most easily determined for X-linked genes, because if two genes both follow an X-linked pattern of inheritance, it follows automatically that these loci are on the same chromosome. Determining linkage for autosomal genes can be much harder. Sometimes, however, a pedigree of two autosomal traits shows an obvious linkage, as in the following example.

The nail–patella syndrome is a rare autosomal dominant trait that causes malformed nails and kneecaps. A pedigree of this trait also records ABO blood group genotypes (Figure 5–9a). Can we conclude from this pedigree that the locus for the ABO alleles is linked to the locus that causes the nail–patella syndrome? If you study the pedigree closely, you will find that with one exception, each person who has the syndrome also has blood type B. We can reasonably infer that the two loci are linked.

A genetic interpretation of the same pedigree (Figure 5–9b) shows that individuals II5, II8, II14, and III3 resulted from crossing over in either the grandmother (I2) or the father (II3). Out of 16 offspring recorded, 4 (25%) show recombination through crossing over, so we can suggest that the two loci are linked, with approximately 25 map units separating them.

A linkage analysis for X-linked loci is done the same way. Consider the following recessive traits:

1 Hemophilia A: "bleeder disease"; missing a blood clotting factor called the antihemophilic factor.
2 Hemophilia B: also a "bleeder disease"; missing a different clotting factor called the "Christmas factor."
3 Red–green color blindness: missing the green retinal pigment (deuteranopia, or **deutan color blindness**).
4 Red–green color blindness, protan: missing the red retinal pigment (protanopia, or **protan color blindness**).

A man who has hemophilia A and deutan color blindness marries a woman who has neither trait, and they produce a daughter who is free of both afflictions but is a carrier of both (Figure 5–10a). The daughter marries a normal man, and they have three sons: one is normal, and two have both hemophilia and colorblindness. They

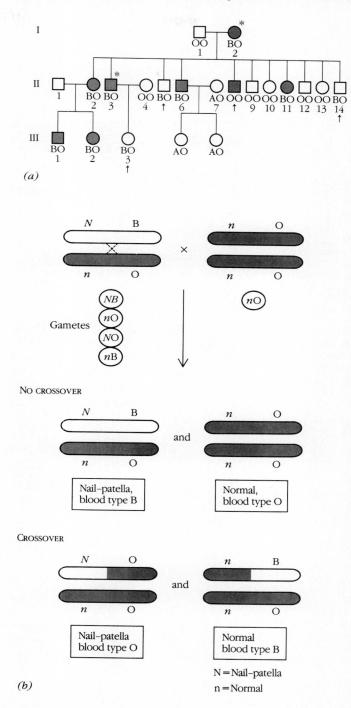

Figure 5–9

(*a*) A pedigree for the nail–patella syndrome, with ABO blood type noted. An arrow indicates recombinant offspring. An asterisk indicates an individual at the point where a crossover occurred. (*b*) A genetic interpretation of the pedigree.

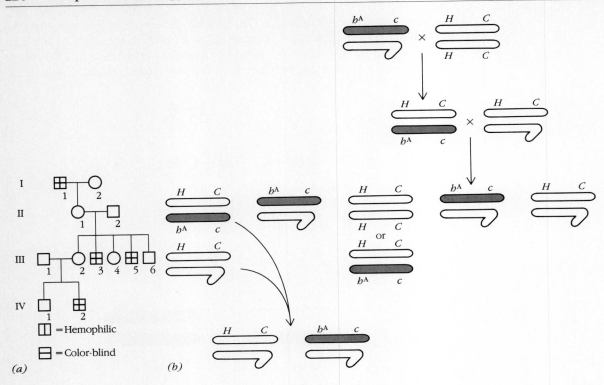

Figure 5–10
(*a*) Pedigree showing transmission of deutan color blindness and hemophilia A together. (*b*) A genetic interpretation of the pedigree. *C* = non-color-blind; *c* = color-blind; *H* = normal; *h*^A = hemophilia A.

also have a daughter, who produces a son with both traits. A genetic interpretation of this pedigree, in Figure 5–10b, shows that the two loci not only are linked but are so close together that no crossing over between them has occurred.

Contrast these results with those obtained from a cross between a man who has both hemophilia B and deutan color blindness and a woman who does not have these traits and is not a carrier. They have a daughter who must be a carrier, and she

Figure 5–11
(*a*) Pedigree for transmission of deutan color blindness and hemophilia B together. (*b*) A genetic interpretation of the pedigree.

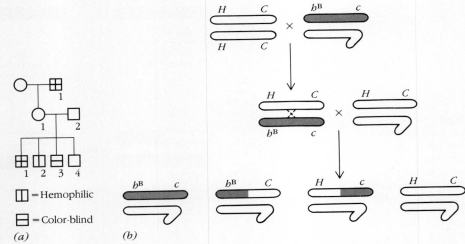

Figure 5–12
A tentative map of the human X chromosome. The numbers show approximate map unit distances.

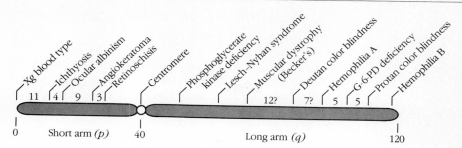

produces four genetically different sons (Figure 5–11a). The only way we can interpret this result is to suggest that there has been extensive crossing over between the two loci to generate all four genotypes in a sample of four. The two loci must be very far apart (Figure 5–11b).

We can conclude from these data that the three gene loci (hemophilia A, hemophilia B, and deutan color blindness) are X-linked, and that the hemophilia A gene locus is closer to the deutan color blindness locus than is the hemophilia B locus. Pedigree analysis of this sort has provided enough information to construct a tentative map of the human X chromosome (Figure 5–12).

Statistical approaches

In the direct approach to pedigree analysis, linkage is established by reference to cases that cannot easily be explained in any other way. Unfortunately, a clear-cut demonstration of autosomal linkage is often impossible, because we are unable to determine the genotypes of all the individuals in a pedigree. We must therefore commonly resort to indirect, or statistical, methods of evaluating possible linkage relationships. With the use of such methods, the probability of obtaining a particular pedigree if two loci are linked can be compared with the probability of obtaining the same pedigree if the two loci are not linked. But further discussion of these methods is beyond the scope of this book.

Human Gene Mapping

We can usually tell from the analysis of pedigrees whether genes are sex-linked or autosomal. And through the application of direct or indirect methods of pedigree analysis, we can also usually say with some certainty whether particular gene loci are linked or not. But we want to know more about the location of genes on chromosomes. On which specific autosome is a particular gene located? Where on that autosome is the gene? Is there any pattern to the arrangement of genes on chromosomes?

In peas, fruit flies, and even mice, we are able to carry out carefully designed crosses that provide answers to these questions. In humans, we cannot perform such crosses, and even when a crucial mating occurs by chance, the number of offspring produced is so small that it is difficult to draw any statistically valid conclusions. To map genes in humans, we must rely on some very special techniques.

Associating a gene with a specific chromosome: Somatic cell genetics

In recent years, genetic analysis of somatic (body) cells grown in a culture medium has provided us with vital new information. Our expertise in growing mammalian cells took a quantum leap forward in 1960 when cytologists made an astonishing

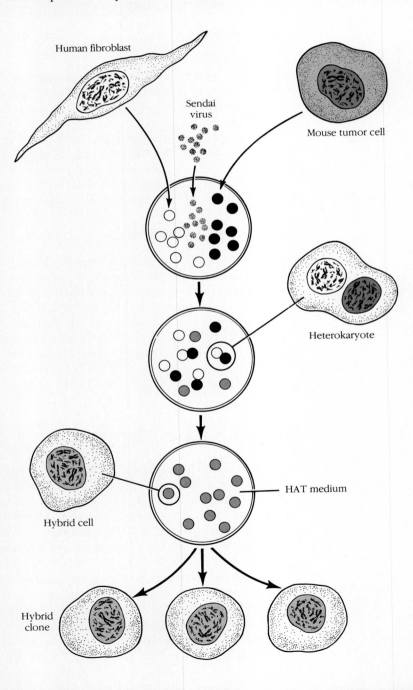

Figure 5–13
Somatic cell hybridization. Human fibroblasts are fused to mouse tumor cells deficient in either HGPRT or TK enzymes. Sendai viruses (inactive) promote the fusion. Some of the cells fuse to produce, first, cells with two nuclei (heterokaryons), then true hybrids. The hybrid cells are grown on a selective medium (HAT), where the hybrids thrive at the expense of the nonfused cells. As the hybrid cells divide, they lose human chromosomes randomly, then stabilize into different clones containing different human chromosomes.

discovery. Two cell lines from different mouse strains were growing in a single culture flask. Soon a new cell type appeared: each cell had a single nucleus and a chromosome number equal to the sum of the chromosome numbers in the two original cell lines. This new cell type was a **hybrid cell**, the product of the **fusion** of cells from different cell lines.

The discovery of somatic cell hybrids led cytologists to ask whether cells from different species would also fuse. The answer was soon provided: human cells and mouse cells could be induced to form hybrids! The technique, outlined in Figure 5–13, requires the presence of a specific but inactivated virus or some other agent, which stimulates some of the cells to fuse.

How do mouse–human hybrid cells apply to the problem of gene mapping in humans? First, for reasons not well understood, hybrids formed by the fusion of mouse and human cells tend to lose the human chromosomes. The loss is random, which means that when the hybrid cell line stabilizes, the cells can carry any one or more of the human chromosomes. Second, hybrid cells can be selected on the basis of the presence or absence of certain gene products, such as enzymes. By combining these two ideas, we can tie the activity of a specific gene product to the presence of a specific chromosome and can thus demonstrate that the gene is on that chromosome. The following example illustrates this approach.

Both the mouse and the human carry a gene that codes for the production of an enzyme called thymidine kinase (TK). In one experiment, mutant mouse cells lacking TK were fused with human cells that produced this enzyme (Figure 5–14). Cells lacking TK are unable to grow on a special culture medium called HAT (containing *h*ypoxanthine, *a*minopterin, and *t*hymidylate). After being cultured on a standard medium for a few days so that hybrid cells could form, the two cell lines were transferred to a HAT medium, where the TK-deficient mouse cells died and the human and hybrid cells persisted. Hybrid cells were detected and isolated. These contained the mouse chromosomes and anywhere from 2 to 12 human chromosomes each.

There was something very interesting about these hybrid cells. Though there was considerable variation as to which of the human chromosomes persisted in the hybrid cells, one human chromosome was always present in all cells growing on the HAT medium—chromosome 17. This meant that only hybrid cells carrying human chromosome 17 could survive on HAT; and since TK-deficient cells perish on this medium, chromosome 17 must carry the TK gene.

Similar experiments have succeeded in localizing other human genes. The techniques of somatic cell genetics are among the most important in our efforts to associate genes with specific chromosomes.

Determining the location of a gene on a chromosome

We are interested in knowing not only which chromosome a gene is on, but also where on the chromosome it is located. Several special approaches have been used for localizing human genes, including somatic cell genetics.

Somatic cell genetics and translocations We just described how a hybrid cell experiment was used to pinpoint the TK gene to chromosome 17. This experiment did not provide any information that would locate the TK gene in a specific region of

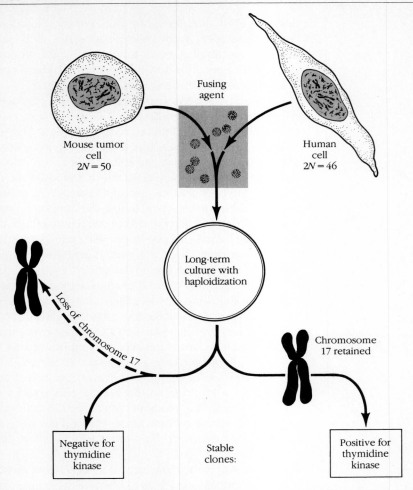

Figure 5–14
Gene mapping using cell
fusion. The enzyme thymi-
dine kinase, coded for by
the TK gene, is present
only when chromosome 17
is present.

this chromosome. However, by combining the techniques of somatic cell genetics
with those of classic cytogenetics, we have been able to locate the TK gene more
precisely.

In one study, mouse–human hybrid cells were formed and the hybrids examined for the presence of human TK. The experimenters found an intriguing cell line
in which all the human chromosomes were eliminated, yet human TK continued to
be produced. This perplexing problem was resolved when they discovered that this
cell line carried a **translocation** of part of human chromosome 17 on one of the
mouse chromosomes. This discovery was remarkable in itself, since it showed that
chromosomes from remotely related species could successfully participate in a
translocation. When the chromosomes were stained so that their banding pattern
was clear, the experimenters found that the long arm of 17 (the q arm) was translocated to a mouse chromosome. The human TK gene was thus shown to be located
on the long arm of chromosome 17.

Geneticists have further used translocations in hybrid cells to show that a group
of genes is located on a particular chromosome segment. In this case, a human cell
line carried a translocation of the long arm of the X chromosome on chromosome
14. In mouse–human hybrid cells formed from this cell line, a hybrid line emerged

in which all the human chromosomes had been eliminated, except this chromosome 14 with the long arm of X. The hybrid cell line produced three human gene products known to be X-linked:

HGPRT (hypoxanthine-guanine-phosphoribosyltransferase)
PGK   (phosphoglycerate kinase)
G6PD   (glucose-6-phosphate dehydrogenase)

The same cell line was also producing a product known to be coded for by an autosomal gene: nucleoside phosphorylase (NP). From these facts, we are able to draw the following conclusions: The HGPRT, PGK, and G6PD genes are on the long arm of the X chromosome, and the NP gene is on chromosome 14.

Deletion mapping   Translocations are not the only form of chromosome abnormality that provides us with information about the location of human genes. Sometimes a segment of a chromosome is simply **deleted** (lost), and the effects of the deletion on the unfortunate person who carries such a chromosome can tell us which genes are in the missing segment. Consider the following case. A boy with multiple congenital abnormalities (Figure 5–15a) was shown by cytological examination to have a deletion in chromosome 2. **Deletion mapping** showed that the breakpoint was in the portion of band 2*p*23 farthest from the centromere (Figure

Figure 5–15
(*a*) Boy with a partial deletion of chromosome 2. (*b*) Chromosome 2 of the boy. The solid line is the centromere position. The dashed line is the distal end of 2P23. (*c*) A summary of the different break points in chromosome 2. Point a is the breakpoint in the patient pictured. Points b, c, and d are breakpoints identified from other studies. Only a break at point a leaves the ACP locus on the chromosome. The other breaks include the ACP locus in the deleted segment.

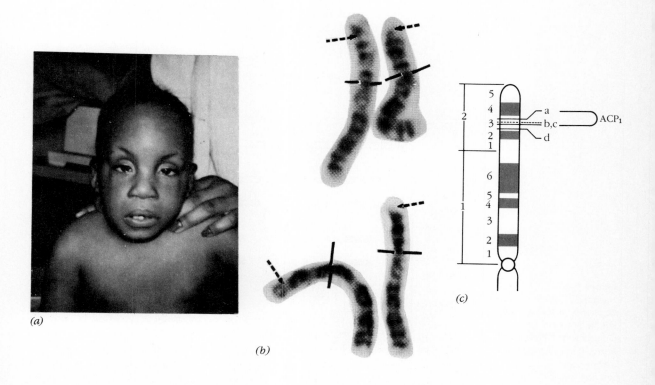

(*a*)

(*b*)

(*c*)

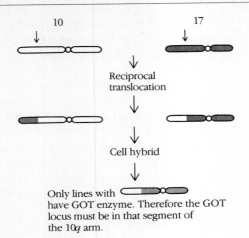

Figure 5–16
Using translocated chromosomes, we can localize the GOT locus to a segment of the 10*q* arm.

Only lines with ⬭🔲◖◗ have GOT enzyme. Therefore the GOT locus must be in that segment of the 10*q* arm.

5–15b), with the deleted segment extending to the terminus, or end of the chromosome. Earlier studies had indicated that the gene for acid phosphatase (ACP₁) in the human red cell is located in this region of chromosome 2, and the deletion found in this boy allowed investigators to locate the ACP₁ gene locus more precisely. Of the three codominant ACP₁ alleles (*A*, *B*, and *C*), the child carried alleles *A* and *B*, which tells us that since two alleles were present, the deletion did *not* encompass the ACP₁ locus. A study of another child with a deletion in 2*p*23, also extending to the terminus, had shown that one of the two copies of the ACP₁ locus was missing—the child

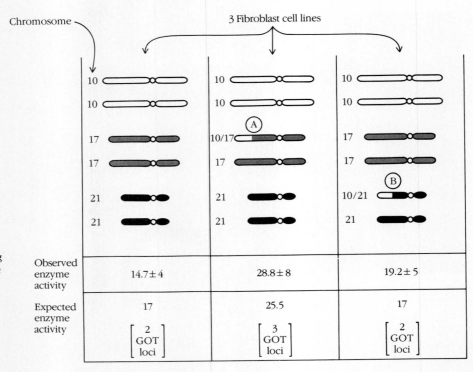

Figure 5–17
Mapping a gene locus using duplication mapping. In the middle fibroblast cell line, the segment of chromosome 10 translocated onto chromosome 17 must be carrying the GOT locus because of the increased level of enzyme activity.

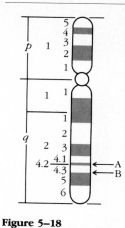

Figure 5–18
Diagrammatic representation of bands in chromosome 10 as observed after G-banding. Arrow A indicates the presumed point of breakage leading to t(10;17)(q24;p13) and arrow B, that leading to t(10;21)(q24;q22). This suggests that GOT maps in the region between A and B, within 10q24.3.

had only one allele for that locus. Since the break in 2p23 involved the ACP$_1$ locus in one child and not in the other, researchers could conclude that the ACP$_1$ locus lies between the two breakpoints—that is, in band 2p23 (Figure 5–15c).

Duplication mapping Translocations can be used to generate duplicated gene loci, and such duplications can often tell us something about the location of that particular locus. For example, suppose that we have determined through somatic cell genetics that the gene locus for glutamic oxaloacetic transaminase (GOT) is on chromosome 10.

We want to know where on chromosome 10 the GOT locus is. A reciprocal translocation of chromosome segments occurs between chromosomes 10 and 17 (Figure 5–16). Only cell lines carrying one of the translocated chromosomes (the line with the tip of the 10q arm) produces GOT, so we conclude that the GOT locus is in that segment of chromosome 10.

Can we be more specific about the location of the GOT locus? There are cell lines known to carry 10q segments originating from different breakpoints (Figure 5–17). One cell line, set up as a control, has two copies of the GOT locus. A second cell line (A) has two chromosomes 10 and a segment of 10 translocated onto chromosome 17. A third line (B) has two chromosomes 10 and a segment of 10 translocated onto chromosome 21. The control line and the line with segment B produce about the same amount of GOT. We infer from this that the GOT locus is not included in segment B. In cell line A, there is about a 33% increase in GOT activity, which is what we would expect if there were three functional copies of the GOT locus instead of two. We infer that segment A carries the GOT locus.

By looking at the breakpoints for the A and B segments, we can pinpoint the GOT locus very closely (Figure 5–18). Using specific staining procedures, we can show that the breakpoints for the A and B segments are different. Since A has the locus and B does not, the locus must be between the A and B breakpoints shown in the figure. This is a very clever and important technique for gene localization.

How Many Genes?

Using somatic cell genetics, deletion mapping, and other techniques, we are developing a human chromosome map that grows in detail almost daily (Figure 5–19). As the human genetic map takes shape, we are prompted to ask how many gene loci there are in humans. It is currently estimated that there are between 50,000 and 100,000 single gene loci in humans and that each codes for the production of a unique polypeptide chain. So far, we have identified about 1200 gene loci, a mere 2% of the total. So even though we are making rapid progress on the human gene map, we still have a long road to travel before it is anywhere near complete.

Gene Clustering

Careful scrutiny of the human gene map will probably lead you to conclude that the gene loci are distributed randomly in regard to their function. In many respects, this is a justified conclusion, but there appear to be some exceptions to this randomness. For example:

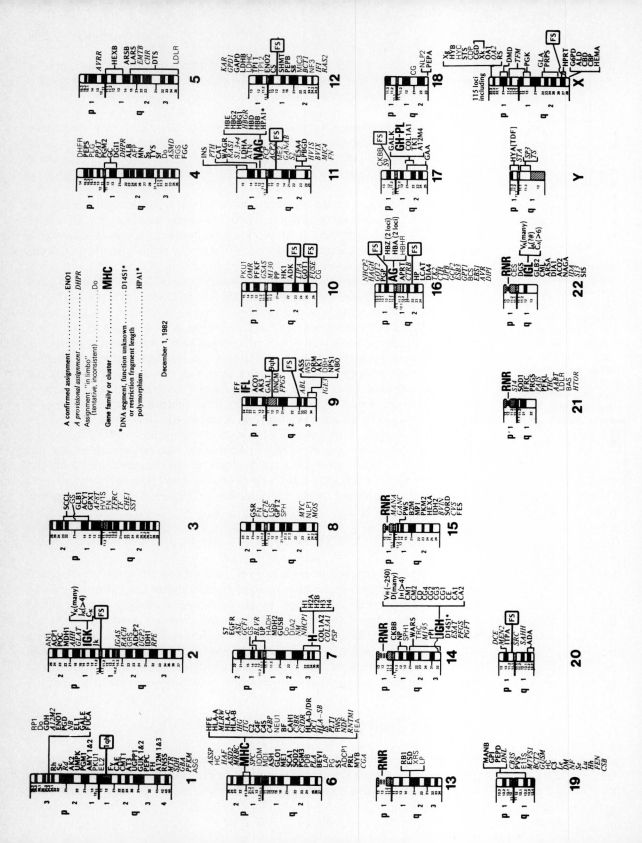

Figure 5–19
(Facing page.)
The human gene map. The abbreviations refer to some of the gene loci that have been identified so far. For instance, "ABO" on chromosome 9 refers to the ABO blood group, "CB" on the X chromosome refers to color blindness, and HEM$_A$ on the X chromosome refers to classic hemophilia.

1 The gene loci for the different human hemoglobin proteins are closely linked on chromosomes 11 and 16.

2 Four loci that code for enzymes involved in glycolysis (the initial breakdown of sugar in cellular respiration) are linked on chromosome 1.

3 The two loci for the two forms of red–green color blindness (deutan and protan) are linked on the X chromosome.

4 Certain gene loci involved in the immune response (the histocompatability loci) are closely linked on chromosome 6.

These and other examples show that in some cases, gene loci that are functionally related are physically **clustered**.

Why do such clusters of functionally related genes exist? And if some functionally related gene loci are clustered, why do we not observe all functionally related loci in clusters? The answers to these questions are not at all certain. We think that clustering involves the long-term accumulation both of chromosome aberrations that can cause identical segments to lie next to each other (duplications) and of mutations that alter the functions of genes in these segments. On the other hand, other kinds of chromosome aberrations (translocations and inversions) can break up existing clusters of genes. We will return to the question of how chromosome abnormalities have influenced genetic evolution after we discuss the more immediate effects of such abnormalities. In the next chapter we will deal with the effects of variation in chromosome number, and also structural variations in chromosomes.

Box 5–1   *Linked Genes and Branching Ancestry*

One generalization in biology overshadows all others: the concept that all living forms are genetically interrelated because of common ancestry. Through this concept, biologists combine the many data of biology into a generally coherent and unified whole.

This is not to say that we understand all the details of inheritance or the history of life; we don't. But the basic pattern, like the theory of gravity, is backed up by solid and rigorously tested evidence. Let's consider just one tiny, still-developing example: the preservation of groups of linked genes on individual chromosomes.

It is evident that if some species are derived from others, or if two groups come from common but now extinct ancestry, any genes that were linked in the ancestor should likely show some evidences of linkage in the descendants. We will learn in the next chapter that genes *can* change their linkage relationships, either by changing their order on a given chromosome or by moving to completely different chromosomes. Over long periods of evolutionary time, basic sets of ancestrally linked genes (linkage groups) will be broken up and rearranged as chromosomes are restructured in the descendant species. The important point here is that linked genes do not always remain linked. But traces of ancestral linkages can be found. The more we learn of linkage groups in different species, the more insight we gain into what ancestral patterns likely were.

No major research efforts have been aimed at comparing linkage relationships among species. But as our knowledge of the genetics of different organisms has increased, a number of patterns have been revealed.

For example, of all the mammals, the genetic map of the mouse is best known. (Thanks to somatic cell genetics, the human genetic map is rapidly catching up. But beyond the mouse and the human, we really know little about the genetics of mammals.) It would not be a great surprise, if the

| Man | | | | Mouse | |
Chromosome	Gene Symbol	Chemical Produced by Gene		Chromosome	Gene Symbol
1p	Pgm-1	Phosphoglucomutase		4	Pgm-2
	Ak-2	Adenylate kinase			Ak-2
	Eno-1	Enolase			Eno-1
	Pgd	Phosphogluconate dehydrogenase			Pgd
	Gdh	Glucose dehydrogenase (Hexose-6-phosphate dehydrogenase)			Gpd-1
4	Pgm-2	Phosphoglucomutase		5	Pgm-1
	Pep-S	Peptidase			Pep-7
	Alb	Albumin			Alb-1
6p	HLA	Major histocompatibility region		17	H-2
	Glo-1	Glyoxylase			Glo-1
10	Hk-1	Hexokinase		10	Hk-1
	Pp	Inorganic pyrophosphatase			Pp
11	Ldh-A	Lactate dehydrogenase		7	Ldh-1
	Hbβ	Hemoglobin, beta chain			Hbb
17q	Gk	Galactokinase		11	Glk
	Tk-S	Thymidine kinase, soluble			Tk-1

Source: Adapted from L. G. Lundin, *Clin. Gen.* 16:72–81, 1979.

gene map for rats resembled that already known for the mouse. It is known, for example, that on chromosome 7 of the mouse are four genes in the following order: pink eyes–albinism–hemoglobin beta chain–Warfarin® resistance (Warfarin® is a rodenticide, most popularly known as the active ingredient in d-Con®). It is interesting, but not especially remarkable, that the same four genes are found, and in exactly the same order, in the rat.

But what about humans and rodents? They are *so* different—should their genes show any common linkage relationships? Under the proposal of common ancestry, *possibly* so. So with all apologies to Steinbeck, let's see what kind of linkage tale can be told of mice and men.

The accompanying table indicates that many of the chromosomally linked groups of genes in mice find rather direct counterparts in humans.

But these genes are all autosomal. What of the X chromosome? Many genes are now known to be X-linked in both humans and mice, and the two groups show perfect agreement. No gene is known that is X-linked in one species but not in the other. Indeed, the generalization holds for all the mammals; the X-chromosome, for reasons relating to X inactivation (see Chapter 4), has been extraordinarily "conservative" in evolution. It has undergone very little structural change.

Another highly conservative chromosome is the human chromosome 1. Data are now available for several genes, and to date, whatever genes occur on the chromosome 1 in humans have also been found on the chromosome 1 in the great apes and Old World monkeys.

But we needn't expect all such comparisons to fit—indeed, we know they don't. Genes carried on chromosome 5 of the mouse appear on both chromosomes 4 and 7 of man (see table). Obviously, either mouse or man had undergone a translocation since the time of common ancestry. Other problems also exist. But beyond all the differences lie all the similarities of pattern—beckoning future researchers.

Summary

1 Many traits do not follow Mendel's law of independent assortment. Genes that tend to be inherited together are called **linked**, and constitute a **linkage group**.

2 Linked genes are on the same chromosome, but genes on the same chromosome do not appear to be linked if they are sufficiently far apart. Genes on the same chromosome are called **syntenic**.

3 **Crossing over**, the exchange of parts of nonsister chromatids during meiosis I, accounts for the **incomplete linkage** of genes on the same chromosome. When a crossover occurs, the resulting gametes are called **recombinant** types. The farther apart two loci are on a chromosome, the higher the frequency of crossing over between them. Therefore, very widely separated syntenic loci assort independently.

4 We can **map** the position of genes in organisms such as fruit flies by observing the frequency of recombinant offspring in experimental crosses. This frequency tells us the relative distance between various pairs of loci, and by comparing distances we can determine the sequence of the loci. One **map unit** represents a recombination frequency of 1%.

5 **Three-point crosses**, which establish the position of three linked loci, produce more accurate maps than **two-point crosses**, because they detect most **double crossovers**.

6 In humans, direct analysis of pedigrees tells us whether genes are X-linked or autosomal. It may also tell us whether genes are on the same autosome and, on occasion, how far apart they are. When there is too little information for direct analysis, we use indirect, or statistical, methods to determine the probability that two gene loci are linked.

7 A revolutionary advance in human gene localization came about with the development of **somatic cell genetics**, or cell hybridization. The formation of human–mouse **hybrid cells**, in which human chromosomes are randomly eliminated, allows us to associate specific gene products with specific chromosomes.

8 Somatic cell genetics, coupled with chromosome anomalies such as translocations, sometimes allows us to assign a gene to a specific region of a human chromosome.

9 **Deletion** and **duplication** mapping can pinpoint the location of some human gene loci to specific chromosome bands by comparing the effects of deletion (congenital loss) and duplication of different parts of the same chromosome.

10 The human gene map shows that many loci appear to be scattered randomly with respect to function. However, some functionally related gene loci form clusters on particular chromosomes.

Key Terms and Concepts

Linked genes, linkage groups

Syntenic

Chiasma, chiasmata

Nonrecombinant (parental) gamete

Recombinant gamete

Complete linkage

Incomplete linkage

Mapping

Map unit

Two-point cross

Double crossover

Three-point cross

Deutan color blindness

Protan color blindness

Hybrid cell

Cell fusion

Translocation

Deletion

Deletion mapping

Duplication mapping

Gene cluster

Problems

5–1   A woman gives birth to six sons. Two are normal, three are both color-blind and hemophilic, and one is just color-blind. These traits are sex-linked. Draw the most likely chromosomal constitution of this woman.

5–2   A woman gives birth to six sons. Two are hemophilic; three are color-blind, and one is normal. How would you draw her chromosomal constitution?

5–3   A woman with the genetic constitution

marries a man with the genetic constitution

If uppercase letters are dominant, and if *A* and *B* are 20 map units apart, what type of offspring would this couple produce and in what approximate proportion?

5–4   A woman is heterozygous for three X-linked loci: *Aa*, *Bb*, and *Cc*. She has seven sons with the following genotypes: 2 *ABc*; 3 *abC*; 1 *ABC*; 1 *aBc*
(a) Draw the mother's genotype (the most probable one).
(b) Which of the sons are the result of recombination?

5–5   A population of human cells has a gene that codes for a specific enzyme called E. Mouse cells do not carry this enzyme. Mouse–human hybrids are formed, and the following stable lines are studied for the human chromosomes present and the presence or absence of E:

	Hybrid Cell Population					
	①	②	③	④	⑤	⑥
Human chromosomes present	21, 18, 16 7, 5, 2	18, 16 8, 3	21, 18 1	23, 19, 17 15, 14, 11 9, 8, 3, 1	18, 6, 2	21, 5
E activity present	Yes	Yes	Yes	No	Yes	No

From these data, where is the gene for E located?

5–6   The following translocation occurred between chromosome 14 and X:

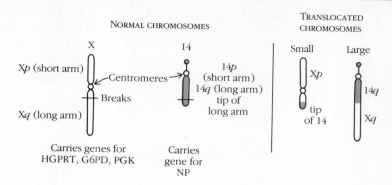

A cell carrying this translocation was fused with a mouse cell line, and the following hybrid cell lines were isolated:

TRANSLOCATED CHROMOSOMES IN HYBRID CELL LINES

		Both	Large One	Small One	Neither
ENZYMES CONTAINED IN HYBRID CELL LINES	G6PD	+	+	−	−
	HGPRT	+	+	−	−
	PGK	+	+	−	−
	NP	+	+	−	−

We know that the genes for G6PD, HGPRT, and PGK are X-linked, and that the gene for NP is autosomal. What else can you deduce from these data?

5–7   A child was born with the karyotype (chromosomal constitution) shown below. The mother carried a reciprocal translocation and was homozygous for an acid phosphatase allele (ACP_a). The father's karyotype was normal, and he was homozygous for a different ACP allele (ACP_b). The ACP alleles are codominant, but the child expressed only the ACP_b allele. What can you conclude about the ACP locus from this study?

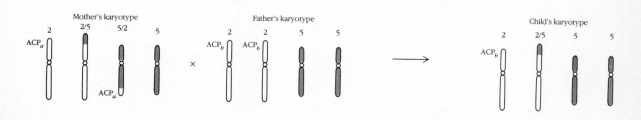

5–8 There is a fragile site (see page 173) on the long arm of chromosome 16. From somatic cell hybrids, we also can show that the Hpα gene (alpha polypeptide of the haptoglobin protein) is on chromosome 16. Among the offspring produced by persons heterozygous for both the fragile site and the Hpα gene, about 8% were recombinant. What do you conclude from this?

5–9 A mouse–human cell line carries only human chromosome 17. Another hybrid line carries $17q^-$ (del q^{21} qter where "del" indicates "deletion" and "ter" indicates "terminus."). Name the products that would be present in the former line that would be absent in the latter (see human gene map, Figure 5–19).

5–10 The X-linked genes *A* and *B* show 10% recombination. Genes *A* and *C* show 16% recombination. Draw the gene arrangement(s) compatible with these data.

5–11 Which of the models suggested in Problem 5–11 would be most likely if *B* and *C* showed about 6% recombination?

Answers

See Appendix A.

References

Cavalli-Sforza, L. L., and W. F. Bodmer. 1971. *The Genetics of Human Populations.* Freeman, San Francisco.

Conneally, M., and M. Rivas. 1980. Linkage analysis in man. *Adv. Hum. Genet.* 10:209–266.

Creagan, R. P., and F. H. Ruddle. 1977. New approaches to human gene mapping by somatic cell genetics. In Yunis, J. J., ed. 1977. *Molecular Structure of Human Chromosomes.* Academic Press, New York.

Emanuel, B. S., et al. 1979. Deletion mapping: Further evidence for the location of the acid phosphatase (ACP$_1$) locus within 2p23. *Am. J. Med. Genet.* 4:167–172.

Jenkins, J. B. 1979. *Genetics,* 2nd ed. Houghton Mifflin, Boston.

Kucherlapati, R. S., and F. H. Ruddle. 1976. Advances in human gene mapping by parasexual procedures. *Prog. Med. Genet.,* New Ser., 1:121–144.

Lebo, R. V., et al. 1979. Assignment of human b, g, and d-globin genes to the short arm of chromosome 11 by chromosome sorting and DNA restriction enzyme analysis. *Proc. Natl. Acad. Sci. USA* 76:5804–5808.

Lin, M. S., J. Oizumi, W. G. Ng, O. S. Alfi, and G. N. Donnell. 1979. Regional mapping of the gene for human UDP Gal-4-epimerase on chromosome 1 in mouse/human hybrids. *Cytogenet. Cell Genet.* 24:217–223.

McKusick, V. A., and F. H. Ruddle. 1977. The status of the gene map of the human chromosomes. *Science* 196:390–404.

Philip, T., et al. 1979. Etat actuel de la carte chromosomique de l'homme (Deuxième partie: Résultats et applications). *J. Genet. Hum.* 27:81–108.

Rao, D. C., et al. 1979. A maximum likelihood map of chromosome 1. *Am. J. Hum. Genet.* 31:680–696.

Renwick, J. H. 1971. Mapping of human chromosomes. *Annu. Rev. Genet.* 5:81–120.

Ringertz, N. R., and R. E. Savage. 1976. *Cell Hybrids.* Academic Press, New York.

Ruddle, F. H., and R. P. Creagan. 1975. Parasexual approaches to the genetics of man. *Annu. Rev. Genet.* 9:407–486.

Ruddle, F. H., and R. S. Kucherlapati. 1974. Hybrid cells and human genes. *Sci. Am.* 231:36–44 (July).

Shows, T. B. Gene mapping, *Adv. Hum. Genet.* (in preparation).

Siminovitch, L. Somatic cell genetics. *Adv. Hum. Genet.* (in preparation).

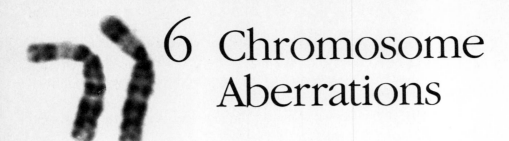

6 Chromosome Aberrations

Words cannot describe the profound sense of shock and tragedy felt by parents who produce a chromosomally abnormal child. Along with their anger and resentment, most people in this situation experience guilt and an altered self-image. It should not be so, since most chromosomal anomalies are spontaneous events that could happen to anyone, but people seldom see things in this light. They may understand the facts intellectually, but deep down they still question their own genetic worth. Genetic counselors must bring great skill and understanding to their efforts to help parents who have had abnormal children.

If mitosis, meiosis, and fertilization were accurate 100% of the time, our species would be free of some terrible genetic tragedies. But mistakes do happen, causing the birth of infants with extra sets of chromosomes, additional or deleted individual chromosomes, structurally altered chromosomes, or bodies composed of two or more chromosomally different cell types. Chromosome anomalies can be divided into two broad categories: (1) aberrations of chromosome number and (2) structural aberrations in individual chromosomes. Interestingly, the two types sometimes produce similar effects. Down syndrome, probably the best known of chromosomally caused disorders, may involve either type of aberration. In this chapter we will consider both categories of chromosome anomalies in some detail.

Aberrations of Chromosome Number

Human zygotes are sometimes produced with chromosome numbers other than 46. Many of these anomalies are spontaneously aborted, some are therapeutically aborted, some are born only to perish shortly after birth, and some survive, the most unfortunate being the severely handicapped and the most fortunate leading normal lives. There are two ways in which chromosomes can deviate from the normal number: **polyploidy**, the addition of whole sets of chromosomes, and **aneuploidy**, the addition or deletion of individual chromosomes.

Polyploidy: Adding whole sets of chromosomes

Possession of any whole-number multiple of the haploid chromosome number characteristic of a species is called **euploidy**. The Greek *eu* means "good," or "true," and a euploid cell has a "good," or complete, set of chromosomes. Any euploid condition higher than diploidy is called polyploidy (Table 6–1).

Table 6–1	**Euploidy nomenclature**	
Symbol	Chromosome Number (Human)	Ploidy (Euploid States)
n	23	Haploid
2*n*	46	Diploid
3*n*	69	Triploid
4*n*	92	Tetraploid
> 2*n*	—	Polyploid

Polyploidy is not detrimental to all organisms. In plants and certain lower animals, it is common and has played a significant evolutionary role. Many of our agricultural crops have been developed through a combination of hybridization and deliberate selection for polyploidy, and modern plant breeders can manipulate dividing cells so as to produce polyploids. Because the cells of a polyploid plant are ordinarily larger than those of its diploid counterpart, the entire plant is larger and more vigorous than a diploid one. In higher animals, however, polyploidy is usually a disaster: most polyploid embryos perish, and human embryos are no exception. There are probably several reasons for this difference between plants and people. One is that humans, like other higher animals, are more delicately balanced physiologically and sexually. The changes polyploidy causes thus disrupt sensitive physiological functioning.

The origin of polyploidy: Mitotic and meiotic failure   How does a polyploid condition arise in a normally diploid organism? A diploid somatic cell may give rise to a **tetraploid** one (four sets of chromosomes) if it experiences **mitotic failure**, that is, if the chromosomes replicate and the sister chromatids separate from each other but cell division does not occur (Figure 6–1). This is what eventually happens, for instance, to diploid cells treated with colchicine (Chapter 3). Without a spindle, the chromatids cannot be pulled to opposite poles, but eventu-

Figure 6–1
Mitotic failure.

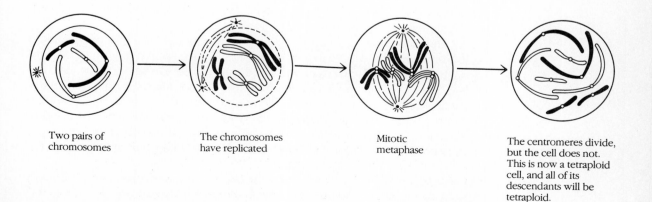

Two pairs of chromosomes

The chromosomes have replicated

Mitotic metaphase

The centromeres divide, but the cell does not. This is now a tetraploid cell, and all of its descendants will be tetraploid.

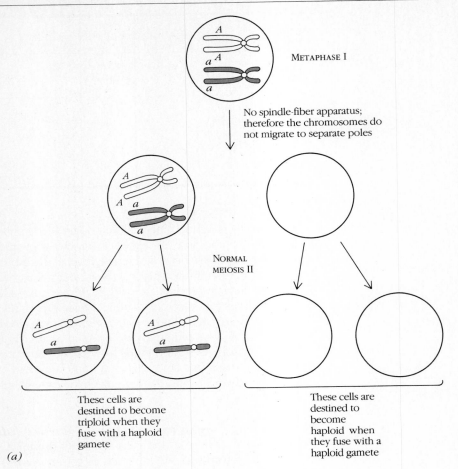

METAPHASE I

No spindle-fiber apparatus; therefore the chromosomes do not migrate to separate poles

NORMAL MEIOSIS II

These cells are destined to become triploid when they fuse with a haploid gamete

These cells are destined to become haploid when they fuse with a haploid gamete

Figure 6–2
Meiotic failure. (*a*) Failure of meiosis I. (*b*) Failure of meiosis II.

(a)

ally they fall away from each other and a nuclear membrane forms around all of the daughter chromosomes, creating a tetraploid nucleus. If the drug is withdrawn, the cell will eventually replicate normally, but its descendants will be tetraploid. Any gametes arising from such a cell line will be diploid instead of haploid, and self-fertilization involving such gametes will produce tetraploid offspring. The colchicine treatment is one method plant breeders use to produce true-breeding tetraploid strains.

In humans, a mitotic failure in cells of the ovary or testis can produce tetraploid cells that ultimately give rise to diploid gametes. But the same result can be brought about by **meiotic failure** in a diploid cell during gamete formation. When meiosis I fails (Figure 6–2a), homologous chromosomes do not migrate to opposite poles, and a nuclear membrane forms around the entire complement of 46 chromosomes. If meiosis II then proceeds normally, the result is two diploid gametes instead of four haploid ones. Alternatively, a failure of meiosis II in one daughter cell from a normal meiosis I produces one diploid and one aneuploid gamete (Figure 6–2b).

Polyploidy in humans: a lethal condition   Trouble arises when a diploid gamete unites with a normal haploid one, producing a **triploid** (3*n*) zygote. Of

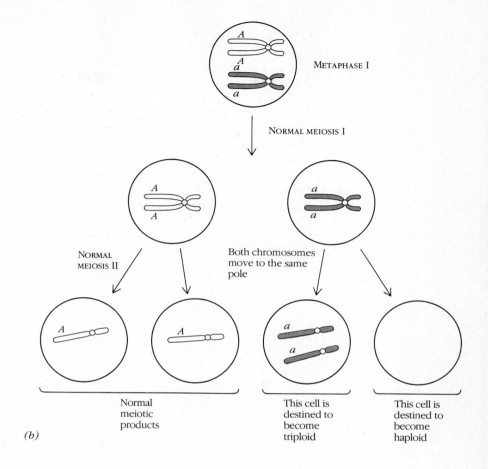

METAPHASE I

NORMAL MEIOSIS I

NORMAL MEIOSIS II

Both chromosomes move to the same pole

Normal meiotic products

This cell is destined to become triploid

This cell is destined to become haploid

(b)

course, a diploid gamete could theoretically unite with another diploid gamete to produce a tetraploid zygote, but in a non-self-fertilizing organism, the odds of two such anomalies meeting are exceedingly small. Human tetraploidy probably arises in most cases from a mitotic failure after a normal fertilization. If such a failure occurred in the zygote's first division, all the embryo's cells would be tetraploid; but if it occurred in a later division, only some of the cells would be tetraploid.

Besides the union of a haploid and a diploid gamete, there is another way that a triploid zygote can arise. On very rare occasions a normal, haploid egg may be fertilized by two normal sperm. Finally, a polar body, instead of being pushed out from the egg, may fuse with the egg nucleus, so that when fusion with the sperm nucleus takes place, the result is a triploid cell.

Polyploidy is the normal condition in certain human tissues: liver, bronchial epithelium, and the amniotic membrane surrounding the fetus. Human polyploidy is rarely observed in other tissues. Most polyploid fetuses are spontaneously aborted early in pregnancy, and the few that do survive to birth are usually a mosaic of diploid and polyploid cells, because the mitotic error occurred either in the second division of the zygote or later (Figure 6–3a). Individuals whose bodies are composed of cells of two or more different karyotypes are called **mosaics** (Figure

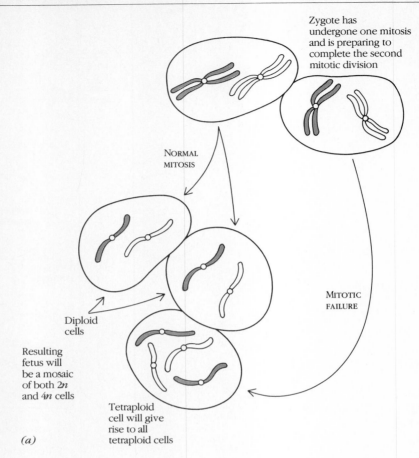

Zygote has undergone one mitosis and is preparing to complete the second mitotic division

NORMAL MITOSIS

MITOTIC FAILURE

Diploid cells

Resulting fetus will be a mosaic of both 2*n* and 4*n* cells

Tetraploid cell will give rise to all tetraploid cells

Figure 6–3
(*a*) The formation of a diploid/tetraploid mosaic. (*b*) A tetraploid child, mosaic for two cell types: tetraploid (XXYY) and diploid (XY). The child died at nine months.

(*a*)

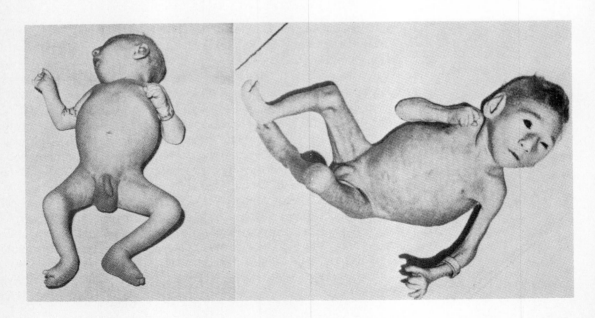

(*b*)

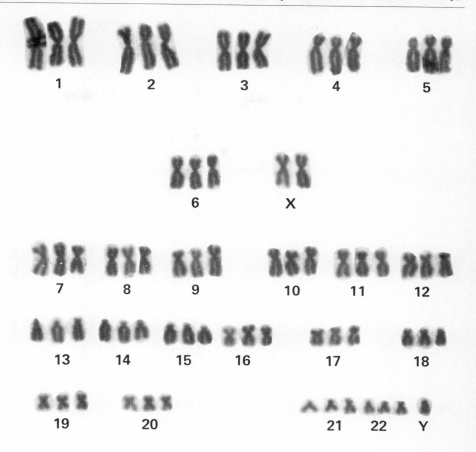

Figure 6–4
Human triploidy (from abortus).

6–3b). Aborted polyploid fetuses are usually triploid or tetraploid (Figure 6–4), although higher levels of ploidy have been observed in very early embryos.

Aneuploidy: Addition or deletion of individual chromosomes

As we noted earlier, a cell that has one or more chromosomes of a set added or missing is called aneuploid. In contrast with a euploid cell, an aneuploid cell contains incomplete chromosome sets. A diploid cell with one extra chromosome ($2n + 1$) is **trisomic** for that particular chromosome, and a diploid cell lacking a chromosome ($2n - 1$) is **monosomic** for that chromosome. The lack of both members of a chromosome pair is invariably lethal.

The origin of aneuploidy: Nondisjunction   Aneuploid cells are the result of **nondisjunction** of the chromosomes during cell division. Nondisjunction (Figure 6–5) means either that sister chromatids fail to separate properly during mitosis (or meiosis II) or that homologous chromosomes fail to separate properly

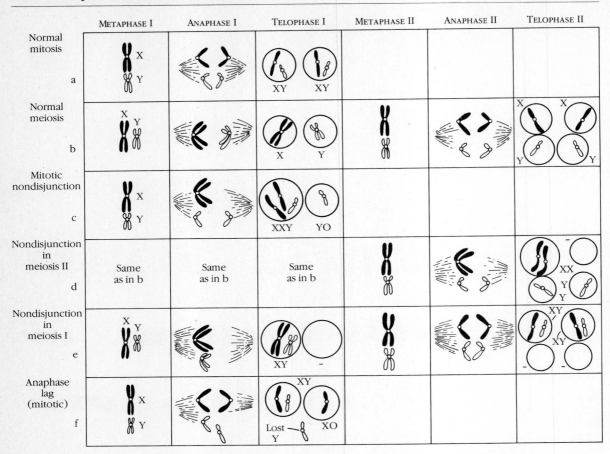

	METAPHASE I	ANAPHASE I	TELOPHASE I	METAPHASE II	ANAPHASE II	TELOPHASE II
Normal mitosis a	X Y		XY XY			
Normal meiosis b	X Y		X Y			X X Y Y
Mitotic nondisjunction c	X Y		XXY YO			
Nondisjunction in meiosis II d	Same as in b	Same as in b	Same as in b			XX Y Y
Nondisjunction in meiosis I e	X Y		XY —			XY XY — —
Anaphase lag (mitotic) f	X Y		XY Lost — Y XO			

Figure 6–5
Nondisjunction.

during meiosis I. If the two chromatids or chromosomes migrate together, one nucleus then lacks a chromosome and the other has an extra one. Such an event may come about because of incorrect attachment of the spindle fibers during metaphase, or because the centromere fails to divide.

If, instead of sticking together all the way to the poles, the sister chromatids or homologous chromosomes separate midway through anaphase, the result is a type of nondisjunction called **anaphase lag**: one chromosome is delayed in its poleward movement. Since it does not arrive at the pole in time, this chromosome is not incorporated into the new nucleus during telophase and is eventually lost in the cytoplasm. Consequently, one cell is normal, and the other is missing a chromosome. When anaphase lag occurs in meiosis I, it is often due to the failure of the homologous chromosomes to synapse properly during prophase I. This failure can cause the two members of a pair to arrive at the metaphase plate separately and then proceed to the poles at different times.

Autosomal aneuploidy: Down syndrome (trisomy 21) In humans, the loss of an autosome is invariably lethal. In fact, autosomal monosomy has not been positively identified even in abortuses (fetuses that are dead or nonviable—incapa-

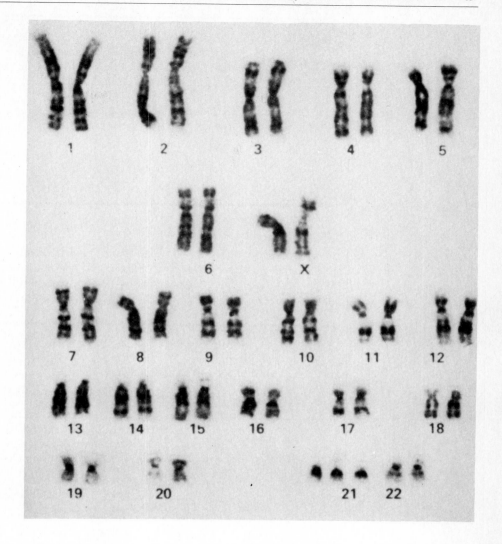

Figure 6–6
Trisomy-21 karyotype
(Down syndrome).

ble of developing or functioning normally), which indicates that development of a monosomic human being does not proceed very far, if at all, beyond the zygote stage. Autosomal aneuploid states greater than trisomy, such as tetrasomy ($2n + 2$) or pentasomy ($2n + 3$), are also fatal.

Most autosomal trisomies are also lethal, but we do find many of them in abortuses (Tables 6–2 and 6–3), so having an extra chromosome is apparently not as disruptive to human development as lacking a chromosome. A few autosomal trisomies are semilethal; that is, they allow the affected person to survive until birth, but not much longer. Some semilethal autosomal trisomies are summarized in Table 6–3. In only one autosomal trisomy do affected individuals commonly survive into adulthood: Down syndrome, or **trisomy 21**.

Down syndrome, usually caused by the presence of an extra chromosome 21 (Figure 6–6), was named for J. Langdon Down, who first systematically described the disorder in 1866. He did not, of course, know its chromosomal basis. It was not until

Table 6–2	Relative frequencies of type of chromosome anomaly of abortuses expressed as percentage of total abortuses with identifiable chromosome anomalies

Type	Percent
Trisomy	52
45,X	18
Triploid	17
Tetraploid	6
Other	7

100 years later, in 1959, that J. Lejeune found that an extra chromosome was responsible for the syndrome. Trisomy 21 was, in fact, the first chromosomal disorder in humans to be explained. This syndrome is the most common and well known of all genetically related malformations. About 1 in every 700 to 800 live births is a tri-

Table 6–3	Some semilethal and lethal autosomal trisomies in humans

Trisomy	Name of Syndrome	Identifying Features; Comments
18	Edward	Malformations in most organ systems. Developmental and mental retardation. Elongated skull, small jaw, low-set ears. Flexion of finger, rocker-bottom feet.
13	Patau	Similar to trisomy 18. Especially common: small head; small, malformed eyes; spasmodic seizures; deafness; extra digits; split tongue.
8		Mental retardation; abnormal skull shape and vertebral column; reduced joint mobility; no kneecaps; cleft palate. Live births may all be mosaics of trisomic and normal tissue.
"22"		Mental and growth retardation; small penis and undescended testicles; small head; abnormal facial structure; low-set and malformed ears; small jaw; congenital heart disease; cleft palate; fingerlike thumb; long, beaked nose. Long regarded as trisomy-22 syndrome, and discussed as such in numerous books and articles, this syndrome is actually a translocation between chromosomes 11 and 22.
9		Deformed brain, heart, and blood vessels; small head; narrow eyelid slits; large nose; small penis and testicle abnormalities; no psychomotor development; severe bone abnormalities. One child lived to the age of nine; all others died in infancy.

Figure 6–7
Trisomy 21 (Down syndrome). (*a*) The face of a child with Down syndrome. (*b*) Handprint showing the simian line, a mid-palmar crease that is found in about two-thirds of Down syndrome victims, but sometimes also appears in other generalized congenital disorders.

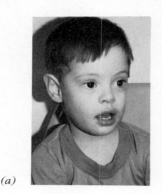

(a)

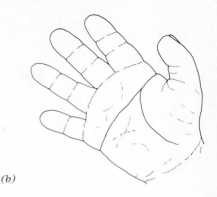

(b)

somy-21 baby, and about 15% of the patients institutionalized for mental deficiency have this syndrome.

The characteristics associated with trisomy 21 are so standard that they form a pattern known as a **clinical syndrome** (Figure 6–7a and b). A person with trisomy 21 has a broad, short skull, flat at the back, and a broad, short neck. The face is quite distinctive, with a protruding and pleated tongue and prominent eyelids that slant downward toward the nose. (The misleading term *mongolism*, formerly used to describe the syndrome, refers to the superficially Oriental appearance of the eyes.) The iris often has spots, called Brushfield spots, arranged in a ring around the pupil. The pinnae (external parts of the ears) are flat and low-set. People with Down syndrome are short, and have short, broad hands with a single crease (called a simian line) across the palm (Figure 6–7b).

Congenital heart defects are common among Down syndrome babies and are a major cause of death during the first 12 months after birth. The other major cause of death in infancy is a reduced capacity to fight respiratory infections. Nearly 20% of trisomy-21 infants die during the first year of life. If, however, they manage to survive that first year, there is a good chance that they can have a nearly normal life span. Nevertheless, when all the data are tabulated, the average life expectancy is about 18 years.

The great majority of persons with Down syndrome have IQs between 25 and 49, but cases have been reported of IQ scores as high as 75. When the afflicted children are raised at home and given conscientious parental attention, their chances for intellectual and social growth are much greater than they are for those who are raised in an institution.

Maternal age is an important factor in the incidence of trisomy 21 (Figure 6–8), as it seems to be in most other trisomies. A woman of 32 runs about three times as great a risk of having a Down syndrome child as does a woman of 19. For a woman of 40, the risk is more than 20 times that for a 19-year-old. Recent data suggest that the age of the father is less relevant in the occurrence of Down syndrome. There has been much speculation as to why the mother's age should matter so much. The most reasonable explanation for the **maternal age effect** is now centered on the potential dangers of holding primary oocytes (immature eggs) in a state of suspended meiosis I for more than 30 years (p. 61). Over such a long period, an increasing number of these cells may experience metabolic problems that are

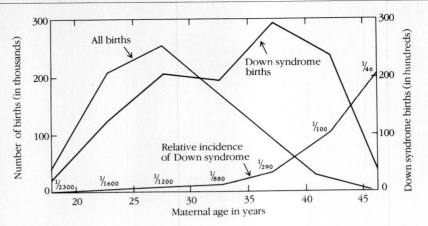

Figure 6–8
Maternal age and chromosome abnormality as demonstrated by the incidence of Down syndrome at birth. More recent data indicate the relative incidence of Down syndrome births may even be higher than indicated here.

reflected in meiotic abnormalities. These problems apparently start becoming quite acute around the age of 35.

Since chromosome 21 has satellites (see Figure 3–18) of varying length, examining the chromosomes of a trisomy-21 child and those of its parents can often reveal the parental origin of the divisional error. We can sometimes even tell whether the error occurred in meiosis I or in meiosis II. In the case diagrammed in Figure 6–9, we know that the nondisjunction occurred in the mother because one of the child's three 21 chromosomes, like the father's, has no visible satellites, while two, like the mother's, have distinct satellites. Thus, one of the child's chromosomes must have come from the father and two from the mother. We can also tell that the meiotic error occurred in meiosis II because the child's two maternal chromosomes are similar. If one of the chromosomes had a large satellite and one had a small one, we would conclude that the error had occurred in meiosis I (see Figure 6–5). Using the approach just described, researchers recently found that 77% of all cases of trisomy 21 are of maternal origin, and that the error usually

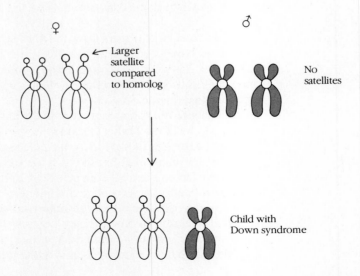

Figure 6–9
This child with Down syndrome has two 21 chromosomes with larger satellites, which must have come from the mother as a result of a failure during meiosis II.

occurs in meiosis I. Paternal nondisjunction is involved but is not so age-dependent.

Trisomy 21 is usually a sporadic event; that is, the tendency for this error to occur does not usually run in families. But, as you will see later in the chapter, Down syndrome can also be caused by a structural chromosome aberration, which can be transmitted from generation to generation like a Mendelian gene. Nevertheless, trisomy 21 is a far more common cause of the syndrome than structural anomalies.

Sex chromosome aneuploidy: Turner syndrome, Klinefelter syndrome, XYY syndrome

The addition or deletion of sex chromosomes does not usually have the disastrous consequences that autosomal aneuploidy does. Sex-chromosome aneuploidy is relatively common, and most types are not so weakening that they commonly prompt spontaneous abortion. Indeed, there are some sex-chromosome aneuploid states that present no clear phenotypic abnormalities.

A human being cannot survive without at least one X chromosome, so Y monosomy (YO) is always lethal. On the other hand, X monosomy (XO) is a viable condition; that is, the person is capable of living despite the lack of the whole chromosome. Trisomy, tetrasomy, and pentasomy of the X and Y chromosomes are viable, though phenotypically abnormal, states. Of the nearly 80 sex chromosome anomalies that have been described, we will discuss only a few. Notice that a male phenotype requires the presence of at least one Y chromosome (or a specific piece of the Y).

Turner syndrome, or X monosomy (XO), was confirmed as a chromosomal anomaly in 1959. Though it is a relatively benign syndrome, presenting no life-threatening problems after birth, it must be a severe handicap to the unborn female since nearly 20% of all abortuses with chromosomal anomalies are 45,XO (Table 6–2). There is no generally acceptable explanation for the apparent high lethality of the syndrome in embryos.

The diagnostic features of Turner syndrome are very regular. The XO person (Figure 6–10) is female, usually short and sexually undeveloped, with webbing of the neck and a reduced carrying angle at the elbow (cubitus valgus). In addition, the phenotype includes a low hairline, wide chest with broadly spaced nipples, narrowing of the aorta (the great trunk artery that carries blood from the heart), and puffy feet in newborns. Apparently no mental retardation is associated with this syndrome, though there is some disagreement on this issue, especially regarding mathematical and spatial ability.

Turner syndrome females are infertile. The ovaries are undeveloped, so the hormones they normally produce are absent. However, with proper hormone treatment, XO females can lead a normal sex life, though they remain sterile.

Current estimates indicate that nearly 75% of the XO females carry their mother's X chromosome, and only 25% carry their father's. This shows that in contrast to trisomy 21, Turner syndrome usually results from a meiotic error in the father, though we cannot tell whether the mistake occurs in meiosis I or meiosis II (see Figure 6–5).

Klinefelter syndrome (XXY) was first described by H. F. Klinefelter in 1942, but its chromosomal basis was not discovered until 1959, when other researchers described the XXY karyotype.

The clinical features of Klinefelter syndrome are not apparent until after puberty. Even then, the phenotype (Figure 6–11) is not strikingly different from that of

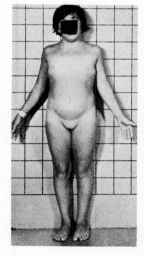

Figure 6–10
Turner syndrome (XO). Characteristically, this person, with 45 chromosomes (an X is missing), is short and has female external genitals, a webbed neck, a shieldlike chest with undeveloped breasts and widely spaced nipples, and very imperfectly developed ovaries.

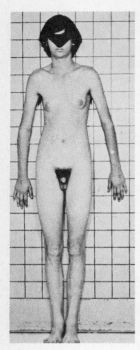

Figure 6–11
Klinefelter syndrome (XXY). This person, with 47 chromosomes (there is an extra X), has normal male external genitals but the testicles are small. His body hair is sparse. He has some female breast development and a rather feminine body conformation in general.

the normal XY male. The lower extremities tend to be longer and the testicles small and firm. XXY males are sterile. About 50% of them develop breasts. Body hair is usually sparse, and its distribution pattern resembles that of typical females. The Klinefelter male has a low sex drive, but this can be corrected by proper hormonal treatment. The XXY male is characteristically somewhat mentally disadvantaged, especially in reading and writing.

The birth frequency of XXY males is about 1.3 per 1000 live male births, and there is a maternal age effect. About 60% of these males result from the fertilization of an XX egg by a Y-bearing sperm, while the remaining 40% result from an X-bearing egg being fertilized by an XY sperm. Therefore, most of the nondisjunctional events occur during oogenesis.

Several variants of the XXY Klinefelter syndrome are known. These include XXYY, XXXY, and XXXXY chromosomal conditions. In all cases, the phenotypes become more abnormal as the number of X chromosomes increases. The XXYY male is taller, more retarded, and more aggressive than the XXY male. The XXXXY has a wide range of rather severe difficulties (Figure 6–12).

The **XYY syndrome**, with its extra Y chromosome, raises some serious issues. In 1968, a series of newspaper articles described a startling discovery: between 2% and 4% of the males in prisons and mental hospitals were XYY. The claim was made by some that the extra Y chromosome predisposes XYY persons to violent, antisocial behavior. Thus emerged the myth of the "criminal chromosome."

The truth of the matter is that the evidence linking XYY to criminal or antisocial behavior is very questionable. About 1 out of every 700 live male births is XYY, and the vast majority of these persons lead normal lives as far as we can tell.

The phenotypic alteration of XYY males is very subtle. They tend to be taller than XY males, but their muscular strength is lower and their coordination poorer. They may be prone to some skin problems. Their sexual development is normal. Their IQ scores are below average.

Psychiatric examination of XYY prisoners has shown that, in the prison setting at least, the XYY male may have a rather explosive personality, with the violent outbursts aimed primarily at property rather than people. Deviant sexual behavior is also particularly common among these prisoners. But one must bear in mind that prisoners are a select population—being in jail has a way of bringing out the worst in anybody. It may be that XYY males, on the average, have a greater tendency than other males toward impulsive behavior, but it is difficult to characterize those who are not institutionalized as anything other than normal. If there are indeed people who are genetically predisposed toward antisocial behavior, their condition would generate some challenging legal and ethical dilemmas. Should such people be monitored? Should we have a screening program in place to identify them?

Sex-chromosome mosaicism   When a mitotic error occurs after fertilization, the result is a mosaic individual, as we saw in the discussion of polyploidy. The following scenario shows how sex-chromosome mosaicism, in this case an XO/XY state, can come about (Figure 6–13):

1 Normal fertilization occurs between an X-carrying egg and a Y-carrying sperm to produce an XY zygote.
2 During the first mitotic division, all the chromosomes except one of the Ys migrate properly; that Y chromosome is lost.

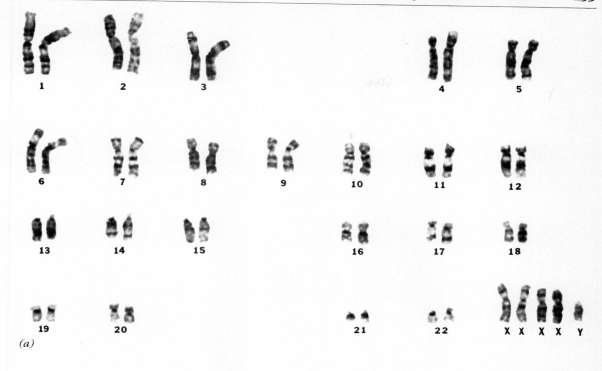

(a)

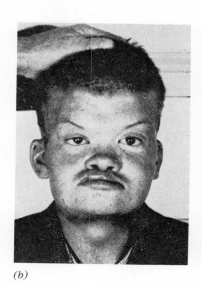

(b)

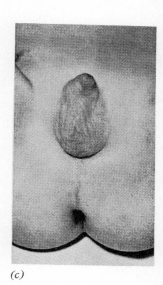

(c)

Figure 6–12
XXXXY Klinefelter syndrome. (*a*) The karyotype of a male child carrying four X chromosomes and a Y. The child carrying this chromosomal constitution presented some of the phenotypic characteristics of a Down syndrome child. (*b*) Facial abnormalities. (*c*) Genital abnormalities.

3 Of the two resulting cells, one is XY and one is XO. All cells replicate faithfully thereafter, and the resulting embryo is an XO/XY mosaic.

The XO/XY mosaic phenotype is extremely variable. About 15% of XO/XY individuals are phenotypic females, and about 5% are phenotypic males. The remaining

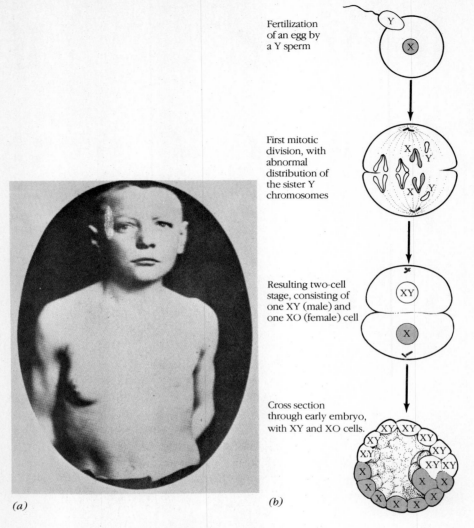

Figure 6–13
The origin and phenotypic consequences of XO–XY mosaicism. (*a*) A young man thought to be an XO-XY mosaic. (*b*) The formation of an XO–XY mosaic.

Fertilization of an egg by a Y sperm

First mitotic division, with abnormal distribution of the sister Y chromosomes

Resulting two-cell stage, consisting of one XY (male) and one XO (female) cell

Cross section through early embryo, with XY and XO cells.

(*a*)

(*b*)

80% are gynandromorphic (having characteristics of both sexes), with genitals consisting of both male and female parts. Most of these gynandromorphs have a uterus, and some exhibit certain Turner syndrome characteristics. There are many other types of sex-chromosome mosaicism, as you can see from Table 6–4.

Aberrations of Chromosome Structure

We can consider structural alterations of chromosomes from several perspectives. On the one hand, these anomalies often cause very serious medical problems. On the other hand, they have played an enormously important role in the evolution of

Table 6–4 **Human sex-chromosomal mosaics**		
Female	Male	Gynandromorphic
XO/XX	XX/XXY	XO/XY
XO/XXX	XY/XXXY	XO/XYY
XX/XXX	XXXY/XXXXY	XO/XXY
XXX/XXXX	XY/XXY/?XXYY	XX/XY
XO/XX/XXX	XXXY/XXXXY/XXXXXY	XX/XXY
XX/XXX/XXXX		XX/XXYY
		XO/XX/XY
		XO/XY/XXY
		XX/XXY/XXYYY

Note: The mosaics may combine two or three chromosomal constitutions. Phenotypically, the mosaics may be female, male, or gynandromorphic.

new species, a role we will consider in the final chapter. Finally, structural changes can tell us something about the arrangement of genes on a chromosome.

Structural changes come about when chromosomes break and are rejoined in the wrong way. Environmental agents such as radiation, viruses, and certain chemicals cause some chromosome breaks, but others occur spontaneously. The cell has mechanisms for repairing broken chromosomes, and no doubt most are rejoined correctly, but errors do occur. These errors include **duplication** (a repeated gene sequence), **deletion** (a missing sequence), **inversion** (a reversed sequence), and **translocation** (a misplaced sequence).

Duplications and deletions

A duplication may come about when a chromosome incorporates a piece of a homologous chromosome, so that it has two segments with genes coding for the same products. A deletion occurs when a fragment of a chromosome breaks off and is not returned to it. When a chromosome or chromatid is involved in a deletion, the result is one deleted chromosome and one **acentric fragment**—a piece of a chromosome without a centromere (Figure 6–14a). Such a fragment is lost when cell division occurs, because without a centromere it cannot be incorporated into either nucleus during mitosis or meiosis I. Most commonly, however, duplications and deletions come about simultaneously, when two chromatids undergo **unequal crossing over** during meiosis.

Unequal crossing over   Normally, a homologous pair of chromosomes shows a precise band for band alignment during synapsis, and a crossover between nonsister chromatids results in the exchange of equivalent chromatid segments (Figure 6–14b). But if the synapsed homologs are not properly lined up when crossing over occurs, the exchanged chromatid segments will not be equivalent (Figure 6–14c). One chromosome will end up with a duplication, the other with a deletion. This is unequal crossing over.

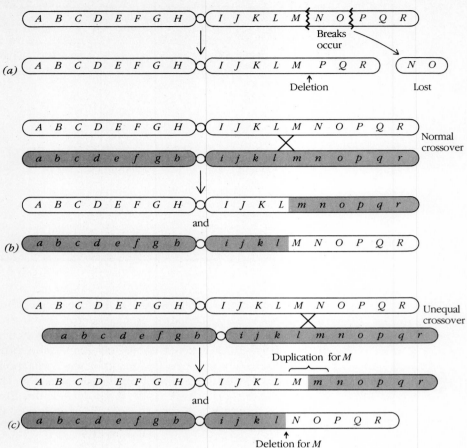

Figure 6–14
(*a*) A deletion occurs when a chromosome segment is lost. (*b–c*) The generation of duplications and deletions through unequal crossing over; (*b*) normal, (*c*) unequal. Only one chromatid for each chromosome is shown.

Although an unequal crossover produces one deletion and one duplication, the deleted and duplicated chromosomes end up in different gametes. When a gamete with a deleted or duplicated chromosome unites with a normal gamete, the result is a zygote with one normal and one abnormal chromosome for the pair in question. Thus, a chromosomally abnormal person will usually have only a duplication or a deletion, not both. The normal chromosome may compensate for the abnormal one, but deletions, in particular, often have unfortunate consequences even when heterozygous.

Observing duplications and deletions   A chromosome with a deletion is usually shorter than its normal homolog, whereas one with a duplication is longer. The deleted or duplicated segment can often be detected by comparing the banding pattern of the affected chromosome with that of the normal one. Duplications and deletions can also be detected by changes in the configuration of the bivalent during synapsis (Figure 6–15). A loop may form in the duplicated chromosome when it is paired with a normal one, because there is no equivalent segment for the duplicated

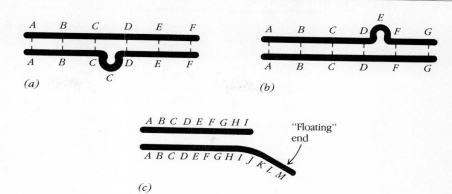

Figure 6–15
The appearance of duplication and deletion chromosomes during synapsis.
(*a*) Duplication pairing.
(*b*) Deletion pairing. (*c*) "Floating" end of nondeleted chromosome in a terminal deletion.

segment to pair with. Similarly, a loop may form in a normal chromosome when it is paired with a deleted one. If a deletion is terminal, the nondeleted chromosome has a "floating" end that does not pair. Terminal duplications are rare, because, unlike the broken ends of chromosomes, the unbroken ends are not "sticky" and therefore do not usually acquire added segments.

Deletions can be extremely valuable in determining a gene's position on the chromosome, as we saw in the discussion of gene mapping in Chapter 5. If a particular gene product is always absent when a particular chromosome band is deleted, the evidence is strong that the gene is located in that band.

Wilms' tumor: A deletion associated with a cancer   Deletions are often lethal, and they cause a great variety of malformations and anomalies. Because a deletion in a particular region of a particular chromosome may vary greatly in length from one person to another, the clinical features also vary widely. Only a karyotype analysis can determine for sure that specific abnormalities are caused by a deletion. Table 6–5 summarizes some of the better-known deletions and the clinical syndromes associated with them. The one we have selected for more detailed discussion is especially interesting because it involves a cancer.

Wilms' tumor is a distinctive type of kidney cancer that affects children. In a recent study of 44 children with Wilms' tumor, 6 turned up who had aniridia (no irises). Aniridia (Figure 6–16b) was thus 1000 times more frequent in these patients than in the general population, suggesting that there is a connection between the lack of irises and this cancer. Other physical defects, primarily in the digestive system and genitals, are also associated with Wilms' tumor, as is mental retardation (Figure 6–16a–c).

The association of Wilms' tumor with specific congenital defects led researchers to look for chromosome anomalies. In 1974, a young boy without the tumor but with the other defects, including aniridia, was found to have a small deletion in the short, or p, arm of chromosome 11. Other such cases soon turned up, and all proved to be $11p^-$ (Figure 6–16d). Significantly, some of these persons later developed Wilms' tumor. Today there are at least 20 patients known to have the $11p$ deletion, aniridia, the other physical and mental defects, *and* Wilms' tumor.

What is the relationship of the $11p$ deletion to the kidney tumor and the various other abnormalities? It may be that there is a specific gene that when absent or defective causes the tumor to develop, and this gene may lie next to others that

Table 6–5	Some chromosome deletions and their consequences	
Deletion	Name of Syndrome	Characteristics
$4p^-$	Wolf–Hirschhorn syndrome	Psychomotor and growth retardation; seizures; small head with asymmetrical features; wide space between eyebrows and eyes; low-set ears; cleft lip or palate and small jaw; absent uterus and streak gonads; congenital heart malformations. About one-third die within first two years of life.
$5p^-$	Cri-du-chat syndrome	Catlike, weak, shrill cry in infancy; small, round head with widely spaced eyes; broad nose; low-set ears; small jaw; severe mental retardation (IQ less than 25); various skeletal anomalies.
$11p^-$	Wilms' tumor	Associated with kidney cancer in children; absence of iris; mental retardation; genitourinary abnormalities; overgrowth of one side or parts of one side of the body.
$11q^-$	—	Cardiac defects; irregular and short digits; psychomotor retardation; prominent forehead; broad nasal bridge; short nose; downturned angles at the mouth. Phenotype is variable.
$13q^-$	—	Sometimes associated with retinoblastoma, an eye cancer.
$18p^-$	—	Mental retardation; slow growth; some Turner syndrome features. Phenotype is variable.
$18q^-$	—	Profound mental retardation; retarded growth; low-pitched voice; long, tapered fingers and dimples around knuckles and knees; rib and heart anomalies; genital abnormalities; deep-set eyes, short nose, and structural ear anomalies.
$21q^-$	Antimongolism	Mental and growth retardation; tense muscle tone; skeletal malformations; undescended testes; abnormal urinary tract; blood cell abnormalities; small head; low ears; antimongoloid eyes; small jaw and cleft palate.

when damaged or deleted cause aniridia, retardation, and so on. A sufficiently long deletion in this region of the chromosome could then cause all these problems.

Wilms' tumor is not the only cancer suspected of having a connection with chromosome aberrations. Children with Down syndrome, for example, have a much greater chance of developing leukemia than do normal children. Many investigators think that chromosome abnormalities may be responsible for several kinds of cancer.

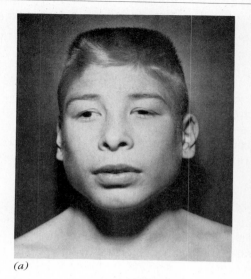

(a)

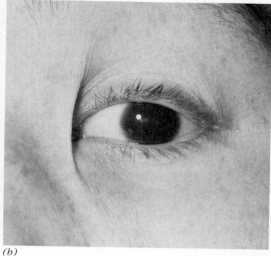

(b)

Figure 6–16

A 15-year-old male with the aniridia–Wilms' tumor syndrome. Note the small cranium (microcephaly) (*a*), absence of the iris (aniridia) (*b*), and scant pubic hair (*c*). As an infant he had a Wilms' tumor of the left kidney, and that kidney was later removed. A similar tumor developed later in the right kidney. The karyotype (*d*) shows the deletion in chromosome 11 that bands in 11p13 and 11p14.

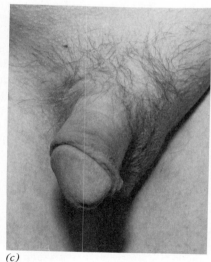

(c)

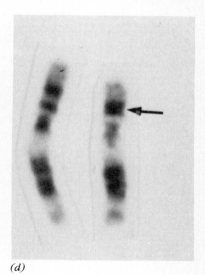

(d)

Clinical problems with duplications: Duplication in 9p   A duplication usually has a much less drastic effect on a person's phenotype than a deletion, and few duplications are lethal. Indeed, it may well be that not all duplications have visible phenotypic effects in humans—many may go undetected.

Nevertheless, some duplications produce serious problems, and we shall consider the duplication in 9p ($9p^+$) as a representative example. Recall that the *p* arm is the small arm of the chromosome. The phenotype associated with this anomaly has very distinctive features. The afflicted person is severely mentally retarded. The head is small and is abnormally short in diameter from front to back. The eyeballs are deeply recessed, and the eyelids have an antimongoloid slant—that is, they slant downward away from the nose. The nose is broad and bulbous.

The literature of human genetics is rich with reports of other duplications and their clinical consequences. But phenotypes are highly variable and are apparently dependent on which genes and how many genes are duplicated.

Inversions

An inversion is a reversal of a chromosome segment. If we were to designate a sequence of chromosome bands as *ABCDEFGH*, a possible inversion would be *AB[FEDC]GH*. Inversions come about when two breaks occur within a chromosome and the broken fragment (in our example, *CDEF*) is turned 180° and reinserted into the gap.

Inversion as a source of duplications and deletions   At first glance it might seem an inversion should have no phenotypic effects, since material is neither added nor deleted. In fact, the individual who carries an inversion is usually not affected by it. But inversions can create enormous meiotic problems, for during synapsis the paired chromosomes must be matched gene for gene. If an inversion chromosome is paired with a normal one, a loop develops in the inverted region, bringing the corresponding alleles into contact (Figure 6–17). The main problems associated with inversions result from a crossover within the inversion loop. Figure 6–18a shows what happens when a crossover occurs within an inversion segment that contains the centromere (**pericentric inversion**). Two of the resulting chromatids are genetically balanced—that is, they contain the correct genetic information—although one of them has an inversion. The other two chromatids, however, have exchanged segments so that each of them contains a duplication and a deletion. When these abnormal chromatids are incorporated into gametes, they can be passed on to the next generation, where they may have a variety of harmful effects.

If the centromere is outside the inverted segment (**paracentric inversion**), crossing over within the inversion leads to even more complex problems (Figure

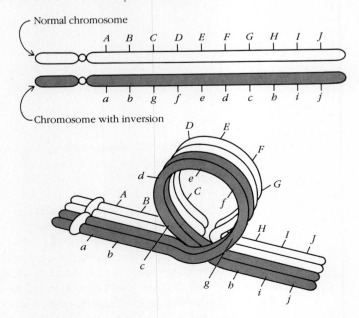

Figure 6–17
An inversion loop forms when a chromosome carrying an inversion pairs with a normal (noninverted) homolog.

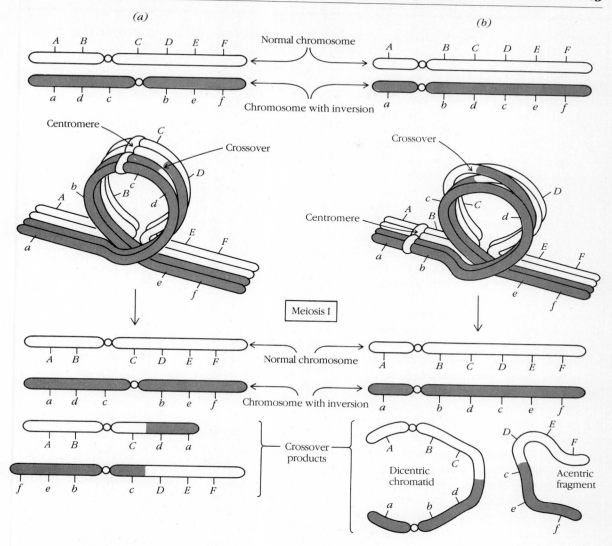

Figure 6–18
(*a*) Crossing over within a pericentric inversion causes duplications and deletions in the crossover products. (*b*) Crossing over within a paracentric inversion results in a dicentric chromatid (two centromeres) and an acentric fragment (no centromere).

6–18b). The end result in this case is two genetically normal chromatids (one inverted), an acentric (centromereless) fragment, and a **dicentric chromatid**—a chromatid with two centromeres. The acentric fragment is not pulled in either direction during cell division and thus is lost. But the dicentric chromatid is often pulled in both directions and either is torn apart, producing chromosomes with duplications and deletions, or disrupts cell division altogether.

Clinical manifestations of inversions Recent studies indicate that inversion chromosomes may occur about once in every 7500 births. Some of these inversions are new; some have been in the family line for generations. One study traced an inversion in chromosome 3 through six generations of a family line. Altogether, 20 infants suffered from severe multiple malformations as a result of the duplications

Box 6–1   *Chromosome Structure and Biological Relationships*

Just as gene linkage patterns can provide clues to the ancestry and relatedness of living species (Box 5–1), so can chromosome banding patterns. Linkage patterns and banding patterns are really just two expressions of a common system; they are both affected by the restructuring of chromosomes by the mechanisms outlined in this chapter.

It has long been known that chromosomes of at least some species will accept certain chemical stains in patterns that produce light and dark bands extending transversely across the chromosomes. The chromosomes of every normal member of a species show precisely the same band sequences. If a piece in the middle of a chromosome is inverted, it can be detected. Translocations and other rearrangements are also readily evident in the microscope.

Such banding patterns have been known for 50 years in some species of *Drosophila* (a group of small flies), and each band has been assigned a specific reference number. The bands can often be used to determine which species are ancestral to others. Assume, for instance, that the chromosomal banding pattern is checked for three rather similar species. One of them has a banding sequence that we will designate *ABCDEFGH,* another has *ABFGCDEH,* and the third has *ABFEDCGH*. It is obvious that the three differ by a pair of inversions and can be arranged in the sequence 1→3→2, as though species 1 gave rise to 3, which in turn gave rise to 2. But in this very simple example, however, the evolutionary sequence could have run in the reverse direction (2→3→1). Or, chromosome 3 could be ancestral and have independently given rise to chromosomes 1 and 2 (1←3→2). Additional data, for example, from biochemical, behavioral, and geographical studies, would be needed to determine the real order. From the banding patterns alone, however, we know that it would be highly likely that chromosome 1 gave rise to chromosome 2, which then gave rise to chromosome 3; that would require a highly unlikely series of double inversions at each step.

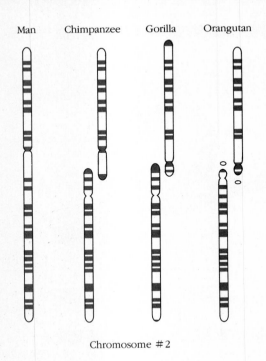

Chromosome #2

A similar but even more fascinating comparison can be made between the chromosomes of humans and the great apes (chimpanzees, gorillas, and orangutans). In the past few years we have developed techniques that can produce about 1000 bands on the basic haploid set of 23 chromosomes. (Humans have a haploid number of 23; all three ape species have 24.) Every major band can be readily identified in all four species, permitting precise comparison.

It turns out that (except for an essentially nongenic substance called heterochromatin), six chromosomes (6, 13, 19, 21, 22, and X) are identical in all four species. Seven more chromosomes are identical in three of the four species, and all the chromosomes can be traced among all the species by a small series of inversions and translocations. But what of the difference in haploid number, 24 versus 23?

The human chromosome 2 seems to have no immediately evident counterpart among the chromosomes of the ape species—and two chromosomes in each ape have no quickly identifiable counterpart in humans. But a second glance easily solves the problem. If the two "unmatched" chromosomes in the chimp, for example, are fused together end to end, their combined banding pattern now perfectly matches the human chromosome 2! The two unmatched chromosomes of the orangutan and the gorilla need only one short inversion plus the fusion to match perfectly the pattern of the human chromosome 2. So the banding patterns match, and the haploid numbers match.

The fusing of two chromosomes in this manner is very common and is found all through the biological world. The comparison between humans and the three great apes provides a classic demonstration of how one basic banding pattern has been modified into four distinct patterns and species. Researchers at the University of Minnesota, who specialize in this type of work, have worked out the apparent ancestral banding pattern from which all four modern species evidently arose (Yunis and Prakash, 1982).

Do the chromosome comparisons completely resolve the question of who among these four species is most closely related to whom? Of course not! The problem must be approached by a wide variety of other techniques, many confirming, but some complicating, the insights from chromosome banding patterns. You can find out more about that fascinating story if you go on to more advanced studies of genetics.

and deletions that occurred when crossing over took place between the inversion chromosome and a normal one.

Though the main danger of inversions lies in the duplications and deletions they generate, inversions themselves do sometimes create clinical problems. It was recently reported, for example, that a severely retarded girl carried a paracentric inversion in the short arm of chromosome 1. Her father also carried the inversion but was normal. Another young girl, who had multiple deformities, was found to have an inversion in chromosome 3. Her father and three siblings also carried the inversion, but all were normal. We do not know what accounts for the difference between the persons who carry the inversion.

Translocations

A translocation occurs when a segment of one chromosome is transferred to another, nonhomologous, chromosome. The most commonly observed type of translocation is the **reciprocal translocation**, which involves the exchange of segments of nonhomologous chromosomes (Figure 6–19a). Two other types of translocations also exist. For instance, an interior segment of one chromosome may be transferred to the interior of a nonhomologous chromosome; or, rarely, the end of one chromosome may be transferred to the unbroken end of a nonhomologous chromosome.

Translocation as a source of unbalanced gametes Since genetic material is neither lost nor gained in a translocation, it often produces no phenotypic abnormalities. The real problem with translocations, as with inversions, is the formation of gametes that carry duplications and deletions, giving rise to aborted fetuses or abnormal children. Since a block of genetic material has been shifted from one chromosome to another, it stands to reason that the chromosome pairing in meiosis will be affected.

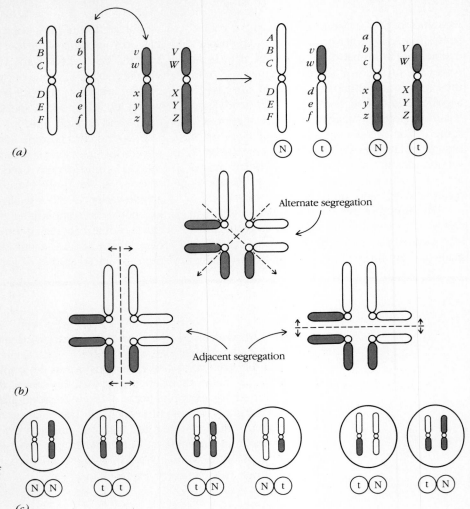

Figure 6–19
The formation and meiotic consequences of reciprocal translocations. (*a*) A reciprocal translocation between two nonhomologous chromosomes. (*b*) Synapsis of translocated chromosomes with their normal homologs. (*c*) Possible results of meiosis in a translocational heterozygote. N = normal, t = translocation.

In a reciprocal-translocation heterozygote, synapsis produces a characteristic cross-shaped configuration, because *two* pairs of chromosomes must synapse to get all the homologous segments together (Figure 6–19b). Studying the diagram should convince you that crossing over can occur in this configuration without affecting the gene sequence; crossovers do not have the disastrous structural consequences here that they do in the case of an inversion. Still, problems can arise as a result of the segregation of the chromosome pairs, depending on how they happen to assort.

If the two translocated chromosomes segregate together and the two normal ones segregate together (alternate segregation), the gametes will be balanced, because each will have a member of every allelic pair. If, however, a translocated chromosome and a normal one segregate together (adjacent segregation), the gametes will be **unbalanced**, because they will have duplications and deletions. When an unbalanced gamete is involved in a fertilization, the result will depend on

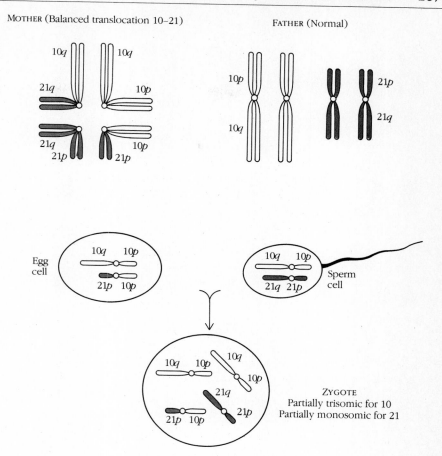

MOTHER (Balanced translocation 10–21)   FATHER (Normal)

ZYGOTE
Partially trisomic for 10
Partially monosomic for 21

Figure 6–20
The case of a mother carrying a translocation and producing an abnormal child.

the number and nature of the affected genes, particularly those that have been deleted.

A translocation may go undetected in a family line for generations if the parents always pass normal chromosomes or balanced-translocation chromosomes to their children. But sooner or later, an unbalanced gamete is likely to be involved in a fertilization, with unhappy results.

Consider the case of a severely retarded 46,XY boy with an abnormal chromosome 21. The banding pattern showed that, in place of a piece of chromosome 21, there was a piece of chromosome 10. Thus, the child was partially trisomic for 10 and partially monosomic for 21. His mother, who was mentally and physically normal, turned out to be a balanced reciprocal-translocation heterozygote (Figure 6–20). Her contribution to her child was a normal chromosome 10 and the translocated 21p10p chromosome; the father contributed a normal 10 and a normal 21. As a consequence of the unbalanced gamete from the mother, the boy had a duplicated segment in 10 and a deleted segment in 21.

The Philadelphia chromosome In 1960, researchers working in Philadelphia discovered that patients who suffered from **chronic granulocytic leukemia**

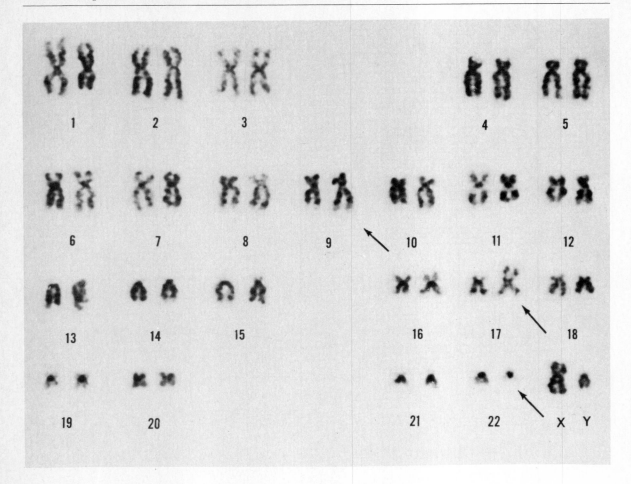

Figure 6–21
Banded karyotype from a patient with chronic granulocytic leukemia. In addition to the typical translocation from chromosome 22 to chromosome 9, t(9q;22q), producing the Philadelphia chromosome, there is an isochromosome for the long arm of 17, iso-17q, replacing a normal 17. This additional abnormality is also observed commonly in patients with chronic granulocytic leukemia.

(CGL) usually carried a specific chromosome anomaly. The long (q) arm on chromosome 22 was shorter than normal in these people. The aberrant chromosome 22, with its short q arm, was named the **Philadelphia chromosome (Ph¹)**. In 1973, after the development of banding techniques, the deleted segment of the Philadelphia chromosome was discovered—attached to the long arm of chromosome 9 (Figure 6–21). Thus, the Philadelphia chromosome was not a simple deletion, but a translocation. Patients who have CGL but not the Ph¹ chromosome do not have any added chromosomal material on their chromosome 9.

The relationship between Ph¹ and CGL is rather mystifying. Some CGL patients have the deletion but do not appear to have the deleted segment translocated to any of their other chromosomes. It may be there, but so far it has successfully resisted detection. The fact that CGL develops in people with the translocation argues that gene position may be crucial to the development of CGL, for all the genes are present, only their arrangement has been altered. As for the CGL patients who do not have the Ph¹ chromosome, it may be that there are two different kinds of CGL, one that involves Ph¹ and one that does not. A profile of the disease does seem to indicate that there are two types.

It appears that the Philadelphia chromosome is not transmitted to offspring by parents but develops in the blood-forming tissues after conception. The mitotic error may occur in a single cell that goes on to produce a clone of Ph[1] cells.

Translocational Down syndrome As we mentioned earlier in the chapter, trisomy 21 is not the only way to produce Down syndrome. About 5% of the afflicted children have 46 chromosomes but carry a translocation that provides them, in effect, with a partial trisomy 21. The translocation is usually between the long arms of chromosomes 21 and 14, but sometimes a chromosome other than 14 is involved.

In about half the cases of translocational Down syndrome, we find that the condition has appeared in several generations of a family line. In most of these cases, one of the parents (almost always the mother) has 45 chromosomes but is phenotypically normal. This situation comes about because of a translocation in which the long arm of a chromosome 21 has fused to the long arm of a 14 or 15 to form a single chromosome. (This type of translocation, in which whole arms of different chromosomes fuse at the centromere, is called a **centric fusion**, or **Robertsonian translocation**.) When this happens, the two short arms may also fuse to form a second, very tiny chromosome, but it is usually lost, leaving a total of 45. However, since the chromosome formed by the long arms contains nearly all the material of the 14 and the 21, the person with this karyotype is essentially normal. In addition to the fused 21–14 chromosome, the 45-chromosome individual has one normal 21 and one normal 14 (Figure 6–22).

The problem, as in most translocations, arises when random segregation produces unbalanced gametes. The possibilities are shown in Figure 6–23. If both normal chromosomes go to one gamete and the translocation chromosome goes to the other, both gametes are genetically balanced. One, on being fertilized, will produce a normal 46-chromosome person, and the other will produce a person who has 45 chromosomes but is phenotypically normal. Each of the other two possibilities, however, leads to one partial monosomy and one partial trisomy. The partial monosomies are lethal, as is the partial trisomy 14, but the partial trisomy 21 produces Down syndrome.

In the cases where translocational Down syndrome arises without a family history of the condition (approximately half of all cases), the translocation presumably occurred during the formation of one of the gametes that gave rise to the affected person.

Isochromosomes and rings

Duplications and deletions may also occur in other ways. An **isochromosome** is an abnormal chromosome with two arms of equal length and bearing the same genes in reverse sequence. The most common explanation of the origin of such a chromosome is that it is the product of a chromosome that split the wrong way during mitosis or meiosis II. That is, instead of the centromere splitting along a line parallel to the long axis of the chromosome, it split along a line perpendicular to that axis (Figure 6–24). There is evidence, however, that isochromosomes may be the result of centric fusion of homologous chromosomes.

An isochromosome carries a partial duplication and a partial deletion. Since most of these are lethal, we do not see many of them in live births, and those that do

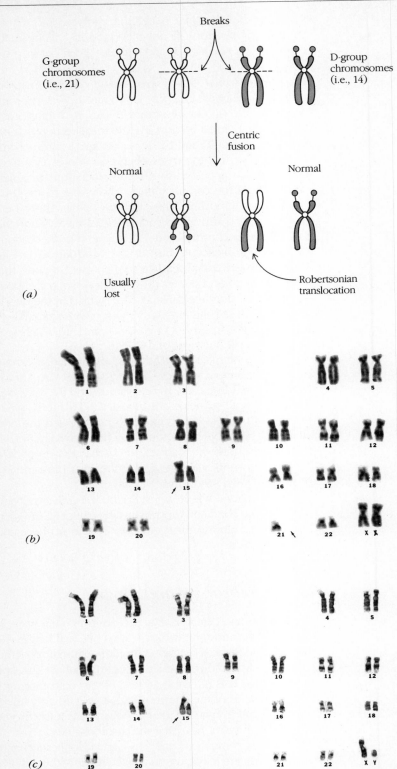

Figure 6–22
(*a*) The formation of a Robertsonian translocation. (*b*) Karyotype of a female carrier of a 15/21 translocation. The missing chromosome 21 is translocated to 15. (*c*) Karyotype of a Down syndrome child with a 15/21 translocation.

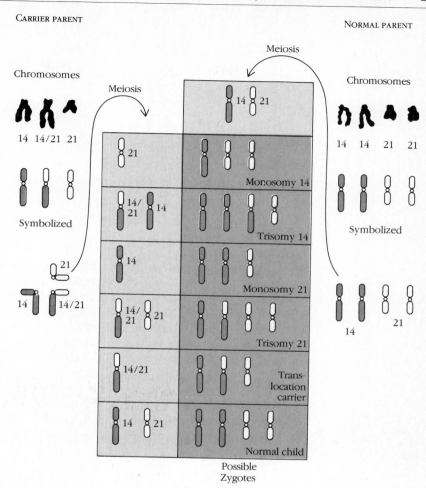

CARRIER PARENT

NORMAL PARENT

Meiosis

Chromosomes

Meiosis

Chromosomes

14 14/21 21

14 14 21 21

Symbolized

14 21

Symbolized

21

Monosomy 14

14/21 14

Trisomy 14

14

Monosomy 21

14/21 21

Trisomy 21

14/21

Translocation carrier

14 21

Normal child

14 14/21

Possible Zygotes

14 21

Figure 6–23
Inheritance of centric fusion, or Robertsonian translocation. The carrier parent appears to have only 45 chromosomes, but this is because of a 14–21 translocation. Usually it is the mother who carries this translocation. The two monosomies and the trisomy 14 are all lethal.

occur usually involve the X chromosome. People with X isochromosomes usually show various features of Turner syndrome. Autosomal isochromosomes usually involve the acrocentric chromosomes in groups D and G.

When breaks occur at both ends of a chromosome, the two ends sometimes fuse to form a **ring chromosome** (Figure 6–25). If the ring carries a centromere, it may replicate. But when it does, it generates interlocking rings that can cause serious cell-division problems. Ring chromosomes are usually associated with phenotypic anomalies.

Genetically determined chromosome breaks

We said earlier that not all chromosome breaks are caused by environmental insults. The causes of spontaneous breaking are not well understood, but two recently investigated phenomena—unstable chromosomes and fragile sites—indicate that certain breakpoints can be inherited like genes.

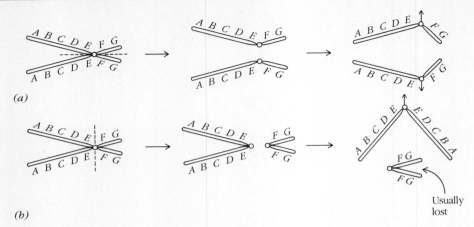

Figure 6–24
Formation of isochromosomes. (*a*) Centromere divides normally in mitosis and meiosis II. (*b*) Centromere "misdivides." This type of split is commonly suggested as the cause of an isochromosome (plus a centric fragment).

Usually lost

Unstable chromosomes   Three well-studied genetically caused diseases show that there can be a genetic cause of **unstable chromosomes**—that is, chromosomes with spontaneous breaks. The three diseases are **Bloom's syndrome**, characterized by low birth weight and dwarfism; **Fanconi's anemia**, characterized by aplastic anemia (an anemia in which red blood cells fail to regenerate), and anomalies of the kidneys, heart, and extremities; and **ataxia telangiectasia** (AT), characterized by defective muscular coordination, telangiectases (red spots caused by dilation of small blood vessels) of the eyes and the skin, and immunity problems that heighten susceptibility to respiratory infections.

Interestingly, the three diseases share features in addition to chromosome breaks: in all of them the patient has sun-sensitive skin spots and/or telangiectases and has a greatly increased chance of developing leukemia. The skin conditions are

Figure 6–25
(*a*) Formation of a ring chromosome. (*b*) Photo of a ring chromosome.

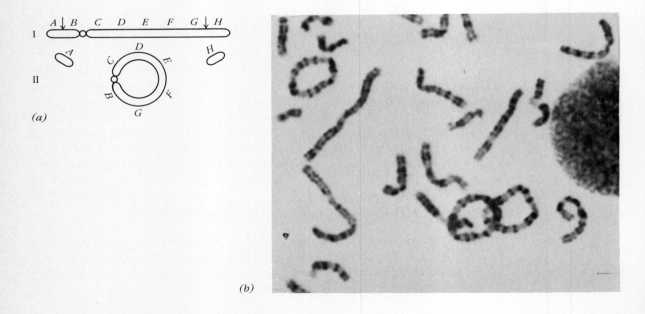

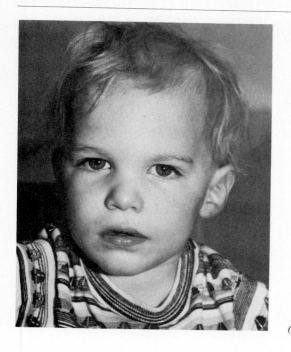

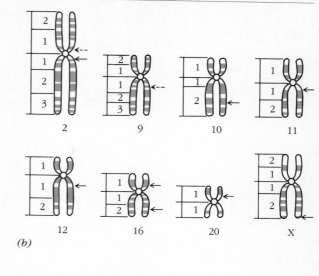

(b)

(a)

Figure 6–26
Fragile sites. (*a*) This child, though normal in appearance, is known to have chromosomal fragile site breakage. (*b*) Well-established fragile sites (solid arrows) and sites that have not yet been absolutely confirmed (broken arrows).

particularly interesting, because we know that sun-sensitive spots and telangiectases are related to the malfunctioning of genetically controlled processes that repair damaged chromosomes.

In patients with all three conditions, the chromosomes are especially unstable and show many breaks. Recent discoveries show that the chromosomal breakpoints are not random; in AT, for example, one of the breakpoints always seems to be in a specific region of the long arm of chromosome 14. We do not yet know the significance of the specific breakpoints.

Fragile sites   A **fragile site** is a specific point on a chromosome that breaks or shows a gap when white blood cells are cultured under specific conditions. The sites are always the same and are inherited as though they were Mendelian genes. At the time of this writing, eight fragile sites are well established and two are probable but unconfirmed (Figure 6–26b).

Fragile sites are truly curious cytogenetic phenomena. A given fragile site does not occur in every cell of a given person, and homozygotes (with breaks occurring at the same site in both chromosomes of a pair) have not been observed. The sites seem to be innocuous (i.e., harmless), since most of them apparently are not associated with phenotypic abnormalities (Figure 6–26a). Though one fragile site on the X chromosome is associated with a specific form of mental retardation, that is probably because the gene causing the retardation is located very close to the fragile site, so that the two are inherited together.

As you may well imagine, geneticists have many questions about fragile sites. So, first of all, we must understand what these sites mean. Then, by studying them, we hope to learn more about chromosome structure in general, the function of particular bands, and the position of loci responsible for genetic diseases.

Personal Aspects of Chromosome Aberration

About one in every 250 live births is a chromosomally abnormal child. It is therefore no wonder that prospective parents are often anxious about the condition of their child-to-be. If the baby does turn out to have a chromosomally caused defect, the couple may well seek genetic counseling before deciding to have another child. Table 6–6 shows the incidence of specific abnormalities.

Genetic counselors have a difficult task. They must understand the precise nature of the genetic defect that has prompted the request for counseling, and they must be able to assess the degree of risk involved in each potential birth. In addition, they must deal effectively with the complex emotional problems created by the birth of an abnormal child.

If a couple has produced a child with Down syndrome, a genetic counselor needs to know the specific chromosome aberration that caused the condition. This

Table 6–6	**Incidence of chromosome abnormalities, newborn surveys (total = 56,952; M = 37,779; F = 19,173)**
Type	Incidence
Sex chromosome:*	
Males: total abnormalities	1:400
47,XYY	1:1100
47,XXY	1:1100
Other	1:1300
Females: total abnormalities	1:700
45,X	1:9500
47,XXX	1:950
Other	1:2700
Autosome:	
Trisomies: total incidence	1:700
Trisomies 13	1:19,000
18	1:8000
21	1:800
Other	1:5500
Autosomal structural abnormalities:	
Unbalanced	1:1700
Balanced	1:500

*Represents 36% of total abnormalities.

Source: E. B. Hook and J. L. Hamerton: The frequency of chromosome abnormalities detected in consecutive newborn studies, in *Population Cytogenetics—Studies in Humans*, ed. by E. B. Hook and I. H. Porter, Academic Press, New York, 1977.

aberration can be determined by cytological studies of parents and child, with additional information provided by a family pedigree. If the cause of the syndrome is trisomy 21, and if the mother is young, the chances of recurrence are small—much less than 1%. If the mother is 35, the chances are much greater—about 5%—but even this probability is low enough that many people facing such a risk would consider having another child.

If the problem is a translocation, however, the parents face a much greater risk. If the mother carries a balanced-heterozygous translocation between chromosomes 21 and 14, she can produce a normal child with normal chromosomes, a normal child with a balanced translocation, or another Down syndrome child—and the risk of Down syndrome does not change much with her age. The factors that determine the proportion of children in each category are quite complex and not entirely understood, but empirical data show that there is an 11% chance of giving birth to a second child with translocational Down syndrome. A woman who becomes pregnant knowing this risk must certainly experience tremendous anxiety.

One way parents can deal with the problem of anticipated chromosome defects is to have a prenatal diagnosis. **Amniocentesis**, a technique developed in the 1960s, enables us to assess the chromosomal constitution of the fetus long before birth, and it is becoming a fairly common procedure for pregnant women with a particularly high risk of bearing an abnormal child—women over 35 and those with a personal or family history of chromosome defects. The procedure is illustrated in Figure 6–27. A sample of the amniotic fluid that surrounds the fetus is withdrawn by

Figure 6–27
The process of amniocentesis.

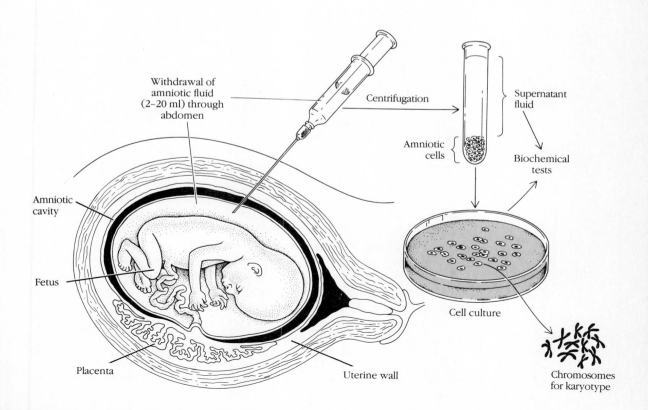

a long hollow needle inserted in the mother's abdomen during the fourteenth to sixteenth week of pregnancy. The fetal skin cells that are routinely shed into the fluid are cultured in a special culture medium for about three weeks, then fixed and stained for chromosome analysis. The procedure is considered to be of low risk, since it causes spontaneous abortion only about once in 200 cases.

Analysis of the fetal chromosomes can detect all known chromosome defects, as well as the sex of the fetus. In addition, biochemical tests of the cultured fetal cells can detect many problems due to genetic aberrations that cause no visible changes in the chromosomes. We are able to detect about 60 kinds of enzyme deficiency and other metabolic defects in the fetus by means of these biochemical procedures. The cytological and biochemical analyses take at least a week, so when the results are finally available the fetus may be 20 weeks old, the usual legal limit for therapeutic abortion.

When a woman and her doctor agree that an amniocentesis should be performed, the usual assumption, explicit or implicit, is that the woman will have an abortion under certain circumstances. But it is not always easy to decide what these circumstances should be. When amniocentesis shows that the fetus has a semilethal condition, most women choose abortion. Most also have abortions if they know the child will have Down syndrome, even though some children with this condition can lead reasonably well-adjusted lives, given special care. But there are conditions that are much less serious, medically, and the choice here is not so clear-cut for many people. What would you decide to do, for instance, if your unborn child were found to be XXY or XO, both fairly benign and treatable conditions? What would you do if the child were diagnosed as XYY, which has no clearly established anomalies associated with it, but which *may* result in social or behavioral problems? And if you elected not to abort an XYY fetus, would you tell the child later on that he is genetically abnormal? These are not easy questions to answer, and as our ability to diagnose problems prenatally improves, more and more such questions will arise.

Assessing the genetic risks of cancer is another area in which scientific progress may lead to some hard choices. Certain inherited diseases, such as Fanconi's anemia, ataxia telangiectasia, and Bloom's syndrome, are chromosomally unstable conditions that entail a high risk of cancer, especially leukemia. Is it appropriate for a physician to tell the immediate family of an AT patient that they themselves are five times more likely than other people to develop cancer? Many people do not want to know of their cancer risk because they believe that there is nothing they can do to prevent the cancer from developing in any case. But who is to decide whether to tell them or not?

We must also ask whether it is appropriate or useful to set up chromosome screening for the general population, so as to spot those who have a high cancer risk. Some would argue that such screening could lead to a slew of new problems. For example, some persons might be denied employment because of their propensity to develop cancer, or they might be denied access to certain jobs within a particular industry. Industry could specifically select people with a low cancer risk to work in special jobs involving toxicologic hazards, then lower the safety levels in those jobs to save money. People who were shown to be at risk for cancer might be denied life insurance. The more we learn about the relation of chromosome defects to cancer—as we surely will—the more we will have to face thorny questions of a practical, ethical, and social nature.

Summary

1 Errors in cell division can result in aberrations of chromosome *number*.

2 **Euploidy** is any whole number multiple of the haploid number. Any euploid state beyond diploidy is called **polyploidy**, and is caused by replication of chromosomes without cell division (mitotic or meiotic failure). Polyploidy is lethal in humans.

3 The loss or addition of individual chromosomes is called **aneuploidy**, and is caused by **nondisjunction**: failure of sister chromatids or homologous chromosomes to separate during mitosis or meiosis.

4 The loss of one chromosome of a pair is called **monosomy**. Loss of an autosome is invariably lethal in humans. The addition of a third chromosome to a pair is called **trisomy**. Most autosomal trisomies are lethal; a few are semilethal. Only trisomy 21 (Down syndrome) is considered viable. Higher autosomal aneuploidies (tetrasomy, etc.) are lethal in humans.

5 There are many kinds of sex-chromosome aneuploidy. Most are not severely disabling, and in some the phenotype is normal. Several extra Xs or Ys may be added. One X is necessary for survival, so YO is lethal, but XO (**Turner syndrome**) is a nearly normal condition.

6 Mitotic nondisjunction in the second or later division of the zygote results in a human **mosaic**: a person with cells of different karyotypes.

7 Aberrations of chromosome *structure* occur when chromosomes break, either spontaneously or as a result of environmental influences, and are not repaired properly.

8 A **duplication** is a repeated sequence of gene loci; a **deletion** is an omitted sequence. A deletion and a duplication may be created simultaneously through **unequal crossing over** (exchange of nonequivalent segments) in meiosis.

9 The clinical manifestations of duplications and deletions depend on the number and nature of the genes involved. Duplications usually have less severe consequences than deletions, and some may not affect the phenotype at all. Deletions are usually lethal; when they are not, they tend to cause severe problems. **Wilms' tumor**, a kidney cancer, is associated with a deletion in 11*p*, as is a lack of irises.

10 An **inversion** is a reversal of a chromosome segment and generally has no phenotypic effect on the carrier. The main problems associated with inversion are caused by crossing over within the inverted segment, which gives rise to chromosomes with deletions and duplications and to **acentric fragments** and **dicentric chromosomes**. These can cause problems in the next generation.

11 A **translocation** is the shift of a chromosome segment to a nonhomologous chromosome. Most translocations are **reciprocal**: two nonhomologous chromosomes exchange segments. A **centric fusion**, or **Robertsonian translocation**, involves the fusion of whole arms of different chromosomes.

12 Translocations usually have no effect on their carriers, but they cause meiotic problems that result in gametes with duplications and deletions (**unbalanced gametes**). A centric fusion of chromosomes 14 (or 15) and 21 causes translocational Down syndrome when a child receives both the translocated chromosome and a normal 21 from one parent. A translocation that apparently causes direct harm is the **Philadelphia chromosome**, associated with a form of leukemia.

13 An **isochromosome** has two arms of equal length that bear the same genes. It is formed by abnormal centromere division or by centric fusion and carries du-

plications and deletions. Isochromosomes are usually lethal. A **ring chromosome** forms when both ends of a chromosome break off and the arms fuse. Its presence creates meiotic problems and clinical symptoms.

14 **Unstable chromosomes**, which break easily at specific sites, are associated with three genetic diseases. The three have overlapping symptoms, including increased susceptibility to cancer. **Fragile sites** are points on a chromosome that break when white blood cells are cultured under specific conditions. These phenomena show that breakpoints on chromosomes can be inherited like genes.

15 **Amniocentesis** is the sampling of fetal cells to be cultured and analyzed for their karyotype and biochemical characteristics. It is particularly valuable for older women and those with family histories of genetic defects. The information it provides can help prospective parents decide whether to continue a pregnancy.

16 Genetic counselors can help people make reasonable decisions about having children when there is a likelihood of chromosome aberrations or other genetic defects. The advice they can give is much more tenuous when it comes to the relationship between chromosome defects and cancer, even though in certain cases high cancer risks can be spotted.

Key Terms and Concepts

Polyploidy

Aneuploidy

Euploidy

Tetraploid

Mitotic failure

Meiotic failure

Triploid

Mosaic

Trisomic

Monosomic

Nondisjunction

Anaphase lag

Down syndrome (trisomy 21)

Clinical syndrome

Maternal age effect

Turner syndrome

Klinefelter syndrome

XYY syndrome

Duplication

Deletion

Inversion

Translocation

Acentric fragment

Unequal crossing over

Wilms' tumor

Pericentric inversion

Paracentric inversion

Dicentric chromosome

Reciprocal translocation

Unbalanced gamete

Chronic myelogenous leukemia

Philadelphia chromosome (Ph[1])

Centric fusion (Robertsonian translocation)

Isochromosome

Ring chromosome

Unstable chromosomes

Bloom's syndrome

Fanconi's anemia

Ataxia telangiectasia

Fragile site

Amniocentesis

Problems

6–1   What is the parental origin of the extra chromosome in the Down syndrome child below, and was the error in meiosis I or meiosis II?

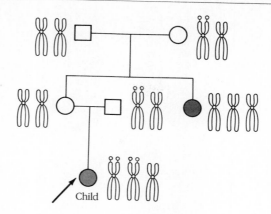

6–2 How many different kinds of gametes can an XYY male form, and what are they?

6–3 On rare occasions, a person with trisomy 21 is fertile. If a trisomy-21 female mates with an XYY male, what types of zygotes could be formed?

6–4 What types of zygotes would you expect from a mating between two people with trisomy 21?

6–5 What proportion of the zygotes formed from the mating in Problem 6–4 would you expect to survive to birth, and why?

6–6 A normal woman and a normal man produce a daughter who suffers from hemophilia, a sex-linked trait. How is this possible?

6–7 through 6–10 Study the human chromosome spreads shown below. For each spread, determine the genetic sex of the person and the chromosomal anomaly, if any, that is present. Refer to pages 67 and 69 to classify the chromosomes.

6–7 (left)
6–8 (right)

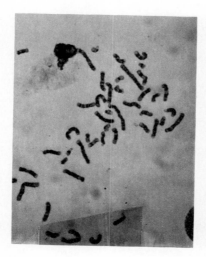
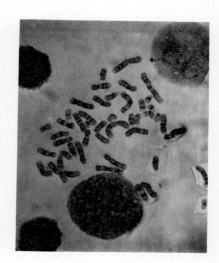

6–9 (left)
6–10 (right)

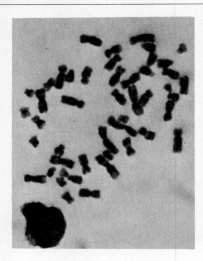 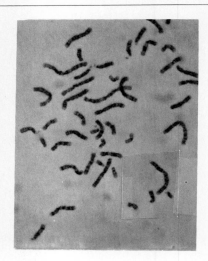

6–11   Using Figure 3–23, point out the following chromosomal regions: 4*p*16; X*q*25; 15*q*22; 2*q*34; 7*p*21.

6–12   Propose a mechanism for the generation of an XXXXX female.

6–13   A pair of chromosomes has the following sequence of regions: *ABCDEFGHIJK* and *abchgfedijk*. Draw this pair of chromosomes as they would appear during prophase of meiosis I.

6–14   Suppose a crossover occurs between E and F while the above pair of chromosomes is synapsed. What would the anaphase of meiosis I look like? Draw a diagram.

6–15   A reciprocal translocation has occurred between the following nonhomologous chromosomes:

ABCDEFGQRSTUV   LMNOPHIJK

Assuming that this is a translocational heterozygote, draw the chromosomes as they would pair in meiosis I.

6–16   What kinds of gametes would be produced from the pairing diagrammed in Problem 6–14?

6–17   Diagram the pairing that would occur between the following pair of chromosomes, one of which carries a deletion:

ABCDEFGHIJKLM   abcdefjklm

6–18   Diagram the pairing that would occur between the following pair of chromosomes, one of which has a duplication:

ABCDEFGHIJKJKLM   *abcdefghijklm*

6–19 A normal woman with 45 chromosomes has a child with Down syndrome. This child has 46 chromosomes. How would you explain this?

Answers

See Appendix B.

References

Allderdice, P. W., N. Browne, and D. P. Murphy. 1975. Chromosome 3 duplication q21 → qter deletion p25 → pter syndrome in children of carriers of a pericentric inversion inv(3)(p25q21). *Am. J. Hum. Genet.* 27:699–718.

Archidiacono, N., M. Rocchi, M. Valente, G. Filippi. 1979. X pentasomy: A case and review. *Hum. Genet.* 52:69–79.

Bartsch-Sandhoff, M., and G. Hieronimi. 1979. Partial duplication of 17p: A new chromosomal syndrome. *Hum. Genet.* 47:217–220.

Carr, D. H. 1971. Chromosomes and abortion. *Adv. Hum. Genet.* 2:201–258.

Deroover, J., J. P. Fryns, J. Haegeman, and H. Van Der Berghe. 1979. Paracentric inversion in the short arm of chromosome 1. *Hum. Genet.* 49:117–121.

Dobzhansky, Th. 1970. *Genetics of the Evolutionary Process.* Columbia University Press, New York.

Erickson, J. D. 1979. Paternal age and Down syndrome. *Am. J. Hum. Genet.* 31:489–497.

Emery, A. 1976. *Methodology in Medical Genetics.* Churchill Livingstone, New York.

Ghosh, P. K., R. Rani, and R. Nand. 1979. Lateral asymmetry of constitutive heterochromatin in human chromosomes. *Hum. Genet.* 52:79–84.

Gorlin, R. J., and W. S. Boggs. 1977. Genetic aspects of facial abnormalities. *Adv. Hum. Genet.* 8:235–346.

Hamerton, J. L. 1971. *Human Cytogenetics,* vol. I: *General Cytogenetics;* vol. II: *Clinical Cytogenetics.* Academic Press, New York.

Hamerton, J. L., N. Canning, M. Ray, and S. Smith. 1975. A cytogenetic survey of 14,069 newborn infants. I. Incidence of chromosome anomalies. *Clin. Genet.* 8:223–243.

Hecht, F., and B. Kaiser-McCaw. 1979. The importance of being a fragile site. *Am. J. Hum. Genet.* 31:223–225.

Hook, E. B., and I. H. Porter, eds. 1977. Population Cytogenetics: Studies in Humans. Academic Press, New York.

Howard-Peebles, P. M., G. R. Stoddard, and M. G. Mims. 1979. Familial X-linked mental retardation, verbal disability, and marker X chromosomes. *Am. J. Hum. Genet.* 31:214–222.

Hsu, T. C. 1979. *Human and Mammalian Cytogenetics—An Historical Perspective.* Springer Verlag, New York.

Magenis, R. E., K. M. Overton, J. Chamberlin, T. Brady, and E. Lovrien. 1977. Parental origin of the extra chromosome in Down syndrome. *Hum. Genet.* 37:7–16.

Miklos, G. L., and B. John. 1979. Heterochromatin and satellite DNA in man: Properties and prospects. *Am. J. Hum. Genet.* 31:264–280.

Nowell, P. C., and D. A. Hungerford. 1960. A minute chromosome in human chronic granulocytic leukemia. *Science* 132:1497.

Nowell, P. C., and D. A. Hungerford. 1961. Chromosome studies in human leukemia. II. Chronic granulocytic leukemia. *J. Natl. Cancer Inst.* 27:1013–1035.

Opitz, J. M., J. Herrmann, J. C. Pettersen, E. T. Bersu, and S. C. Colacino. 1979. Terminological, diagnostic, nosological, and anatomical-developmental aspects of developmental defects in man. *Adv. Hum. Genet.* 9:71–164.

Riccardi, V. M., H. Mintz-Hittner, U. Francke, S. Pippin, G. P. Holmquist, F. L. Kretzer, and R. Ferrell. 1979. Partial triplication and deletion of 13q: Study of a family presenting with bilateral retinoblastomas. *Clin. Genet.* 15:332–345.

Roberts, J. A. F., and M. E. Pembrey. 1978. *An Introduction to Medical Genetics,* 7th ed. Oxford University Press, Oxford.

Robinson, A., H. A. Lubs, and D. Bergsma. 1979. Sex chromosome aneuploidy: Prospective studies on children. (Birth Defects Original Article Ser., vol. 15, no. 1.) New York: Alan R. Liss.

Rowley, J. D. 1973. A new consistent chromosome anomaly in chronic myelogenous leukemia identified by quinacrine fluorescence and giemsa staining. *Nature* 243:290–293.

Rowley, J. D. 1976. Population cytogenetics of leukemia. In Porter, I. H., and E. D. Hook, eds., *Population Cytogenetics: Proceedings of Birth Defects Institute Symposium.* Academic Press, New York.

Sutherland, G. R. 1979. Heritable fragile sites on human chromosomes. I. Factors affecting expression in lymphocyte cultures. *Am. J. Hum. Genet.* 31:125–135.

Sutherland, G. R. 1979. Heritable fragile sites on human chromosomes. II. Distribution, phenotypic effects, and cytogenetics. *Am. J. Hum. Genet.* 31:136–148.

Therman, E. 1981. *Human Cytogenetics.* Springer Verlag, New York.

Tjio, J. H., and A. Levan. 1956. Chromosome number of man. *Hereditas* 42:1–6.

Verma, R. S., and H. Dosik. 1979. Precise identification of human chromosome abnormalities. *Am. J. Hum. Genet.* 31:82–83.

Yunis, J. J., ed. 1974. *Human Chromosome Methodology,* 2nd ed. Academic Press, New York.

Yunis, J. J., ed. 1977. *New Chromosomal Syndromes.* Academic Press, New York.

Yunis, J. J., and O. Prakash. 1982. The origin of man: A chromosomal pictorial legacy. *Science* 215:1525–1530.

7 The Molecular Basis of Inheritance

Consider Michael, the child with cystic fibrosis we met in Chapter 1. Michael's symptoms could all be traced to a malfunctioning of the mucus-secreting glands, which could in turn be traced to the absence of a single enzyme—perhaps the normal form of NADH dehydrogenase. His cells produced only a nonfunctional form of the enzyme because he had two abnormal alleles for the enzyme's production (see Figure 1–1). These alleles had been passed from generation to generation in the families of Michael's parents, their effects masked by normal alleles that coded for a functional form of the enzyme. Their presence was unsuspected until misfortune struck—and Michael inherited them from both his parents.

We have said repeatedly that genes exert their influence through enzymes and other proteins, and that an alteration in a single protein can have widespread effects, as it did in Michael. But we have not yet explained *how* genes cause the production of particular proteins. We have discussed the inheritance patterns of genes and the mechanisms by which copies of genes are passed from cell to cell and generation to generation, but we have not explained how the copies are made. Indeed, we have not yet defined a gene, except in the vaguest of terms. In this chapter we will begin to probe the most fundamental questions of modern genetics: What is a gene? How does it reproduce itself? How does it produce its effects? The answers to all these questions lie in the molecular structure of the nucleic acids.

The Nucleic Acids: DNA and RNA

The genetic material of all living organisms is one of two nucleic acids, **deoxyribonucleic acid (DNA)** or **ribonucleic acid (RNA).** DNA is the genetic material in all cellular organisms and in most of the viruses; only in a few simple viruses does RNA carry the genetic information. Nevertheless, RNA plays a vital role in the heredity of all organisms, for it serves as an intermediary in the translation of genetic information into proteins. In addition, as recently discovered, it assists in the replication of DNA.

Identifying DNA as the genetic material

DNA was discovered in the mid-nineteenth century when the Swiss physician Friedrich Miescher isolated a substance he called "nuclein" from cell nuclei of salmon sperm and the white blood cells from pus he obtained from bandages. However, neither Miescher nor any of his colleagues were in a position to suggest that this substance was the genetic material. By the 1920s, chemical analysis of chromosomes had shown that they consist of DNA, protein, and a very small amount of RNA, so it seemed possible that genes might consist of DNA. But most biologists favored the idea that the protein component of the chromosome carries the genetic information. Proteins were known to be complex molecules, whereas the molecular structure of DNA was thought to be relatively simple. It did not seem to the biologists of that time that the DNA molecule incorporated enough variation to serve as a code for the structure of a complex organism.

The first real clue to the chemical identity of the gene came in 1928, when Frederick Griffith discovered the phenomenon of **bacterial transformation**. Griffith's experiment is summarized in Figure 7–1. *Pneumococcus* bacteria type IIIS (S for smooth) are enclosed in a carbohydrate capsule that is responsible for the pathogenic (disease-causing) properties of this strain. When these bacteria are injected into a healthy mouse, they cause it to develop pneumonia. If, however, the IIIS cells are heat-killed before they are injected, the mouse develops no symptoms, since only living cells can produce the disease. A mutant *Pneumococcus* strain called IIR (R for rough) lacks a carbohydrate capsule and is thus unable to cause pneumonia. When Griffith injected a healthy mouse with heat-killed IIIS cells and living IIR cells, neither type capable of causing pneumonia by itself, the animal contracted pneumonia and died. When he examined the animal for bacteria, he found living IIIS cells! It was well known from earlier experiments that type II *Pneumococcus* never mutate to type III. Where, then, had the type III cells come from?

After much debate, the correct interpretation of Griffith's results emerged: a substance released from the heat-killed IIIS bacteria was able to transform the IIR cells into pathogenic IIIS cells. This unknown substance was named the "transforming principle," and 16 years later other researchers discovered its chemical nature.

Figure 7–1
Scheme of Griffith's transformation experiment.

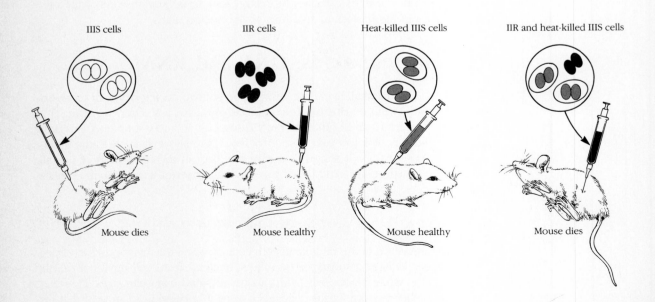

IIIS cells   IIR cells   Heat-killed IIIS cells   IIR and heat-killed IIIS cells

Mouse dies   Mouse healthy   Mouse healthy   Mouse dies

They found that DNA extracted from IIIS pneumococcal cells could transform IIR cells into IIIS, while the proteins, carbohydrates, and lipids that also make up IIIS cells could not cause this transformation. It was evident that DNA can carry genetic information from one cell to another, at least in bacteria. In 1952, still other experimenters showed that DNA is the genetic material in viruses. Since bacteria and viruses are very dissimilar kinds of organisms, biologists inferred that DNA is the universal genetic material.

The chemical composition of nucleic acids

The chemical components of the nucleic acids were analyzed and understood long before DNA was seriously implicated as the genetic material. Molecules of nucleic acids are made of smaller molecules: pentose sugars, phosphates, and nitrogenous bases. Let us briefly examine the meaning of these chemical terms.

A sugar molecule is composed of carbon, hydrogen, and oxygen, with the carbon molecules arranged in a ring. Unlike the more familiar hexoses, or six-carbon sugars (e.g., fructose, glucose), the **pentose sugars** contain five atoms of carbon per ring. DNA contains the pentose sugar **deoxyribose**, and RNA contains **ribose**; deoxyribose is so named because it has one less oxygen atom in its ring than does ribose (Figure 7–2).

A **phosphate** molecule is made up of phosphorous, hydrogen, and oxygen. Phosphates are extremely reactive; that is, they combine readily with many other substances. It is the phosphates that hold the individual sugar molecules together in the long-chain molecule of a nucleic acid.

The most important components of nucleic acids, from the genetic point of view, are the **nitrogenous** (nitrogen-containing) **bases** (Figure 7–2). The simplest chemical definition of a base is that it is a substance that can react with an acid to produce a salt and water. A molecule that is basic by itself can form part of a more complex molecule that is acidic, so it is not surprising that nitrogenous bases are part of nucleic acids. The four nitrogenous bases found in DNA are **adenine, cytosine**, **guanine**, and **thymine**, represented in diagrams as A, C, G, and T, respectively. RNA contains adenine, cytosine, guanine, and **uracil** (U). Notice that adenine and guanine have double carbon–nitrogen rings, while cytosine, thymine, and uracil have single rings. The double-ring bases are called **purines**, and the single-ring bases **pyrimidines**.

Clues to the structure of DNA

Analyzing the chemical makeup of nucleic acids is relatively simple; determining how the molecule is put together is quite a different matter. Merely knowing the ingredients that go into DNA and RNA suggests nothing about the unique properties of these molecules. As biologists began to accept the idea that DNA is the genetic material, they realized that its structure must hold the secret of its function. By 1950, biologists and chemists in laboratories throughout the world were working on the structure of DNA.

More sophisticated chemical analysis during the 1940s and '50s provided some of the necessary clues. Investigation of the DNA of many different organisms showed

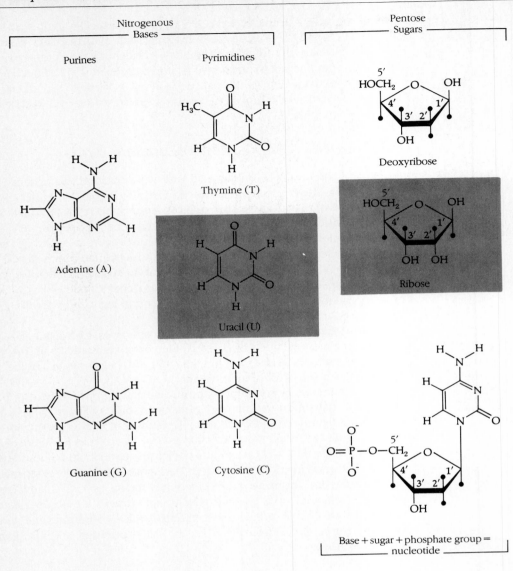

Figure 7–2
Nucleic acid constituents.

that the proportion of nitrogenous bases differs from one species to another. Thus, DNA is not a simple substance of uniform composition, as was previously supposed, but possesses the variability necessary for incorporation of genetic information. Since the variation is in the nitrogenous bases, this finding suggested that the bases might be the key element in the coded information. It was further discovered that in the DNA of any given species the number of adenine units is exactly equal to the number of thymine units, while the number of guanine units is exactly equal to the number of cytosine units. This discovery was an essential precursor to the Watson–Crick model of DNA, for it suggested that molecules of the bases come in pairs: A with T and G with C.

The Double Helix: The Structure and Replication of DNA

To qualify as a genetic material, a substance must satisfy three requirements:

1 It must be able to replicate faithfully.
2 It must be able to store genetic information.
3 It must be able to mutate (produce stable changes in its structure).

In 1953, the British journal *Nature* published the now-famous paper "Molecular Structure of Nucleic Acids," in which James Watson and Francis Crick proposed the **double-helix model for DNA**. So beautifully did this structure meet the requirements of a genetic material that biologists accepted its validity almost at once. In this chapter, we will consider how the DNA double helix replicates and how it stores information; in Chapter 9, we will discuss how it mutates.

Watson and Crick's research consisted largely of trying out various three-dimensional models, and they relied heavily on the discoveries of other researchers in the fields of chemistry and X-ray crystallography. In X-ray crystallography, X rays are bounced off crystalline substances like purified DNA; the patterns these deflected X rays make on photographic film can reveal the shape of the molecules that deflected them. The work of the X-ray crystallographers strongly indicated that the DNA molecule is a helix, and at the time Watson and Crick published their paper several groups of researchers were working on helical models. Watson and Crick relied particularly on the work of Maurice Wilkins and Rosalind Franklin, who had produced the first good X-ray images of DNA.

The genius of Watson and Crick lay in their ability to sort through the rapidly growing mass of information (some of it seemingly contradictory) about DNA and to distill it into a simple, reasonable model. The structure they proposed was extremely plausible chemically and, moreover, suggested the mechanisms both of DNA replication and of the encoding of genetic information. Watson and Crick's paper ushered in the modern age of molecular biology. It stands with the works of Darwin and Mendel as one of the very few single publications that have profoundly altered our understanding of life.

DNA structure

Watson and Crick proposed the following essential features of the DNA molecule (Figure 7–3):

1 It is composed of two parallel helical chains, each made up of subunits called **nucleotides**.
2 Each nucleotide consists of one deoxyribose molecule, one phosphate group, and one nitrogenous base—adenine, cytosine, guanine, or thymine. The deoxyribose and phosphate portions of the nucleotides are on the outside of the molecule, forming the **backbone** of each chain; the bases are on the inside.
3 Each nucleotide is bound to other nucleotides in the same chain by **phosphodiester bonds**, strong chemical bonds between deoxyribose and phosphate (Figure 7–3).

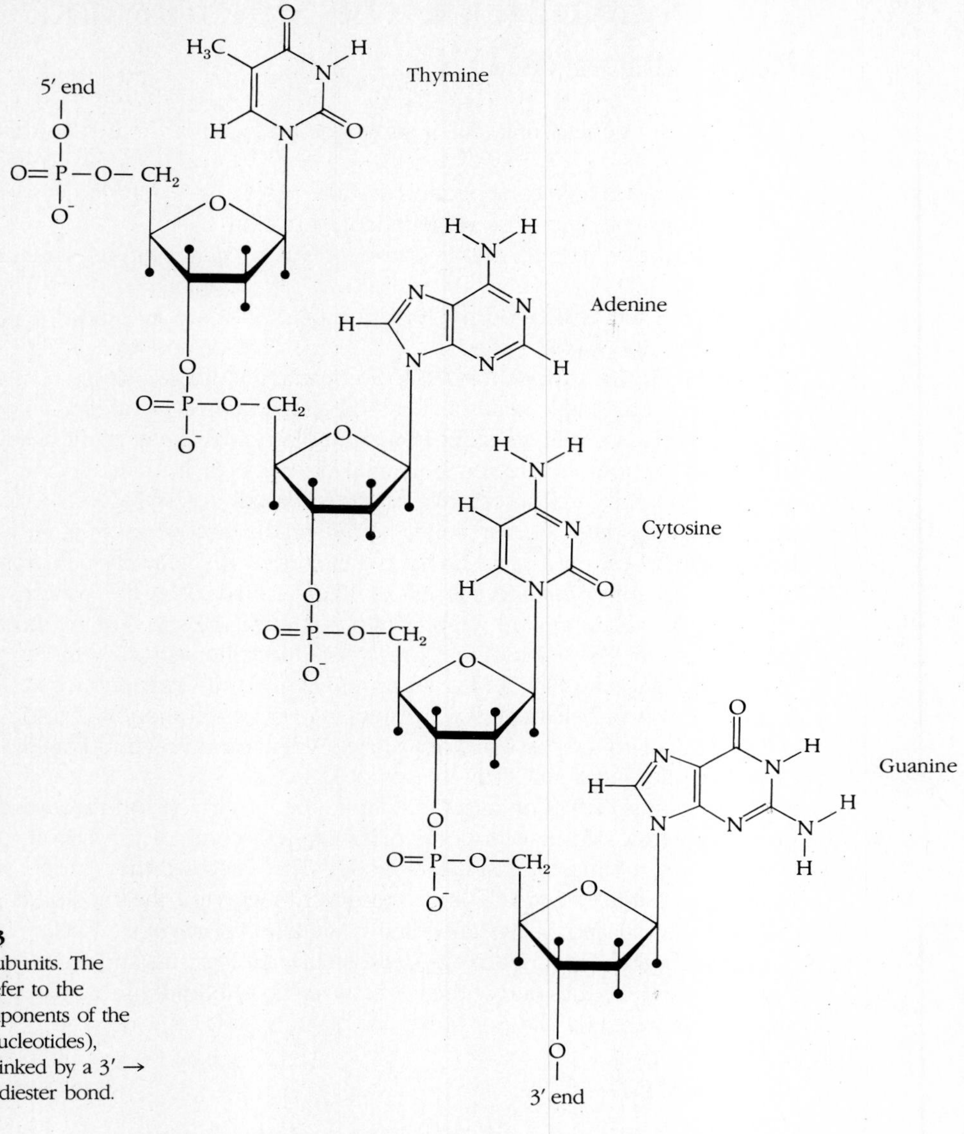

Figure 7–3
The DNA subunits. The numbers refer to the atomic components of the subunits (nucleotides), which are linked by a 3′ → 5′ phosphodiester bond.

4 Each nucleotide is bound to a nucleotide on the other chain by weak chemical bonds called **hydrogen bonds** between specific pairs of bases. Adenine pairs most efficiently with thymine (or uracil) and guanine with cytosine. Thus, the two chains are **complementary** in structure—the sequence of bases on one determines the sequence of bases on the other (Figure 7–4).

Watson and Crick proposed that when DNA replicates, the two chains come apart and each serves as a **template**, or pattern, for the formation of a new chain. They further suggested that the sequence of bases in each chain might constitute a genetic code that is somehow translated into specific proteins. Intensive research

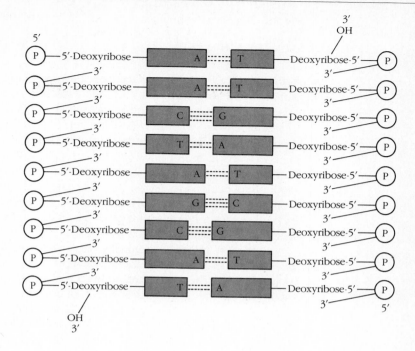

Figure 7–4
The terminus of the DNA molecule has one strand with a free 3′ end and one strand with a free 5′ end. Thus, there is opposite polarity. (P) = phosphate.

during the 1950s and '60s proved beyond a doubt that both these ideas were correct, and we now know the mechanisms of DNA replication and protein synthesis in considerable detail.

DNA replication

The capacity for replication is an essential property of the genetic material, and indeed of life. Billions of years ago, in the virginal environment of the young Earth, there evolved a protobiological system—a "living unit"—that could organize the raw material it confronted into replicas of itself. This primitive capacity for self-replication has developed into an elaborate and efficient process involving cells and multicellular organisms. There are differences in many of the details of reproduction from one species to another, but we find nonetheless that replication of the genetic material is fundamentally similar in organisms as simple as bacteria and as complex as people.

Despite the great diversity of life, DNA comes in only two basic forms. One is the linear molecule we have just described; the other is circular. The linear form is perhaps more common, occurring in some viruses and in all **eukaryotes**—organisms whose cells have true nuclei (i.e., the nucleus is surrounded by a membrane). Circular DNA is more common in some viruses and in the **prokaryotes**—the bacteria and blue-green algae, whose cells contain nuclear material (not bounded by a membrane). Circular DNA is also found in mitochondria (see Chapter 1), which presumably evolved from prokaryotes. Circular DNA is similar in construction to the linear form, being composed of two helical strands of nucleotides. It is not at all unusual in the living world to find linear DNA molecules become circular before replication, then convert back to linear forms after replication is completed.

We will discuss only the linear DNA molecule here, but you should realize that the replication of circular DNA differs only in detail.

Replication is semiconservative   The replication of DNA is called **semiconservative**, because each new molecule conserves or retains half of the material—one strand—of the parental molecule. When a DNA molecule is about to replicate, its strands unwind and the weak hydrogen bonds between them are broken. Each strand serves as a template for the formation of a new strand. Nucleotides, which are preassembled in the nucleus, are linked together by enzymes called **DNA polymerases**. The sequence of nucleotides in the new chain is determined by the sequence of nucleotides on the parental chain, since A normally binds only to T, and C only to G (Figure 7–5). The nucleotides of the newly forming chain are bound to each other by strong phosphodiester bonds and to the nucleotides of the parental chain by the weaker hydrogen bonds.

As Watson and Crick originally conceived it, semiconservative replication begins at one end of the DNA molecule and proceeds continuously to the other. They supposed that as the two parental strands "unzipped," each simultaneously served as a template for the continuous synthesis of a new strand. But there were problems with this scheme. If you examine Figure 7–5 closely, you will see that the strands of the DNA molecule are arranged in an antiparallel fashion. That is, the nucleotides in one chain are all oriented in the opposite direction from those in the other chain.

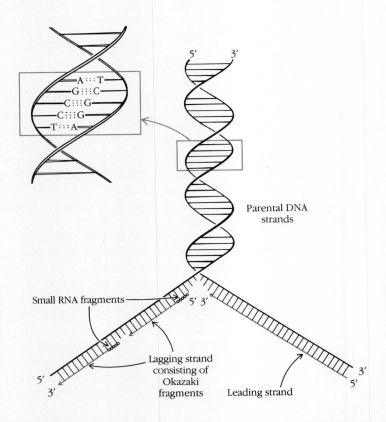

Figure 7–5
Replication of the DNA double helix. This is called semiconservative replication because a parental DNA strand is conserved in each new daughter DNA molecule. The growth of the lagging strand illustrates that DNA replication is discontinuous and uses RNA as a primer.

Parental DNA strands

Small RNA fragments

Lagging strand consisting of Okazaki fragments

Leading strand

Each chain is described as having a **5′ end** and a **3′ end**, a terminology derived from the system of numbering the carbon atoms in the deoxyribose ring (see Figure 7–3), and the 5′ end of one chain lies opposite the 3′ end of the other. For replication to proceed continuously as the parental chains unwind, one newly forming chain would have to add nucleotides in a 5′ → 3′ direction, the other in a 3′ → 5′ direction. But experiments soon showed that DNA strands always grow in a 5′ → 3′ direction, because DNA polymerase must attach to the free 3′ end of a nucleotide.

Replication is discontinuous The dilemma was resolved when it was discovered that replication of one DNA strand is discontinuous and uses short lengths of RNA as primers that supply the required free 3′ end (Figure 7–5). RNA can be synthesized anywhere on the DNA template, because the enzyme that links RNA nucleotides together does *not* require a free 3′ end. Replication of one strand of DNA (the **leading strand**) proceeds continuously in a 5′ → 3′ direction for some distance, but not the full length of the parental strand. Synthesis then switches to the other strand (the **lagging strand**), RNA primers are synthesized, and small chains of DNA are added to their 3′ ends. Thus, each added fragment of the lagging strand is built up in a 5′ → 3′ direction, as required by the enzymes involved, but the strand as a whole grows discontinuously in a 3′ → 5′ direction. The RNA primers are eventually removed with the aid of enzymes, and the small DNA fragments, called **Okazaki fragments** (after their discoverer, Reiji Okazaki), are enzymatically linked together to form a continuous strand.

In the large DNA molecules in humans, replication does not proceed from one end of the molecule to the other, but is initiated at several points. This creates a series of "bubbles" that expand as replication proceeds (Figure 7–6). The bubbles eventually fuse, and the daughter molecules separate.

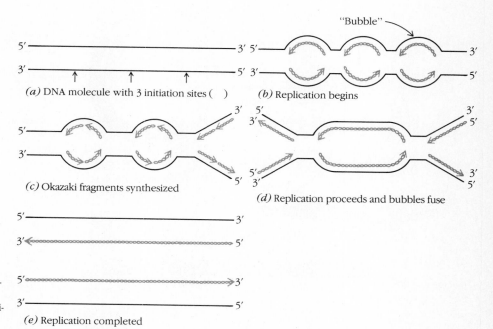

Figure 7–6
DNA replication in eukaryotes, including humans. Notice that there are multiple initiation sites.

Box 7–1   *Recombinant DNA Technology*

One of the most exciting developments in biology today is the manipulation of genes by recombinant DNA technology. This methodology has already contributed significantly to our knowledge of mammalian gene structure and the organization of genes in the eukaryotic chromosome. It may revo- lutionize the treatment of human inherited dis- eases by providing human enzymes or hormones for replacement therapy. At present, the proteins extracted from animals for this purpose differ enough from their human analogs to cause aller- gic reactions in sensitive people.

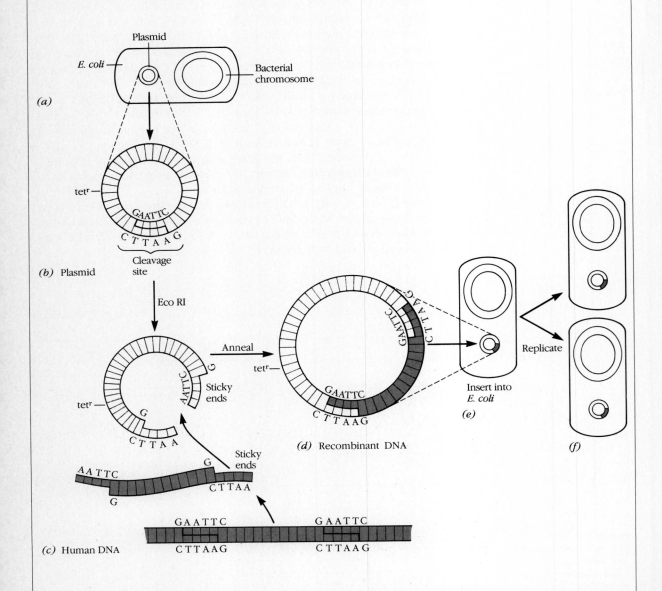

Recombinant DNA technology makes use of small, self-replicating circles of DNA, called plasmids, that occur naturally within bacterial cells. Plasmid-borne genes may provide the host bacterium with resistance to one or more antibiotics, such as tetracycline. The circle of plasmid DNA may be broken at specific sites by enzymes called *restriction endonucleases*. For example, the endonuclease Eco RI cleaves the nucleotide sequence $\frac{\text{CTTAAG}}{\text{GAATTC}}$ between G (guanine) and A (adenine) on each strand. If a plasmid with only one such sequence is chosen for cleavage, the DNA circle is opened up, leaving TTAA unpaired at one end of the molecule and AATT unpaired at the other end. If DNA from a "foreign" source, such as human DNA, is also cleaved by Eco RI, the fragments produced have single-stranded ends that are base-complementary to the ends of the broken circle of plasmid DNA. The complementary ends can join, or anneal, with one another, thus producing a larger circle of DNA, with the human DNA fragment spliced into the plasmid DNA (Figure 1). This hybrid molecule is referred to as *recombinant DNA*. Covalent bonds are formed enzymatically between the sugar-phosphate backbones of the plasmid (vector) DNA and the human (insert) DNA. A bacterial cell may take up the recombinant DNA and thereby become resistant to specific antibiotics. More important, the bacteria carrying the recombinant plasmid may produce the product of the inserted gene, even though the gene is of human origin. This gene product may be an important molecule such as insulin, which would be of great value to diabetics. Individual transformed cells may divide repeatedly, producing a clone of identical cells, all resistant to the antibiotic and producing the product of the inserted gene.

In many cases, it is easier to identify and isolate the messenger RNA for a particular protein than it is to locate the unique fragment of DNA carrying the gene. For example, red blood cells are a rich source of nearly pure mRNA for globin, the protein portion of the red pigment hemoglobin. The purified mRNA can be copied into a double-stranded DNA by the enzyme *reverse transcriptase*. The cDNA (copy DNA) can be inserted into a plasmid and cloned (see Box 8–1) to yield large amounts of globin DNA. The globin DNA can then be used as a probe to hunt for the globin gene within the chromosomal DNA. DNA extracted from human cells is digested with a restriction endonuclease and the fragments are separated by electrophoresis. A radioactively labeled cDNA is denatured (the strands separated) and added to the separated chromosomal DNA fragments under conditions that allow complementary sequences to hybridize with one another. The observation that globin cDNA binds to several chromosomal fragments led to the unexpected discovery that the globin gene is made up of coding regions separated by noncoding (untranslated) regions. Similar studies have revealed that most eukaryotic genes share this type of structure.

Because the genetic code is known for each amino acid, biochemists can deduce the nucleotide sequence coding for any protein whose amino acid sequence has been determined. They can chemically synthesize the gene by linking together nucleotides and can insert this synthetic gene into a plasmid. This procedure was used to manufacture genes that code for biologically active mammalian somatostatin (a hormone) and human insulin. Inserted in bacterial plasmids, such genes can be used to produce these hormones for medical use.

The early fears that this technology might produce some uncontrollable new life form have been calmed by the demonstrated safety of the procedures currently in use. Most investigators feel that the potential benefits of recombinant DNA technology far outweigh any risks involved.

Replication in humans is much slower than in bacteria—about one-sixth the rate. But, perhaps to compensate for this more leisurely rate, multiple origin points of replication exist.

In spite of variation in replication rate and mechanisms, all living organisms share the same basic semiconservative scheme of replication. Time and evolutionary divergence have resulted in the variations as well as the basic similarities.

It is evident that enzymes play a very important role in DNA synthesis. Some of the same enzymes that are involved in replication are also involved in DNA repair and recombination. We will discuss these processes in Chapter 9.

From Gene to Protein

A protein, you will recall, is composed of one or more polypeptides—chains of amino acids. By the time Watson and Crick published their model of DNA structure, it was well established that at least some genes code for polypeptides. It was evident as soon as the Watson–Crick model was proposed that the seemingly random sequence of bases in the DNA molecule might be a sort of ticker-tape code for the manufacture of specific polypeptide chains.

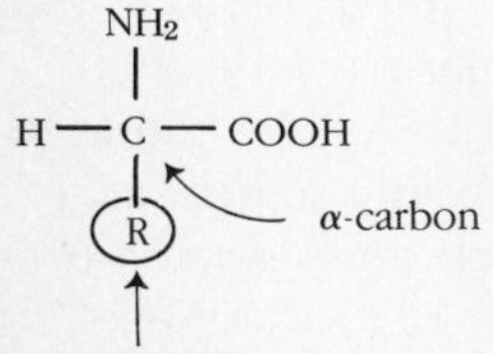

Amino acids, polypeptides, and proteins

The basic unit of a protein is an **amino acid**, a molecule composed of carbon, hydrogen, oxygen, and nitrogen. The chemical structure of an amino acid is shown in Figure 7–7. The COOH group, which is acidic, is called a **carboxyl group**, and the NH_2 group is called an **amino group**—hence the name *amino acid*. R (for radical) represents any of 20 groups called **side chains,** a few of which are shown in Figure 7–8. There are thus 20 different amino acids. Amino acids are linked to one another by **peptide bonds**, bonds between the amino group of one amino acid and the carboxyl group of another. Figure 7–9 shows how such a bond is formed. Two linked amino acids form a dipeptide, three a tripeptide, and four or more a **polypeptide**.

Proteins consist of one or more polypeptides, but they are not simple, linear structures. The R groups of the various amino acids in a protein form bonds with each other, causing the polypeptide chain (or chains) to twist and fold into various three-dimensional configurations (Figure 7–10). The reaction between two specific R groups is determined by the nature of the groups, the distance between them, and the presence of other specific groups in their vicinity, so the folding of a protein is exactly determined by the amino acid sequences of its chains. The three-dimensional configurations thus produced either conceal or expose various parts of the molecule, determining which groups of atoms can react with various substances in

This R group varies, giving 20 different common amino acids

Figure 7–7
The general formula for an amino acid.

Figure 7–8
The structures of five amino acids. The R groups distinctive to each molecule are shaded gray.

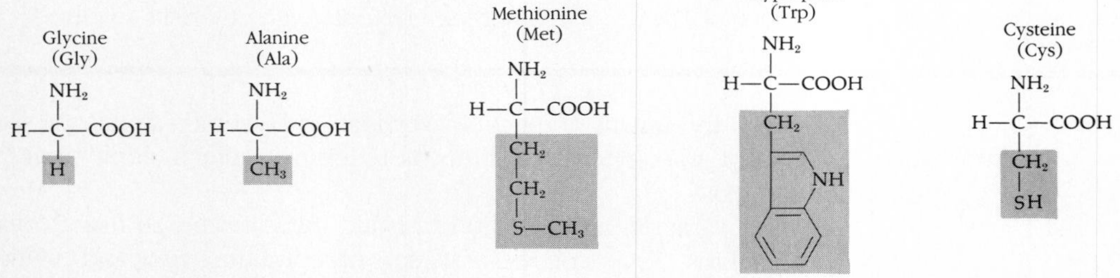

$$\cdots + HO-\underset{\underset{O}{\|}}{C}-\underset{\underset{H}{|}}{\overset{\overset{R}{|}}{C}}-N-H + HO-\underset{\underset{O}{\|}}{C}-\underset{\underset{H}{|}}{\overset{\overset{R'}{|}}{C}}-N-H + \cdots$$

$$\downarrow$$

$$HO-\underset{\underset{O}{\|}}{C}-\underset{\underset{H}{|}}{\overset{\overset{R}{|}}{C}}-\underset{\underset{H}{|}}{\overset{\overset{H}{|}}{N}}-\underset{\underset{O}{\|}}{C}-\underset{\underset{H}{|}}{\overset{\overset{R'}{|}}{C}}-N-H + H_2O$$

Peptide
bond

Figure 7–9
The formation of a peptide bond between two amino acids.

the environment. Thus, the function of a given protein—its reaction with other metabolic substances—is ultimately determined by the amino acid sequence of its constituent polypeptide chains. Sometimes the omission or substitution of a single amino acid causes such a distortion of the molecular configuration that the protein loses its function entirely.

Figure 7–10
Protein structure. (*a*) Linear sequence of amino acids. Note that each amino acid is designated by a three-letter abbreviation. (*b*) The helix and pleated sheet are common examples of three-dimensional configurations that can be assumed by a polypeptide. They are determined by the amino acid sequence of the polypeptide chain. (*c*) Polypeptide helices and pleated sheets can fold up in turn to produce the more complex three-dimensional structures of functional proteins.

Ser—Tyr—Ser—Met—Glu—His—Phe—Arg—Trp—Gly—Lys—Pro—Val—Gly—Lys
Glu—Asp—Glu—Ala—Gly—Asn—Pro—Tyr—Val—Lys—Val—Pro—Arg—Arg—Lys
Ser—Ala—Glu—Ala—Phe—Pro—Leu—Glu—Phe

(*a*)

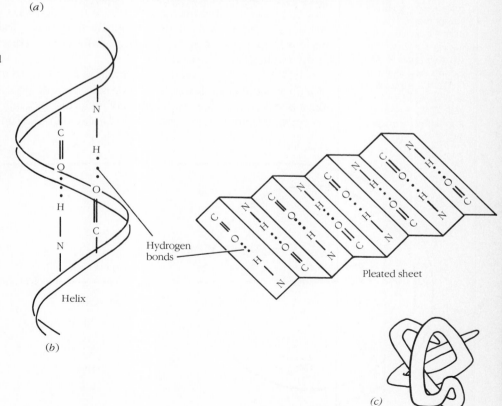

Hydrogen bonds

Pleated sheet

Helix

(*b*)

(*c*)

One-directional information flow

Once they knew the structure of DNA, and that genes coded for polypeptides, biologists predicted that the sequence of nucleotides in a gene determines the sequence of amino acids in a polypeptide. Intensive research during the 1950s and '60s confirmed this prediction and led to the deciphering of the information coded in the DNA nucleotides.

Translating an alphabet of 4 nucleotides into an alphabet of 20 amino acids has been a highly complex biologic process, and unraveling the mechanisms behind this flow of information has required highly sophisticated techniques. A detailed explanation of these techniques is beyond the scope of this book, but as we discuss the action of genes we will mention some of the evidence for our conclusions. Figure 7–11 outlines the steps involved in the transfer of information from gene to protein.

Genes are in the nucleus of eukaryotes, but protein synthesis occurs on the ribosomes, which are located in the cytoplasm (see Figure 3–1). Even before the structure of DNA was known, a great mystery of gene function was how the information gets from nucleus to ribosome, for DNA is not found free in the cytoplasm. When research began on the gene-to-protein information flow, one of the first discoveries was that RNA is involved as an intermediary.

Several observations supported such a conclusion. First, proteins can be synthesized in the complete absence of DNA: cells whose nuclei have been removed can still carry on active protein synthesis. Second, the amount of protein synthesized is directly proportional to the amount of RNA present, and is not correlated with the amount of DNA present. It was eventually shown that RNA is synthesized in the nucleus from a DNA template, then transported to the cytoplasm, where it engages in the synthesis of polypeptides.

In 1958, Crick proposed the generalization that there is a one-directional flow of information from DNA to RNA to protein. So entrenched has this idea become that it is known as the **central dogma** of molecular biology. There is never a flow of information from protein to nucleic acids, although RNA can serve as a template for

Figure 7–11

Scheme for decoding a gene. In a two-step process, a particular sequence of nucleotides in DNA prompts a cell to put together a particular sequence of amino acids to form a protein. RNA molecules are the go-betweens, transporting copies of genetic information across the nuclear membrane. Transcription (DNA → RNA) is the first step, and translation (RNA → protein) is the second step. Sometimes a gene is transcribed into RNA but not translated.

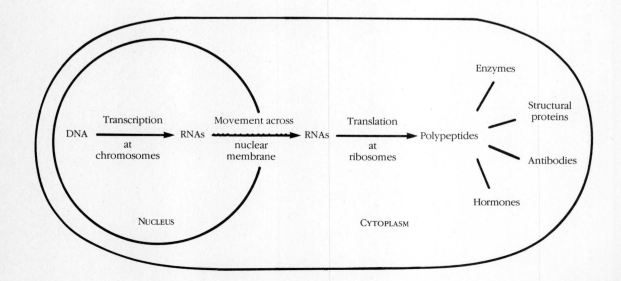

the manufacture of DNA in a few viruses. Some biologists think that in the early stages of the evolution of life, protein was assembled directly from a DNA template. But the mediation of RNA is required in all present-day organisms. The transfer of information from DNA to RNA (or, rarely, from RNA to DNA) is called **transcription,** and the transfer of information from RNA to protein is called **translation.**

RNA structure and species

As we said in the discussion of nucleic acids, RNA differs from DNA in having ribose instead of deoxyribose as its sugar component and in having uracil instead of thymine as one of its bases. The nucleotides of RNA (**ribonucleotides**) are otherwise identical to those of DNA (**deoxyribonucleotides**) and link together in the same manner to form polynucleotide chains (Figure 7–12). However, the difference in

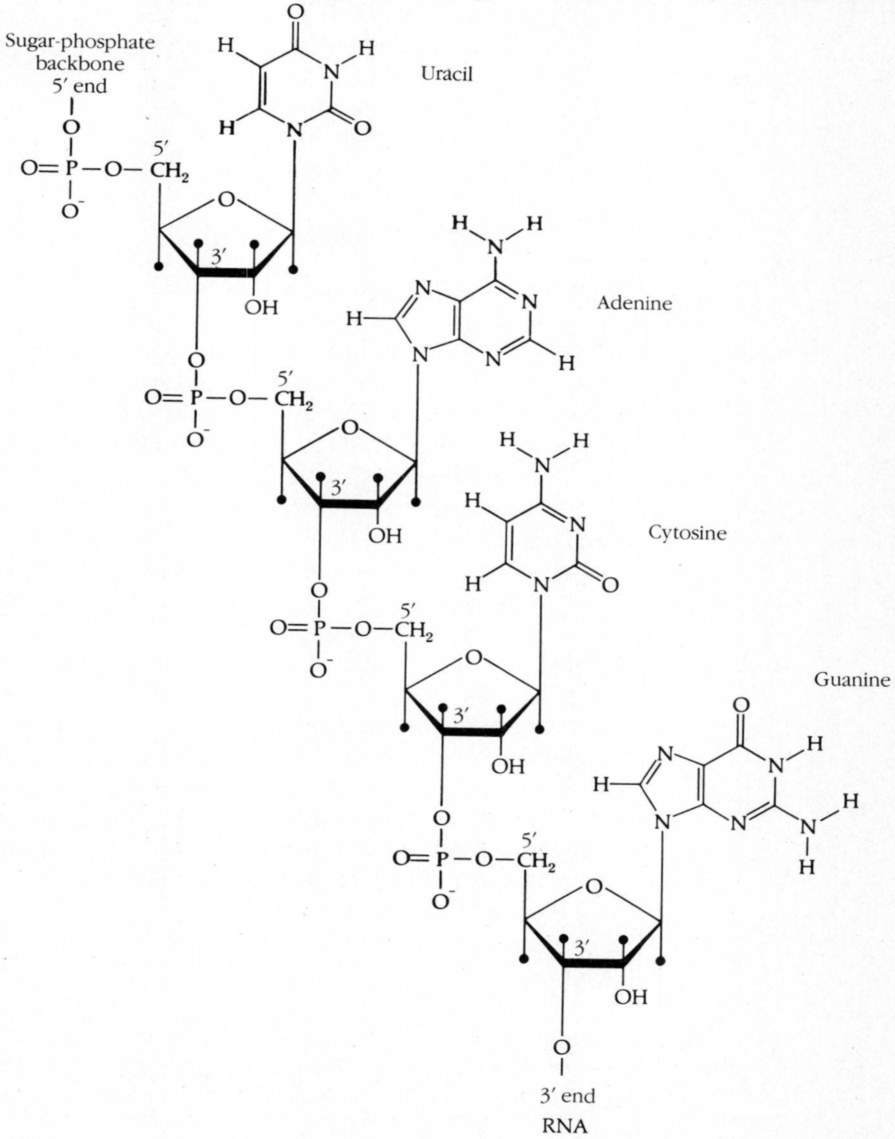

Figure 7–12
The structure of RNA. It is single-stranded, has uracil instead of thymine, and ribose instead of deoxyribose.

sugars between the two nucleic acids has an important structural consequence: RNA chains do not usually pair with each other. Whereas DNA is normally double-stranded, RNA is normally single-stranded.

There are several kinds, or species, of RNA, differing in length, three-dimensional configuration, and function. You have already heard about the small pieces of RNA that aid in the formation of Okazaki fragments, critical elements in DNA replication. Three other species of RNA have essential roles in protein production:

1 **Messenger RNA**, or **mRNA**, carries the genetic message from the nuclear DNA to the ribosome, where it directs the organization of amino acids into a polypeptide chain.
2 **Transfer RNA**, or **tRNA**, attaches to amino acids and lines them up in the order determined by mRNA.
3 **Ribosomal RNA**, or **rRNA**, is a structural component of the ribosome, where polypeptide synthesis takes place.

All RNA species originate in the nucleus, where they are synthesized, or transcribed, on a DNA template. In the following section we will discuss only the transcription of mRNA, but the transcription of other RNA species is virtually the same.

DNA to mRNA: Transcription

RNA transcription requires the presence of a specific enzyme, **DNA-dependent RNA polymerase**, which promotes the bonding of ribonucleotides to each other. Like the DNA strand, the RNA strand grows in a $5' \rightarrow 3'$ direction, using DNA as a template to determine the order of the nucleotides (Figure 7–13). The pairing of the bases is just the same as in DNA replication, except that adenine couples with uracil instead of with thymine. The nucleotide sequence of the RNA strand is thus complementary to that of the DNA strand.

Some early studies of RNA transcription suggested that both DNA strands might be transcribed into RNA. If this were true, one gene would produce two complementary species of mRNA from the complementary DNA strands, and the ultimate products of such a gene would be two different polypeptides. We now know that this is incorrect: *only one strand of each gene* is transcribed into RNA.

The first step in RNA transcription is **initiation**, in which DNA-dependent RNA polymerase binds to a DNA region called a **promoter**—a special nucleotide sequence that indicates the beginning of a gene. The DNA chains must separate to allow initiation, because the bases with which the ribonucleotides must pair are on the inside of the double helix. The second step of transcription is **elongation** of the RNA chain; during elongation the DNA strands continue to unwind along the length of the gene. The third and final step is **termination**, in which RNA polymerase, the RNA strand, and the DNA template part company. Like initiation, termination is signaled by special DNA sequences.

The genetic code

Messenger RNA carries the genetic information coded in DNA in the form of complementary bases. It may be likened to a photographic negative that carries the same image as the print, but in complementary colors. Though all species of RNA are

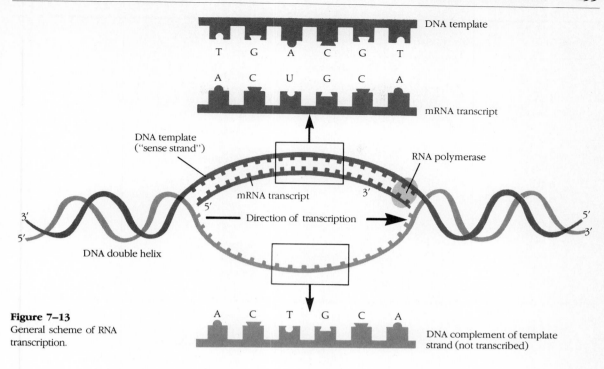

DNA template

mRNA transcript

DNA template ("sense strand")

RNA polymerase

mRNA transcript

Direction of transcription

DNA double helix

Figure 7–13
General scheme of RNA
transcription.

DNA complement of template
strand (not transcribed)

transcribed from genes and have complementary bases, only mRNA can be said to carry the genetic code, for only mRNA is translated into protein. A sequence of three adjacent mRNA nucleotides constitutes a coding unit, or **codon,** and specifies a single amino acid.

When researchers began deciphering the genetic code, it was clear that to specify 20 amino acids the code must have units consisting of at least three nucleotides. With only four kinds of nucleotides (distinguished by their bases), there would be only four coding units, if one nucleotide specified one amino acid. Groups of two nucleotides would provide 16 (4^2) codons, still not enough for one per amino acid. Groups of three would provide 64 (4^3) possible combinations—more than enough for the code.

The hypothesis that the genetic code is triplet in nature was confirmed by a variety of complex experiments that showed which triplets specified which amino acids. One of the techniques employed was test-tube synthesis of polypeptides, using mRNA molecules of known base sequence as templates. The sequence of amino acids in the resulting polypeptides was determined and compared with the sequence of bases in the mRNA. No matter what species contributed the mRNA for such experiments, the outcome was always the same: the genetic code is universal.

With 64 possible codons and only 20 amino acids, a major question for the code-crackers was whether only 20 of the 64 codons actually specify amino acids. The answer turned out to be *no*. All 64 codons are functional, with 61 specifying amino acids and the other three serving to terminate polypeptide synthesis (they are "stop signals"). Most amino acids are specified by more than one codon, but any one codon specifies no more than one amino acid. The genetic code is presented in

Table 7–1   The RNA genetic code words

UUU ⎤ Phenylalanine UUC ⎦	UCU ⎤ UCC ⎥ Serine UCA ⎥ UCG ⎦	UAU ⎤ Tyrosine UAC ⎦	UGU ⎤ Cysteine UGC ⎦
UUA ⎤ Leucine UUG ⎦		UAA ⎤ *Terminate* UAG ⎦	UGA *Terminate* UGG Tryptophan
CUU ⎤ CUC ⎥ Leucine CUA ⎥ CUG ⎦	CCU ⎤ CCC ⎥ Proline CCA ⎥ CCG ⎦	CAU ⎤ Histidine CAC ⎦ CAA ⎤ Glutamine CAG ⎦	CGU ⎤ CGC ⎥ Arginine CGA ⎥ CGG ⎦
AUU ⎤ AUC ⎥ Isoleucine AUA ⎦ AUG Methionine	ACU ⎤ ACC ⎥ Threonine ACA ⎥ ACG ⎦	AAU ⎤ Asparagine AAC ⎦ AAA ⎤ Lysine AAG ⎦	AGU ⎤ Serine AGC ⎦ AGA ⎤ Arginine AGG ⎦
GUU ⎤ GUC ⎥ Valine GUA ⎥ GUG ⎦	GCU ⎤ GCC ⎥ Alanine GCA ⎥ GCG ⎦	GAU ⎤ Aspartic acid GAC ⎦ GAA ⎤ Glutamic acid GAG ⎦	GGU ⎤ GGC ⎥ Glycine GGA ⎥ GGG ⎦

Table 7–1. Our knowledge of this code is the culmination of some of the most creative experiments ever designed in modern biology.

RNA to protein: Translation

An mRNA molecule cannot serve as a direct template for a polypeptide, because chemical bonds do not form between amino acids and the bases of RNA. Other molecules must serve as adapters to fit amino acids into the proper codons, and tRNA molecules play this role.

tRNA   Compared with mRNA, which varies greatly in size, the tRNA molecule is very small, being composed of only 75 to 85 nucleotides. It is folded into a cloverleaf configuration held together by hydrogen bonds between its bases, so that the 5′ end is near the 3′ end (Figure 7–14). The 3′ end of the tRNA molecule attaches to a specific amino acid. Opposite the two ends is a region called the **anticodon**, a sequence of three nucleotides complementary to a specific mRNA codon. The anticodon can bind to its complementary codon on the RNA molecule with the assistance of proteins in the ribosome.

As there are 20 amino acids, there must be at least 20 distinct tRNA molecules, each kind adapted for attachment to a specific amino acid. Where more than one codon codes for the same amino acid, the specific tRNA anticodon is commonly able to recognize two or more of the codons because the base in its third slot often "wobbles" and can pair with more than one kind of base.

A tRNA molecule bound to an amino acid is called **charged tRNA**. The attachment of amino acids to their specific tRNA molecules depends on the activity of enzymes called **aminoacyl-tRNA synthetases**. At least 20 such enzymes exist—one for each kind of amino acid.

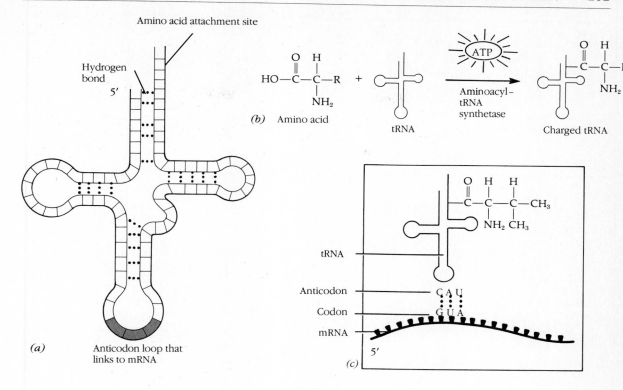

Figure 7–14
(*a*) Cloverleaf structure of transfer RNA. (*b*) Amino acid activation for attachment to tRNA. The enzyme aminoacyl-tRNA synthetase catalyzes the reaction. (*c*) The tRNA anticodon pairs with a complementary codon on the mRNA strand.

Ribosomes and rRNA The human ribosome is composed of two subunits, called **40S** and **60S**, or small and large. The small and large subunits exist independently in the cytoplasm—thousands of each per cell—and associate only during polypeptide formation.

Earlier we compared mRNA to a photographic negative, but in discussing the activity of mRNA it is more useful to think of it as a blueprint containing instructions for the assembling of a polypeptide. To continue the analogy, amino acids are the building materials, and tRNA molecules are the vises that hold the pieces in place on the workbench. The ribosome is the building site, or workbench, where the blueprint is read and materials are organized according to instructions encoded in the mRNA.

This building site is not a simple one. Ribosomes do all of the following:

1 Recognize and bind to a polypeptide initiation site on an mRNA molecule.
2 Bind charged tRNA molecules to their surface so that codon–anticodon recognition can take place.
3 Move along the mRNA molecule during polypeptide synthesis, allowing successive codons to interact with charged tRNA molecules.
4 Influence the accuracy of the codon–anticodon recognition and interaction.
5 Catalyze the formation of peptide bonds between adjacent amino acids.
6 Contain in their structure enzymes involved in the three states of polypeptide synthesis—initiation, transfer, and release.

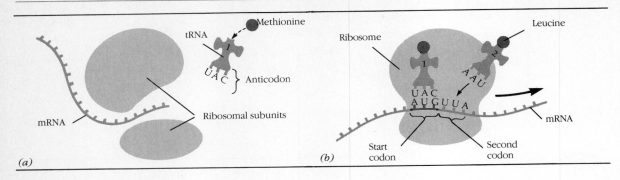

(a)   *(b)*

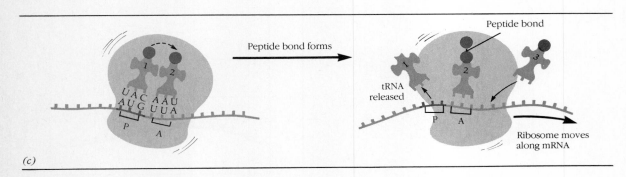

(c)

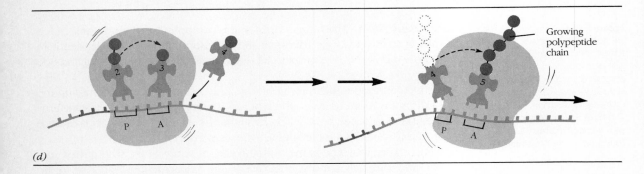

(d)

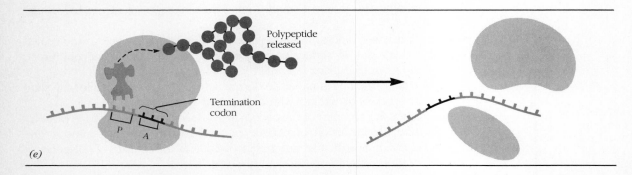

(e)

Polypeptide synthesis The list of ribosomal functions in the preceding section can serve as an outline of the process of polypeptide synthesis (Figure 7–15). A polypeptide is formed on the surface of a ribosome to which an mRNA molecule is attached. Charged tRNA molecules undergo codon–anticodon interactions, locking into the appropriate slots of mRNA, and peptide bonds form between their amino acids. As each new amino acid is added to the chain, the bond that holds it to tRNA is broken, and the tRNA molecule is set free to pick up another amino acid in the cytoplasm.

Let us consider the stages of polypeptide synthesis in more detail. There are two steps in the initiation of polypeptide synthesis. The first is the formation of an **initiation complex,** consisting of the small ribosomal subunit (40S), mRNA, and a tRNA bearing the initiating amino acid of the polypeptide. But polypeptide initiation cannot be performed by just any amino acid—it must always be methionine (met), which is coded for by the initiating codon AUG and attaches to a special tRNA molecule bearing the anticodon UAC.

The second step in initiation is the linking of the initiation complex to the large ribosomal subunit. The 60S subunit has two critical sites: the **P site** (peptidyl-tRNA site), which attaches to Met-tRNA and holds the growing polypeptide, and the **A site** (aminoacyl-tRNA site), which attaches to each incoming amino acid in turn. When the complete ribosome is assembled and Met-tRNA is attached both to mRNA and to the P site, initiation is complete.

Polypeptide elongation (Figure 7–15) begins with the occupation of the A site by an incoming charged tRNA. The specific tRNA, with its associated amino acid, is determined by the mRNA codon opposite the A site, for no other tRNA will fit that codon. In our illustration, the first codon after the initiating AUG is UUA, which codes for leucine (Leu). The A site is therefore occupied by a tRNA molecule with the anticodon CAG, to which valine is attached.

The next step in elongation is formation of a peptide bond between Met and the second amino acid—leucine in our example. This reaction, which also breaks the bond between Met and its tRNA, is catalyzed by an enzyme in the ribosome. The P site now holds an uncharged tRNA and the A site holds Met-Val-tRNA—a dipeptide attached to the second tRNA molecule. The first tRNA molecule is released from the P site.

The third, and perhaps most remarkable, step in elongation is the movement of the ribosome along the mRNA molecule the distance of one codon, so that the complex formerly occupying the A site now occupies the P site. A third charged tRNA can now move into the A site; its amino acid (asparagine in Figure 7–15) will form a peptide bond with leucine, and the tRNA attached to leucine will be released. The ribosome will then move again, so that the last-added tRNA and its attached tripeptide are in the P site, and the A site is free to receive a fourth charged tRNA.

Elongation continues until an mRNA termination signal arrives opposite the A site. The termination signals, you will recall, are the three codons that do not code for any amino acid. Termination results in the dissociation of ribosome, mRNA, and polypeptide. The completed polypeptide then enters the cytoplasm to perform its function.

A single mRNA usually has several ribosomes complexed with it during translation, forming what we call a **polysome**, or **polyribosome**, (Figure 7–16). This allows more polypeptides to be produced per unit time.

Figure 7–15
(Facing page.)
The components and process of mRNA translation into protein.

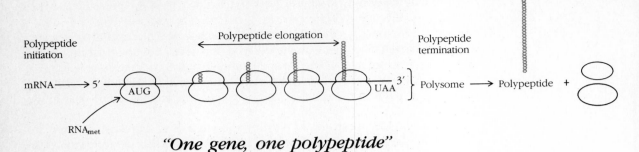

"One gene, one polypeptide"

Figure 7–16
A general view of protein synthesis.

As early as the first decade of this century, biologists had some idea that enzymes might be the products of genes. Garrod's investigations of alkaptonuria (see Chapter 1) suggested that this recessively inherited abnormality might be caused by a defective enzyme. During the decades that followed, many other Mendelian traits were shown to be due to enzyme variations—strong evidence that individual genes produce specific enzymes.

In the 1940s, George Beadle and Edward Tatum did extensive research on the genetics of the bread mold *Neurospora crassa*. They found many simply inherited conditions in which the mold lost its ability to produce some specific chemical: the mold would not grow normally, or at all, except on a special nutrient medium that supplied the missing substance. They further determined that in every case the mutant organism lacked a single enzyme, and that this lack was responsible for blocking production of the missing substance at some stage of its synthesis. This well-documented correlation between Mendelian inheritance and enzymes led Beadle and Tatum to propose the "one gene–one enzyme" theory: a single gene gives rise to one and only one enzyme.

This theory was extremely important as a stimulus to biochemical investigation of gene function, but we know today that it was too simple. Genes code not only for enzymes, but for other proteins, such as antibodies, hormones, hemoglobins, and structural components of the cells. Furthermore, a protein (e.g., hemoglobin), may be composed of two or more dissimilar polypeptide chains, and each of these is coded for by a different gene. We could define all genes that code for protein production by saying "one gene, one polypeptide," but as a general definition of a gene even this would not be adequate, for there are genes that code for RNA alone but not protein, and still others that have regulatory functions. Still, we will be describing the great majority of genes if we say that a gene is that segment of a DNA molecule that codes for a single polypeptide chain.

The Eukaryote Gene

A definition broader than "one gene, one polypeptide" might describe a gene as that segment of the DNA molecule that codes for a single RNA molecule, whether or not the RNA molecule gives rise to a polypeptide. This definition may be adequate for some simple organisms, but it is not inclusive enough for humans. Perhaps the best way we can describe a gene is to say it is a unit of genetic function. This description covers those genes that code for polypeptides, those that code just for RNA molecules, and those that function as controlling sites.

The DNA of prokaryotes is organized in a relatively conservative fashion. There is very little DNA that is not transcribed into RNA, and, aside from rRNA and tRNA, very little RNA that is not translated into protein. The prokaryote gene is usually a sequence of DNA that is completely transcribed into an RNA molecule. The eukaryote gene, on the other hand, often contains an astonishing excess of material. Recent discoveries have revealed, for example, that some genes such as those coding for the protein part of hemoglobin (globin) contain nontranslated, intervening nucleotide sequences within the gene's coding region. Apparently a gene of this type is transcribed into a large molecule of RNA of a class called **heterogeneous nuclear RNA (hnRNA)**. Before hnRNA is transported to the cytoplasm to be translated, it is "tailored": the noncoding sequences are snipped out by enzymes in the nucleus, and the coding pieces are joined together to form a continuous translational sequence (Figure 7–17).

What is the function of the nontranslated genetic material? Though we are not yet sure, we think that the extra DNA has regulatory activity, giving cells control over gene function at different stages of their development. Regulation of gene function is the subject of much current research. We now know what genes are made of, how they replicate, and how they give rise to proteins, yet none of this, by itself, tells us how genes give rise to a complex organism like a human being. We must also know what makes a particular gene function in a particular cell, or at a particular stage in a person's development. These are the questions we will probe in the next chapter.

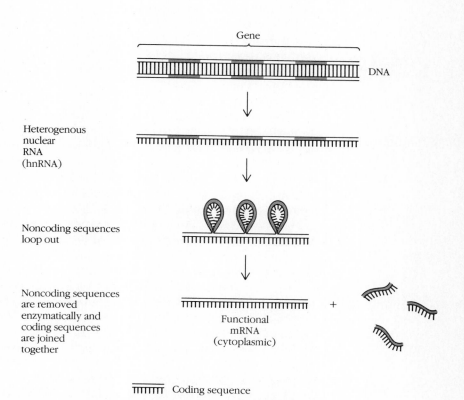

Gene

DNA

Heterogenous
nuclear
RNA
(hnRNA)

Noncoding sequences
loop out

Noncoding sequences
are removed
enzymatically and
coding sequences
are joined
together

Functional
mRNA
(cytoplasmic)

Figure 7–17
General scheme of RNA transcription.

⊓⊓⊓⊓⊓ Coding sequence

⊓⊓⊓⊓⊓ Noncoding sequence

Summary

1 The nucleic acids are **DNA (deoxyribonucleic acid)** and **RNA (ribonucleic acid)**. RNA is the genetic material in a few viruses; DNA composes the genes in all other organisms. RNA's function in cellular organisms is to assist in the translation of the DNA-coded information into protein.

2 Nucleic acid molecules are made of units called **nucleotides**, each made up of a **pentose sugar**, a **phosphate** group, and one of five **nitrogenous bases**. The sugar is **deoxyribose** in DNA, **ribose** in RNA. The bases are **adenine, guanine, cytosine**, and **thymine** in DNA; RNA has **uracil** instead of thymine. Each nucleic acid molecule thus has four kinds of nucleotides.

3 RNA molecules consist of a single strand of nucleotides, DNA molecules of two antiparallel helical strands. The two DNA strands are held together by hydrogen bonds between the bases, which are on the inside of the molecule; the sugar–phosphate **backbones** of the strands are on the outside.

4 The **double-helix model for DNA** was proposed by Watson and Crick in 1953. Its structure provides insights into the mode of replication and the encoding of genetic information, two essential features of the genetic material.

5 DNA replication is **semiconservative**: each new molecule contains one new strand and one old one. During replication the strands of the parent molecule separate, and each serves as a **template** for formation of a new strand from pre-existing nucleotides. The order of the nucleotides in the new strand is **complementary** to that in the parent strand and is determined by the bonding of the bases: adenine pairs with thymine, guanine with cytosine.

6 Each nucleotide and each strand is described as having a **5′ end** and a **3′ end**, and the two strands of DNA are oppositely oriented, or antiparallel. The enzyme **DNA polymerase**, which links the nucleotides of a forming strand to each other, must attach nucleotides to the 3′ end of the last-added nucleotide, so synthesis is always in a 5′ → 3′ direction. Synthesis proceeds discontinuously as the parental strands unwind.

7 The flow of genetic information is from DNA to RNA to protein; it includes the processes of **transcription** (RNA and DNA synthesis) and **translation** (polypeptide synthesis).

8 RNA is synthesized in the nucleus of eukaryotes, using DNA strands as templates. Adenine pairs with uracil, guanine with cytosine. The nucleotide sequence of an RNA molecule is complementary to that of the parental DNA strand.

9 Proteins consist of one or more **polypeptides**—chains of amino acids held together by **peptide bonds**. There are 20 different amino acids. The function of a protein is determined by its three-dimensional configuration, which is determined by the amino acid sequence of the polypeptide(s). Polypeptides are synthesized on the surface of cytoplasmic structures called **ribosomes**.

10 **Messenger RNA (mRNA)** is an RNA molecule that carries information from the nucleus to the ribosomes. mRNA contains the **genetic code**: the order of the bases in mRNA determines the order of amino acids in a polypeptide.

11 A series of three mRNA bases constitutes a coding unit, or **codon**. There are 64 codons (4^3), of which 61 specify amino acids and 3 are stop signals (terminate polypeptide synthesis). Most amino acids are coded for by more than one codon.

12 **Transfer RNA (tRNA)** is a small, folded RNA molecule that attaches to a specific amino acid at one end and has a sequence of three bases called an **anticodon** at the other. The anticodon bases are complementary to those of a specific mRNA

codon or group of related codons; the codon and the anticodon pair during protein synthesis. A tRNA molecule bearing an amino acid is called **charged tRNA.**

13 The ribosome consists of a small and a large subunit which come together during protein synthesis. **Ribosomal RNA (rRNA)** is a structural component of the ribosome.

14 The ribosome has sites for attachment of tRNA. During polypeptide synthesis, the ribosome moves along the surface of an mRNA molecule, allowing successive mRNA codons to react with successive tRNA anticodons. As peptide bonds form between adjacent amino acids, tRNA molecules are released. Synthesis proceeds until the ribosome arrives at a termination codon or stop signal on the mRNA.

15 We may define genes that code for proteins by saying that a gene is the segment of a DNA molecule that gives rise to one polypeptide. But some genes code for tRNA or rRNA molecules, which are not translated into protein. In a general way, a gene is a unit of function.

16 Many human genes have intervening nucleotide sequences that are not translated into protein. The intervening sequences of the mRNA molecule are enzymatically removed before translation. The extra DNA may have a regulatory function.

Key Terms and Concepts

Dexyribonucleic acid (DNA)

Ribonucleic acid (RNA)

Bacterial transformation

Pentose sugars

Phosphates

Nitrogenous bases

Deoxyribose

Ribose

Adenine

Cytosine

Guanine

Thymine

Uracil

Purines

Pyrimidines

Double-helix model for DNA

Nucleotide

Backbone

Phosphodiester bond

Hydrogen bond

Complementary

Template

Eukaryotes

Prokaryotes

Semiconservative replication

DNA polymerase

5' end

3' end

Leading strand

Lagging strand

Okazaki fragments

Amino acid

Carboxyl group

Amino group

Side chains

Peptide bond

Polypeptide

Central dogma

Transcription

Translation

Ribonucleotide

Deoxyribonucleotide

Messenger RNA (mRNA)

Transfer RNA (tRNA)

Ribosome RNA (rRNA)

DNA-dependent RNA polymerase

Initiation

Promoter (region)

Elongation

Termination

Codon

Anticodon

Charged tRNA

Aminoacyl-tRNA synthetases

Initiation complex

P site

A site

Polysome (polyribosome)

Heterogeneous nuclear RNA (hnRNA)

Problems

7–1 What is meant by antiparallel structure?

7–2 What is semiconservative replication?

7–3 Do DNA strands elongate in a $5' \rightarrow 3'$ direction or a $3' \rightarrow 5'$ direction?

7–4 What do Okazaki fragments have to do with DNA replication?

7–5 What is a peptide bond?

7–6 Evaluate the one gene–one enzyme theory.

7–7 What problems would occur if both strands of the gene's DNA were transcribed into RNA?

7–8 If both strands were transcribed into RNA, how could this problem be resolved?

7–9 In light of your knowledge of the different classes of RNA, evaluate the one gene–one enzyme theory and the one gene–one polypeptide theory.

7–10 On theoretical grounds, why was a doublet genetic code (a sequence of two bases coding for one amino acid) rejected?

7–11 What is the function of the promoter region?

7–12 Do RNA chains elongate in a $5' \rightarrow 3'$ direction or a $3' \rightarrow 5'$ direction?

7–13 What is a codon?

7–14 Why is tRNA so crucial to polypeptide synthesis?

7–15 What is an anticodon?

7–16 What is the primary function of aminoacyl tRNA synthetase?

7–17 What is the significance of Met?

7–18 What functions does the ribosome have?

7–19 What is the difference between the A site and the P site on the ribosome?

7–20 How does the prokaryotic gene differ from the eukaryotic gene?

Answers

See Appendix A.

References

Ayala, F. J., and J. A. Kiger. 1980. *Modern Genetics*. Benjamin/Cummings, Menlo Park, CA.

Jenkins, J. B. 1979. *Genetics,* 2nd ed. Houghton Mifflin, Boston.

Lewin, B. 1980. *Gene Expression—Eukaryote Chromosomes,* 2nd ed. Wiley, New York.

Suzuki, D. T., A. J. F. Griffiths, and R. C. Lewontin. 1981. *An Introduction to Genetic Analysis,* 2nd ed. Freeman, San Francisco.

Watson, J. D. 1984. *Molecular Biology of the Gene,* 4th ed. Benjamin/Cummings, Menlo Park, CA.

8 The Regulation of Gene Expression

Human development is nothing less than awe-inspiring. Each of us is a vast array of cells that have a wide variety of specialized functions: nerve cells, muscle cells, light-sensitive cells, ciliated cells, secretory cells, red blood cells, white blood cells, and many more. Yet all of these are derived from a single cell, the zygote, by a progression of mitotic divisions, and, as a general rule, they contain the same genetic information. The differences among cell types are due primarily to **differential gene activity**: certain genes are active in some cells but not in others. Though only a red blood cell can produce hemoglobin, a nerve cell must have an inactive gene for hemoglobin production. The structural and functional differences between a red blood cell and a nerve cell are the result of regulatory messages to their genes, received early in embryonic life.

Differential Gene Activity in Humans

In this chapter we will consider the evidence for differential gene activity and the cellular mechanisms that may account for it. Much of our understanding of gene regulation comes from the study of bacteria. And, since bacteria are one-celled organisms, research has necessarily focused on the question, How can a single cell regulate its gene activity in response to different environmental stresses? Now that we understand some of the mechanisms involved (in bacteria, at any rate), we can begin to tackle the more difficult questions that apply to ourselves: How do cells with the same genome come to be different in form and function—to differentiate? How do these different cells produce different gene products? These questions are the subject of much current biologic research, but the answers are still very tentative. To get some idea of how our own physiology may be regulated, we must consider the mechanisms that have been thoroughly confirmed in bacteria. But before we do, let us take a look at some examples of differential gene activity in humans.

Different activity at different times: Hemoglobin synthesis

One of the most dramatic examples of differential gene activity in human development is the changing pattern of hemoglobin synthesis. **Hemoglobin (Hb)**, which we will discuss at length in Chapter 10, is the oxygen-binding protein in red blood cells; its function is to carry oxygen from the lungs to other tissues. The human hemoglobin molecule varies in composition at different stages of a person's life, but it always includes four polypeptide chains, two of one type and two of another. Adult hemoglobin (HbA) has the formula $\alpha_2\beta_2$, with α (alpha) and β (beta) representing the two types of chains. At all stages of human development, most of the hemoglobin molecules contain the α chains, although at one very early stage a ζ (zeta) chain replaces α in some of the hemoglobin molecules. The non-α chains show a regular pattern of variation during fetal and early postnatal life. There are, in fact, four types of non-α (or ζ) chains: ϵ (epsilon), γ (gamma), β, and δ (delta), each coded for by a different gene. The first of these occurs mainly during early embryonic life, the second during later fetal development, and the last two mainly after birth. The different chains combine to form different kinds of hemoglobin, each appearing at a specific stage of development, as shown in Table 8–1. These different hemoglobins are not all products of the same cells, for the life of an individual red blood cell is very short. Though each newly formed precursor red blood cell must have genes for all six kinds of chains, it produces only the kinds needed to make particular hemoglobins. These hemoglobins appear and disappear in a highly regular and predictable fashion (Figure 8–1).

Somehow, the genes that code for the different polypeptide chains are being regulated with great precision. We do not understand how this regulation occurs, nor do we understand the significance of all the changes. As you will see in Chapter 10, fetal hemoglobin sometimes persists in the adult, but this anomaly seems to be quite harmless, so you may question the reasons for the shift from fetal to adult hemoglobin. Fetal hemoglobin does bind to oxygen at a lower oxygen tension than does adult hemoglobin, easing the transfer of oxygen from mother to fetus. This is a primary advantage of fetal hemoglobin. Regardless of its physiologic significance, hemoglobin synthesis is a striking illustration of how gene activity is regulated.

Different activity in different tissues: LDH synthesis

Our second example of differential gene activity in humans concerns the enzyme **lactate dehydrogenase (LDH)**, which is involved in **cellular respiration**, the

Table 8–1	Human hemoglobin types and the developmental stage at which they appear	
Developmental Stage	Characteristic Chains	Hb Composition
Early embryo	α (alpha) ζ (zeta) ϵ (epsilon)	α_ϵ^4 $\zeta_2\epsilon_2$
Late fetus	α, γ (gamma)	$\alpha^2\gamma^2$
Postnatal	α, β (beta) δ (delta)	$\alpha_2\beta_2$ $\alpha_2\delta_2$

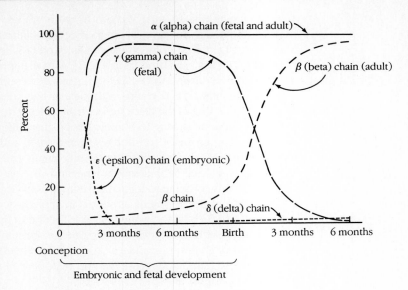

Figure 8–1
Regulation of production of hemoglobin polypeptide chains during human development.

oxidation of organic molecules in cells to produce energy. Two species of polypeptide, designated A and B, respectively, and coded for by separate genes, make up the four-chain enzyme. The two species of polypeptides combine randomly in cells to make LDH, so there are five possible enzyme variants, or **isozymes**:

Isozyme	Composition
LDH-1	A_4
LDH-2	AB_3
LDH-3	A_2B_2
LDH-4	A_3B
LDH-5	B_4

Each isozyme has its own distinctive properties, though all perform the same basic function.

A puzzling feature of the LDH isozymes is that their distribution in the body is not uniform. If equal amounts of the A and B chains are mixed together in a test tube, random association produces the five isozymes in a 1:4:6:4:1 ratio. In the body, however, every tissue has a different mixture of isozymes (Figure 8–2). The A_4 isozyme is found in very high concentrations in the heart and the B_4 isozyme predominates in the muscle tissue. Other isozymes are concentrated elsewhere; AB_3, for example, occurs in large quantities in the kidney, the liver, and red blood cells. The unequal distribution must mean that the A and B chains are produced in different proportions in the different tissues, which suggests a subtle quantitative regulation of gene activity. It seems, in other words, that the transcription and translation of either the A or the B gene into polypeptide are increased or decreased in different kinds of cells to maintain specific A/B polypeptide ratios.

We do not know what control mechanism regulates the differential production of A and B chains in different tissues, any more than we know the mechanism that

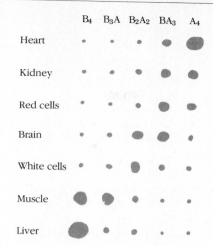

Figure 8–2
Relative content of lactate dehydrogenase isozymes in human tissues. The isozymes are separated by electrophoresis (see p. 389) and then stained.

allows production of different hemoglobin chains at different stages of human development. But we do have an idea of the *kinds* of mechanisms involved in such differential gene activity. Now that we have hinted at the complexity of gene regulation in humans, let us consider some of the regulatory processes themselves.

Regulatory Mechanisms

The life of every organism depends on the organism's ability to maintain **homeostasis** (Greek for "staying the same"). Homeostasis means the balancing of the various processes going on in the organism, so that its structure and internal conditions remain essentially unchanged from one moment to the next. An example of homeostasis in humans is temperature regulation. If the elaborate homeostatic mechanisms that keep our body temperature constant are overwhelmed by extreme external conditions or break down because of a metabolic disorder, we cannot survive very long. Temperature regulation is a homeostatic system involving the interaction of many cells and organs. But every cell, whether it exists independently or as part of a multicellular organism, must also maintain a reasonably constant internal environment. This means, among other things, that cells must vary their output of gene products according to changing internal and external conditions. This variable gene expression may be regulated by genetic or nongenetic mechanisms.

Nongenetic regulatory mechanisms

If a certain protein is active only in certain cells or only under certain circumstances, this does not necessarily mean that the gene or genes coding for that protein are differentially active. *Protein activity* (as distinct from *protein synthesis*) is often controlled by **nongenetic regulatory mechanisms**. These are mechanisms that do not control the transcription or translation of a gene, but affect the expression of the gene farther along some **metabolic pathway**. A metabolic pathway is any series of biochemical events that results in one substance being converted into another in the

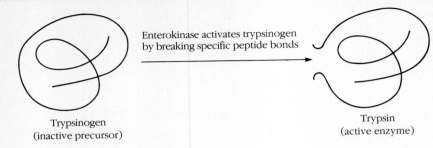

Enterokinase activates trypsinogen
by breaking specific peptide bonds

Trypsinogen
(inactive precursor)

Trypsin
(active enzyme)

Figure 8–3
The changing of an inactive
precursor into an active enzyme.

living organism; our functioning depends on the interaction of the thousands of
such pathways that exist in every cell. The protein initially produced by a gene is not
necessarily the final expression of that gene's function—it may have to engage in a
series of biochemical reactions to produce a final product or a specific effect. The
network of cause and effect in the metabolism of a complex organism is so vast that
we can barely touch on it in this book.

In one simple form of nongenetic regulation, an enzyme is synthesized in an
inactive form and is activated only under a specific set of circumstances. For example, the protein **trypsinogen**, which has no biological activity, is synthesized in the
human pancreas and is activated in the small intestine, forming the digestive enzyme
trypsin. Activation occurs when **enterokinase**, an enzyme found only in the small
intestine, cleaves certain peptide bonds (Figure 8–3). As a result, trypsin is produced
only where there is something for it to digest. Other regulatory mechanisms see to
it that enterokinase is produced only in the small intestine.

A second type of nongenetic regulatory mechanism allows cells to vary the rate
of synthesis of a specific product so that it will always be present in the same
concentration. This mechanism, called **feedback inhibition**, is especially common
in the synthesis of small molecules such as amino acids. Feedback inhibition occurs
when the enzyme that catalyzes the first step in a metabolic pathway is inhibited by
the end product of that pathway. Biosynthesis of the amino acid isoleucine (Figure
8–4) illustrates this type of control. Threonine is converted to isoleucine in five
steps, the first of which is catalyzed by the enzyme threonine deaminase. This enzyme is inhibited when the concentration of isoleucine reaches a certain level,
shutting down the entire synthetic process. When the isoleucine level drops again,
threonine deaminase becomes active again and isoleucine production resumes.

The first of the mechanisms we have described (the activation of trypsinogen)
is an example of a **positive-control system**: the regulatory substance *starts* the
activity in question. The second (the conversion of threonine to isoleucine) is an
example of a **negative-control system**: the regulatory substance *stops* an activity
that would otherwise proceed. *Feedback* (seen in the second example) means that
the end product of a series of reactions serves as the regulatory substance, by affecting the first step in the series. In **negative feedback** (i.e., feedback inhibition) the
end product slows or stops the reaction, whereas in **positive feedback** the end
product increases the reaction rate, resulting in an accelerating effect. Since homeostasis requires that effects be kept within definite limits, positive feedback does not
normally occur in biologic systems, but negative feedback is very common.

Nongenetic regulatory mechanisms are extremely important in maintaining
homeostasis. In this chapter, though, we will focus on the **genetic regulatory**

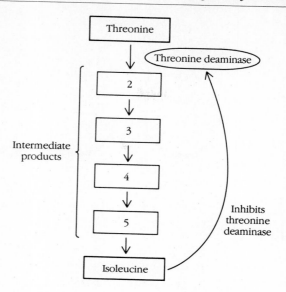

Figure 8–4
Feedback inhibition in the
isoleucine pathway.

mechanisms—those that control gene activity. These mechanisms are particularly interesting to students of human genetics, because we think they are primarily responsible for cell differentiation. Ironically, though, we began to know about them chiefly from research on unicellular organisms—bacteria.

Genetic regulatory mechanisms

Gene activity can be regulated at the point of transcription into RNA, and the regulation may be either qualitative or quantitative. For example, a gene may not be transcribed into RNA unless a specific chemical is present, or the rate of transcription may vary according to specific chemical signals. Another control point is translation: an mRNA molecule may not be translated until it is somehow modified, or it may be more or less readily translated according to the amount of some protein-synthesizing components in the cell. For example, a particular species of tRNA may be a limiting factor in the rate of translation.

It turns out that most regulation of gene activity, at least in bacteria, occurs at the level of transcription: the synthesis of protein is controlled by regulating the manufacture of the appropriate kind of mRNA. This is not surprising, since transcriptional control is undoubtedly more energy-efficient than translational control—the organism does not waste energy making mRNA that is not translated. In eukaryotes we find regulation occurring commonly after RNA has been synthesized.

Regulation of transcription in prokaryotes: The operon   Our first understanding of a genetic regulatory mechanism came from studies of **lactose metabolism** in the bacterium *Escherichia coli*, or ***E. coli***, a normal inhabitant of the human intestine. *E. coli* can utilize various sugars for its energy needs, including lactose (milk sugar). When lactose is present in the environment, *E. coli* synthesizes the enzymes necessary to metabolize it—that is, to transport it into the cell and break it down into simpler substances—but when lactose is absent, the enzymes are

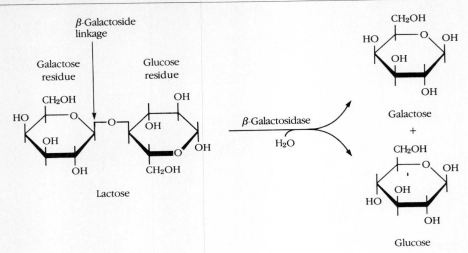

Figure 8–5
Breakdown of lactose by
β-galactosidase.

too. In other words, the bacterium does not waste energy and materials synthesizing unneeded enzymes. The presence or absence of the **substrate** (the substance acted upon by an enzyme) regulates transcription of the appropriate genes.

Two enzymes have known roles in lactose metabolism. One, a **permease,** helps transport lactose across the cell membrane into the cell. The second, β-**galactosidase,** helps break down lactose into simpler substances. The lactose molecule is made up of one molecule each of two simple sugars, **glucose** and **galactose;** to utilize lactose, the bacterium must break it down into these components. β-galactosidase cleaves the bond between glucose and galactose (Figure 8–5).

The enzymes just described are coded for by adjacent genes: *Z*, which codes for β-galactosidase, and *Y*, which codes for permease. A third gene, *A*, is adjacent to *Y* and codes for an enzyme called **transacetylase**. This enzyme catalyzes a chemical reaction called acetylation, but its role in the transport and breakdown of lactose is unclear. It probably modifies lactose to a form that is readily transported into the cell. The three enzyme-specifying genes—*Z, Y,* and *A*—are called **structural genes;** together with two adjacent **regulatory genes** called the **operator** (*O*) and the **promoter** (*P*), they constitute a regulatory unit called an **operon.** The operon codes for a single molecule of mRNA, a multigenic mRNA, which carries the code from all three structural genes at once. This particular operon is called the *lac* **operon**, since it regulates lactose metabolism.

Figure 8–6 shows how the *lac* operon works. For the operon's genes to be transcribed, RNA polymerase must bind first to the promoter region that is a specific sequence of bases recognized by the RNA polymerase as signaling "attach here." Actual transcription begins in the *O* region, then continues through *Z, Y,* and *A*. But another gene, called the **repressor** (*I*), codes for a protein that regulates the operon. When lactose is absent, this repressor protein binds to the *O* region, so that RNA polymerase cannot bind to *P* properly and transcription of the *lac* operon cannot proceed. Thus, in the absence of lactose, *Z, Y,* and *A* genes are not transcribed, and no lactose-metabolizing enzymes are produced. When lactose is present, however, it serves as an **inducer** of gene activity by blocking the action of the repressor. To be more precise, lactose is enzymatically converted to a related substance, **allolactose,** that is the actual inducer. Allolactose inactivates the repressor

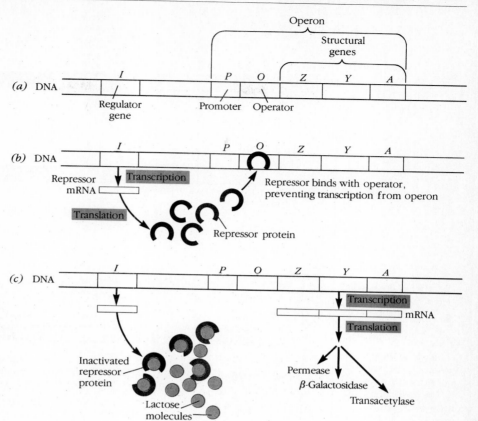

Figure 8–6
(*a*) Segment of an *E. coli* DNA molecule showing the components of the *lac* operon. (*b*) In the absence of inducer, the repressor binds to the operator, excluding RNA polymerase from the promoter. (*c*) The inducer binds to the repressor, inactivating it and permitting RNA polymerase to initiate transcription at the promoter.

protein by binding to it, so that the protein cannot attach to the *O* region. With the repressor no longer in the way, RNA polymerase can bind successively to the *P* and *O* regions of the operon, and transcription can proceed.

The result of operon transcription is a molecule of **polycistronic mRNA**. *Cistron* is a synonym for gene, and polycistronic mRNA means a single molecule of mRNA that codes for more than one gene. Polycistronic mRNA transcribed from the *lac* operon gives rise to permease, β-galactosidase, and transacetylase. Since the three structural genes are regulated by one substance—lactose—these genes are said to be **coordinately regulated**.

If the transport of lactose into the cell depends on the presence of permease, and if permease production depends on the presence of lactose in the cell, how does lactose get into the cell in the first place? The answer is that the repressor protein does not block the *O* region perfectly. In the absence of lactose, a very few molecules of polycistronic mRNA do manage to get made, so there are always a few molecules of lactose-metabolizing enzymes in the cell. Thus, when lactose appears in the environment, enough permease is present to permit a little of the lactose to get into the cell. Once this happens, lactose starts binding the repressor proteins, more mRNA and enzymes are produced, and more lactose can get in. Eventually, all the repressor protein is tied up by lactose, and enzyme production proceeds at full speed.

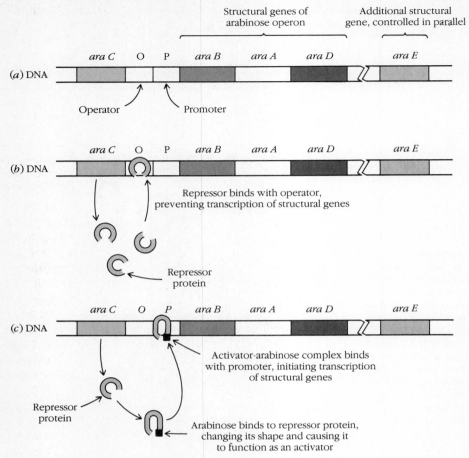

Figure 8–7
(*a*) The arabinose operon. The *ara C* gene product can function in two ways. (*b*) When it is not complexed with arabinose, it is a repressor. (*c*) When complexed with arabinose, it activates the *B*, *A*, *D*, and *E* structural genes.

The *lac* operon is an example of a negative-control system, in which a regulatory substance, the repressor protein, "turns off" a gene that would otherwise be active. (It is not a negative-feedback system, however, because the regulatory substance is not a product of the process it controls.) Positive-control systems also exist in bacteria. An example of such a system is the coordinately controlled gene complex known as the *arabinose operon*. This operon consists of four structural genes (*ara B*, *ara A*, *ara D*, and *ara E*) that code for enzymes involved in arabinose metabolism, a regulator gene (*ara C*), and the operator and promoter regions (Figure 8–7). In this operon, the *ara C* gene product has both positive and negative qualities, depending on the signals it receives. In the absence of arabinose sugar, the *ara C* gene product is a repressor, binding to the *O* region and preventing transcription of the structural genes. When arabinose is present, it is an *activator*, or *expressor*. That is, it binds to arabinose and is required to turn on the structural genes. This latter function is an example of a positive-control system.

There are several other known bacterial operons, in addition to the *lac* operon and the arabinose operon, and all regulate gene activity so that enzymes are produced only when their substrates are present. Any system that accomplishes this allows optimal use of available resources and minimizes energy waste. In prokaryotic operon systems, genes that are functionally related are often contiguous (i.e., in contact) with one another on the chromosome, which simplifies the problem of regulating them in a coordinated fashion.

Regulation of transcription in eukaryotes   The regulation of gene activity in humans and other eukaryotes generally involves systems that are more complicated than the bacterial operons. Nevertheless, our understanding of the operon gives us insight into the coordinate regulation of genes, which certainly occurs in humans.

A major difference between prokaryotes and eukaryotes in coordinate regulation is that functionally related genes are usually not arranged contiguously in eukaryotes; in fact, coordinately regulated genes are often on different chromosomes. Consider, too, the circuitous route for producing a gene-regulating protein in a eukaryote. An mRNA molecule must be synthesized on a DNA template in the nucleus and then transported to the cytoplasm, where it is translated into a regulatory protein; this protein is then transported back to the nucleus, where it regulates the transcription of RNA. Prokaryotes do not have nuclei to contend with.

There is little hard evidence as yet about the nature of genetic control systems in humans, though geneticists have come up with various hypothetical models. We will not present these here, because they are quite complex and will undoubtedly be greatly modified or replaced in the next few years. None of the current models solves the problem of cell differentiation; but, as we learn more about differentiation, we will be able to develop new and better models of eukaryote gene regulation.

Regulation of translation   Though not as common a regulatory mechanism as transcriptional control, translational control is important nonetheless. For example, after polycistronic mRNA is transcribed from the *lac* operon in *E. coli*, the three genes are not translated equally. The Z gene produces about 4 times as much product as the Y gene, which produces about $2\frac{1}{2}$ times as much product as the a gene. We do not know the mechanism of this differential translation. It may be that there is preferential initiation and translation of Z-gene base sequences, or that the rate of translation of certain genes is limited by the availability of particular species of tRNA.

Sometimes mRNA is simply rendered unavailable for translation. This "masked," or stored, mRNA has been transcribed and stored for translation at a later time. The storage process involves complexing the mRNA with proteins or otherwise modifying it so that it cannot be translated. This masking phenomenon occurs, for example, in the mammalian egg cell, where RNA is stored until fertilization triggers translational activity. We will discuss the egg's stored RNA later in the chapter, when we talk about cell differentiation in humans; we mention it here as an example of regulation at the level of translation.

Hormones as regulatory molecules   The control of gene expression, whether at the level of transcription or at the level of translation, involves regulatory substances. We have already described the role of a regulatory protein in our discussion of the *lac* operon, so we will not discuss such proteins further, except to say that they are exceedingly important in both prokaryote and eukaryote control systems. Among the other molecules that play a part in gene regulation are carbohydrates and hormones. Let us take hormones in humans as examples.

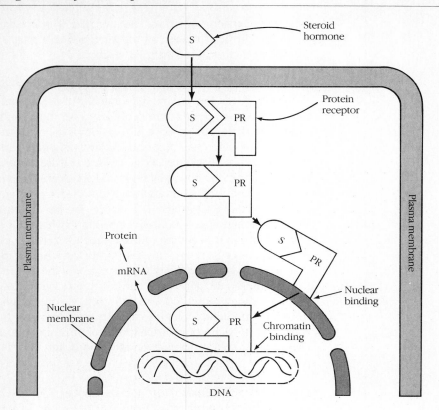

Figure 8–8
A model for the control of gene expression by steroid hormones. The binding of the hormone to the receptor produces a complex that is able to bind to DNA and activate transcription.

For the purpose of discussion, we will consider **hormones** to be chemical messengers that carry information from one type of cell to another in the body. The two main classes of hormones are protein and steroid. The *protein hormones*, such as insulin, act by binding to receptor sites on the surface of cells and triggering cell activity. The *steroid hormones*, such as estrogen, act by complexing with the nuclear DNA and regulating its activity, and it is these we will take a look at here.

The steroid hormone diffuses across the cell membrane and forms an activated unit only when it complexes with specific cytoplasmic *protein receptors*. The steroid–protein complex then moves to the nucleus, where it binds to the chromatin and causes a selective stimulation of certain genes. Only cells that have the specific protein receptors can be stimulated by the hormone. These cells comprise the hormone's *target tissue*. A model of the control of gene expression by steroid hormones is presented in Figure 8–8.

In humans, the development of the mammary gland is a good example of regulation by hormones, though we do not know the specific mechanisms involved. The development of the mammary gland is initiated by increased estrogen secretion at the time a female begins sexual maturation. The function of the mammary gland is to secrete milk, a function that is vital to the survival of the human species. But milk secretion, or *lactation*, does not occur continuously. It occurs only at certain periods, and these are determined by complex hormone signals.

In a pregnant woman, the circulating level of *progesterone* and *lactogenic hor-*

mones such as *prolactin* (from the pituitary gland) and *lactogen* (from the placenta) increases. The hormones stimulate the development of milk-secreting tissue, but not the synthesis of milk proteins such as casein. Progesterone in combination with estrogen inhibits milk-protein synthesis. At birth there is a dramatic decrease in the levels of circulating sex steroids, especially those supplied by the placenta, and this decrease stimulates the synthesis of milk protein. Progesterone clearly seems to hinder milk-protein synthesis by inhibiting both the transcription and the translation of mRNA.

Milk production continues after birth as long as the pituitary hormone prolactin is secreted. Prolactin is inhibited by another hormone, called *prolactin-inhibiting hormone* (PIH), which is produced in the hypothalamus, a small structure located in the floor of the brain. But PIH is in turn inhibited by an infant's suckling at the nipple. The suckling sets up a complex sequence of events that not only inhibits PIH but also stimulates the production of *oxytocin*, another pituitary hormone that triggers the release of milk from the mammary gland. Clearly, one can raise a multitude of questions about precise controls in this series of events, but the answers are not yet available.

RNA processing: post-transcriptional regulation in eukaryotes

For geneticists comfortable with the relatively straightforward mechanisms of transcriptional and translational regulation in *E. coli*, the discoveries now being made about **posttranscriptional gene regulation** in eukaryotes are mind-boggling. We have recently found that eukaryotic RNA is processed before it leaves the nucleus—after transcription and before translation—and that this **RNA processing** may be critical to the selective functioning of genes in different cells.

You will recall from Chapter 7 that heterogeneous RNA (hnRNA) consists of very long molecules with specific nucleotide sequences that are removed before translation. We now have quite a lot of information about hnRNA, and it all supports the idea that this processing has a critical influence on gene expression. A key discovery concerns the nature of the eliminated sequences.

In different human tissues, and at different developmental stages in the human embryo, we find tremendous variation in the kinds of mRNA in the cytoplasm. This is what we would expect, given the variation in the proteins that are synthesized in different cells and at different stages of development. Each kind of mRNA contains only the nucleotide sequences that code for a specific protein. What was quite unexpected was the discovery that in the *nucleus* of a specialized cell there are RNA molecules—hnRNA—with coding sequences for genes not normally expressed in those cells. Evidently, each cell produces hnRNA transcripts for most of its structural genes, then selectively eliminates the RNA sequences it does not use.

This is a most remarkable discovery, and it seems to fly in the face of our earlier statement that the synthesis of nontranslated, or nonfunctional, RNA is energy-inefficient. But a generalization that is true for prokaryotes may not be true for more complex organisms. We cannot discuss the advantages and disadvantages of the posttranscriptional control seen in eukaryotes because we do not yet fully understand the process. The most challenging question that remains to be answered is how the cell "knows" which hnRNA sequences to eliminate. Eukaryotic gene regulation is still far from understood, and each new discovery seems to generate more questions than answers.

The Differentiation of Cells

So far we have discussed how a single cell can vary its gene expression and how different kinds of cells carrying the same genome can manufacture different gene products. In this section we will present the most mystifying problem of all: how cells with the same genes come to be different in the first place.

The role of the cytoplasm

Over the years, there have been many attempts to explain how the zygote proliferates (i.e., grows by rapid production) into a population of diverse cell types. One of the more important of the early ideas was the hypothesis of selective gene loss, according to which genes sort out as the zygote divides, so that different cells receive different genes. Not only was this hypothesis countered by many observations, but the mechanism of mitosis itself strongly suggested that every cell receives a copy of every gene. However, the idea remained that unexpressed genes, though not physically lost, might be permanently inactivated during cell division. One of the most dramatic refutations of this permanent-inactivation concept was an experiment carried out by J. B. Gurdon in 1968 (Figure 8–9).

Gurdon took unfertilized eggs of the African clawed toad (*Xenopus laevis*) and destroyed their nuclei with ultraviolet (UV) irradiation—a procedure called **enucleation.** He then removed nuclei from young but differentiated clawed-toad cells, such as epithelial cells from the intestine of a tadpole, and inserted them into the enucleated eggs. In some cases an unfertilized egg carrying an implanted nucleus from a differentiated cell developed into a normal mature toad. This could not have happened if the specialized cell nucleus had lost genes either through physical sorting out or through permanent inactivation. Only a full set of potentially active genes would support the development of a complete individual from the zygote.

The Gurdon experiment, though still the subject of some controversy, appears to indicate that cell nuclei derived from a single zygote normally carry the same genetic potential, and that cell differentiation is due to differential gene activity, not differential gene *loss*, during development. But what causes differential gene activity? In further experiments with clawed-toad cells, Gurdon and a colleague, D. Brown, demonstrated that cytoplasmic factors are involved.

You will recall that ribosomes contain a form of RNA known as rRNA. In vertebrates, rRNA is synthesized in the developing oocyte, but not in the mature egg. In fact, the embryo does not begin synthesizing rRNA until it reaches the gastrula stage, that is, the first stage of egg development. All of the ribosomes necessary for protein synthesis from the zygote to the gastrula stage are the product of rRNA synthesis in the egg *before* fertilization. Gurdon and Brown took a nucleus from a postgastrula cell that was synthesizing rRNA and inserted it into an enucleated egg in which no rRNA synthesis was occurring. The nucleus promptly stopped synthesizing rRNA and resumed synthesis only when the embryo reached the gastrula stage. This experiment showed that the rRNA genes in the nucleus "know where they are." The only possible conclusion seems to be that they are regulated by a message from the cytoplasm, though we still do not know the nature of the message.

Another point is that protein synthesis up to gastrulation takes place entirely from maternal mRNA molecules synthesized before fertilization. It is only after gastrulation that new mRNA synthesis occurs from both maternal and paternal chromosomes. The mRNA that is stored in the egg cytoplasm and translated during the early stages of embryonic cell division is stabilized or masked by complexing with proteins so that it can be translated in proper sequence and at proper rates.

You will have realized by now that the cytoplasm has a profound influence on gene expression. It is in the cytoplasm that the regulator molecules that affect transcription and translation are synthesized. And it is in the cytoplasm that genetic messages are translated into proteins, which enter into a variety of metabolic pathways that ultimately determine the organism's phenotype. Gurdon and Brown's experiments demonstrated that cytoplasmic influence on gene expression begins

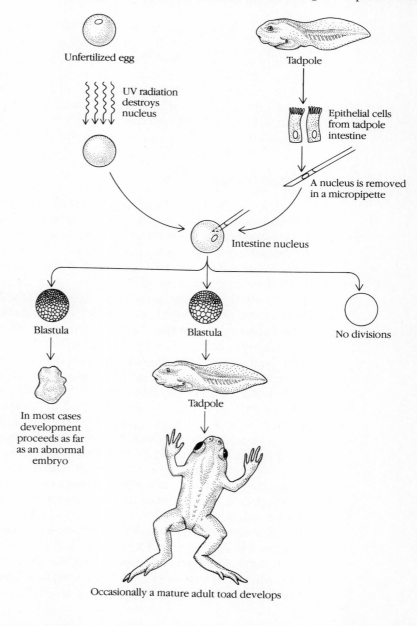

Figure 8–9
Procedure employed to demonstrate that the nucleus from a differentiated cell retains the genetic information required to direct the development of a mature toad from a single cell, the zygote.

early in embryonic life—in fact, in the zygote—which strongly implies that it is involved in all cell differentiation.

Maternal influence

Classical genetics has always insisted that only the genetic material—the "germ plasm" of earlier terminology—is passed from one generation to the next, and that an individual receives an equal amount of material from each parent. We know now that this is not precisely true. The zygote receives cytoplasmic components as well as chromosomes from its parental cells, and these, too, affect the development of the new individual.

But the cytoplasmic contributions of the two parents are far from equal. The egg, with its stored RNA and its great cytoplasmic volume, has much more influence on early development than does the sperm, which is really nothing more than a flagellated nucleus. Since much of a person's early development is guided by the maternally derived "masked" mRNA of the cytoplasm, the maternal genotype may be the dominant force in determining the phenotype of the child. There is good evidence that the maternal genotype is more influential than the paternal genotype in certain nonhuman mammals, but the evidence in humans is not conclusive.

Gene inactivation

In discussing the Gurdon and Brown experiments, we suggested that there is no selective loss or permanent inactivation of genes during cell division. We must modify that statement somewhat, because there are specific but limited instances in which genes are irreversibly inactivated or even lost. One known instance where gene function is indeed regulated by permanent inactivation involves an entire chromosome, the X chromosome.

In female mammals, including humans, one of the two X chromosomes in each somatic cell is inactivated (there is no such inactivation in the sex cells). This **chromosome inactivation**, which occurs at an early stage of embryonic development, is essentially random: the paternal X chromosome is inactivated in some cells and the maternal X chromosome in others. These cells give rise to other cells like themselves, that is, with the same chromosome inactivated. Thus, a normal female human is a mosaic of two different tissue types with respect to the X chromosome. The inactivated chromosome is a distinctly dark-staining body called the **Barr body**, found just inside the nuclear membrane of nondividing cells (Figure 8–10). The Barr body replicates much later in the cell cycle than does the active X chromosome. Mary F. Lyon first identified the Barr body as an inactive X chromosome, so the idea that one X chromosome is randomly inactivated in female mammals is called the **Lyon hypothesis.** The inactivation seems to be a way of achieving **dosage compensation** in females, since they would otherwise have twice as many X-linked alleles as males.

The number of inactivated X chromosomes in a given somatic cell is always one less than the total number of X chromosomes. Thus, a female with Turner syndrome (XO) has no inactivated chromosome; a male with Klinefelter syndrome (XXY) has one inactivated chromosome per cell; and an XXX female has two inactivated chromosomes per cell.

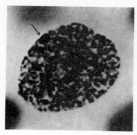

(a)

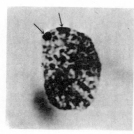

(b)

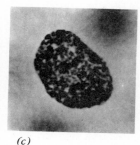

(c)

Figure 8–10
Sex chromatin (or Barr body). The nucleus of a cell with (*a*) one Barr body, in an XX human female, and (*b*) two Barr bodies, as found in individuals with three X chromosomes. In general, the number of Barr bodies is equal to the number of X chromosomes minus one. (*c*) There is no sex chromatin in an XO female (the same result as would be found in a normal male).

Geneticists demonstrated that X-chromosome inactivation is random by examining the cells of females heterozygous for a pair of X-linked genes. Consider, for example, a female who is $G6PD^-/G6PD^+$. She will have some cells that produce glucose-6-phosphate dehydrogenase (those with the $G6PD^-$ gene on the inactive X) and some cells that do not produce this enzyme (those with the $G6PD^+$ on the inactive X).

There is some evidence, though, that the inactivation of an X chromosome is not complete. A small number of X-linked loci seem to express both the maternal and the paternal alleles in a single cell, suggesting that a small segment of the inactivated X chromosome does remain active.

X-chromosome inactivation is not a mechanism for cellular differentiation, although it is an example of a regulatory mechanism. We do not yet know how X-chromosome genes are inactivated, but the process may someday explain selected gene inactivation in differentiating cells.

Genes not on the X chromosome may also be permanently inactivated. For example, cells that make up the webbing between the toes of a frog are differentiated. They produce large amounts of keratin, a protein found in skin cells. When nuclei from keratin-producing cells are transplanted into an enucleated egg, the egg will develop into a tadpole, but not into an adult. What does this tell us? For one thing, it tells us that the nucleus from the differentiated skin cell carries the genetic information to produce heart, liver, lung, nerve, stomach, and other highly specialized cell types. It also tells us that the skin-cell nucleus must contain some irreversibly inactivated genes, because development does not proceed beyond the tadpole stage.

Genes may also be lost when a cell becomes differentiated. Some of the genes that code for certain species of antibodies in white blood cells are actually excised (cut out) as the cell differentiates into a mature antibody-producing cell. We will discuss this phenomenon in more detail in Chapter 11. Now we want to take a more general look at cell differentiation.

Cell differentiation: An overview

Cell differentiation in a developing embryo involves two distinct steps. The first is called **genetic determination**, a step in which a cell becomes genetically committed to a specific differentiated phenotype. This step occurs *before* the cell becomes morphologically or biochemically differentiated. A cell may be genetically committed to become a skin cell before it actually takes on the appearance of a skin cell and synthesizes skin proteins. The second step is called **differentiation**, which is the actual expression of the specific differentiated phenotype. A cell that is not expressing a differentiated phenotype may still be genetically determined to express only that phenotype and no other.

Determination is a gradual, step-by-step process. In some animals it occurs very early in development, so that at the two- or four-cell stage, the cells are already genetically committed to some phenotype. This usually happens because of the way the cytoplasm is partitioned off in the early divisions. In higher animals, including humans, determination occurs only later in development. But as development proceeds, the developmental potential of a cell becomes more and more restricted. In humans, for example, a cell may undergo a sequential series of determinations:

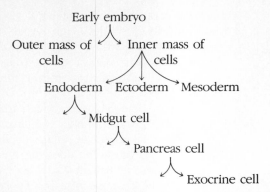

Notice that the genetically determined condition occurs in a step-wise fashion during which the cell's developmental fate becomes progressively more restricted until it forms a specific terminally differentiated cell type. Determination and differentiation require sequential gene activation and inactivation via some of the mechanisms we have discussed.

Aging

Aging is an uninterrupted continuation of normal development. It is, in effect, a kind of programmed death in which birth is followed by maturation, then postmaturation, and finally death.

What is the nature of the aging process? What accounts for more or less predetermined life spans? What causes premature aging? These are the questions we will take up here.

The aging process

Aging persons experience a gradual decline in their ability to adapt to the environment. Their homeostatic mechanisms go into a state of general decline so that such things as heat regulation, water regulation, and disease resistance become defective. There is no simple explanation of this process. Of the several models that attempt to explain aging, none is really adequate. In all likelihood, aging is the result of a combination of mechanisms that interact to cause cell dysfunction, cell death, tissue dysfunction, organismal decline, and finally organismal death.

A general view of aging is presented in Figure 8–11. Following birth, there is a period of growth and development, culminating in the attainment of reproductive maturity. During this process, which is programmed in the genetic material, changes in the synthesis of gene products occur. Patterns of hormones, enzymes, and antibodies change as a person matures and then enters the postmature period. The altered patterns of protein synthesis create new cell environments, and these altered environments lead to deteriorating functions as we enter the postreproductive period of our lives. The final phases of these deteriorating functions are senescence (old age) and death.

A recent discovery about the capacity of cells to divide may help us understand aging better. The **Hayflick limit** (named for its discoverer, Leonard Hayflick) refers to the fact that some cell lines undergo a specific number of divisions in culture and,

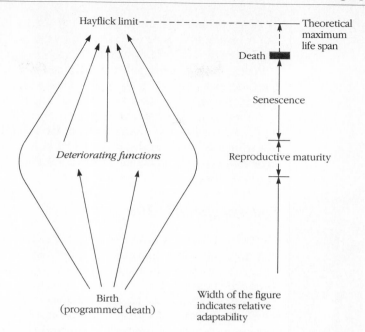

Hayflick limit- - - - - - - - - - - - - - - - - -

Theoretical maximum life span

Death

Senescence

Reproductive maturity

Deteriorating functions

Birth
(programmed death)

Width of the figure indicates relative adaptability

Figure 8–11
General scheme of the aging process.

we infer, in the body. Once that number has been reached, the cells deteriorate and die. A human embryonic connective tissue cell divides 50 times; a connective tissue cell cultured after birth divides about 30 times. In general, the species with the longest life spans have the largest number of cell divisions. Many geneticists believe the Hayflick limit is related to aging, even though some cell types do not appear to have this limit. It may not be necessary to argue that all cells have a division limit, for if certain groups of cells have such a limit, they may be able to trigger the events that characterize the aging process. The Hayflick limit represents the theoretical maximum life span attainable by a single cell line—and hence by an organism. Should life-threatening components of our environment such as cancer and heart disease be eliminated, we could theoretically reach that point of maximum life span.

Many investigators think that aging is controlled by a developmental "clock." One such developmental model is called the **codon-restriction model of aging**. The impetus for this model was the observation that, at different developmental stages, the proportions of different tRNA molecules and the synthetase enzymes that link them to amino acids undergo a series of changes. The model proposes that this pattern change reflects sequential gene expression controlled by **translational languages**—alterations in the codon-producing capacities of the genes. A specific set of codons, called *language one*, leads to the synthesis of a series of gene products that includes specific gene repressors and activators. The activators and repressors trigger the formation of a new set of codons, *language two*, that defines a new set of activators and repressors. These, in turn, give rise to *language three*, and so on. Aging, according to this model, is the sequential activation and repression of selected gene activity. The inference here is that the scheduled appearance and disappearance of certain gene products control the aging process.

If a genetic program determines aging, what factors cause individual variation in the onset and severity of senescence? Mutations and repair processes may play a critical role in this aspect of the aging process. **Somatic mutations**—changes in the genes of nonreproductive cells—may accumulate, giving rise to clones of abnormal cells that have a negative impact on the individual such as cancer. Further, a growing body of information points to an ever-diminishing capacity of cells to repair DNA that has been damaged by environmental insults or normal cellular processes. The older person is much less successful at repairing damaged DNA than a younger person, and this may explain why cancer is predominantly a disease of old age.

There is another idea about aging that may be integrated into the "programmed death" idea, called the **error-catastrophe model of aging**. This model suggests that as errors occur in the synthesis of polypeptides, these altered polypeptides trigger other errors, generating a cascading effect. The accumulation of errors leads to cell dysfunction and ultimately death.

Species-specific life spans

The fact that life spans are more or less fixed for each species is strong evidence that they are genetically controlled. That they vary a bit from one individual to another within a species is undoubtedly due both to genetic variation among individuals and to environmental influences on the aging process.

The species-specific life span has always been difficult to explain. We know that the Hayflick limit is different for different species, but we do not know why. One suggestion focuses on **genetic redundancy**, the redundant, or repetitive, DNA sequences found in nearly all species. Repetitive genes, which are simply multiple copies of a single gene, may function in a protective way: a mutation that renders one copy defective does not eliminate the gene's function if other copies are left unaffected. Perhaps life spans are influenced by the loss of these redundant genes—when all the copies of a critical gene are inactivated, senescence and death result. If this is so, the more redundancy a species has, the longer its life span should be, and there is some indication that there is, in fact, such a correlation. Further, studies clearly show that as an organism ages, it loses some of its genetic redundancy.

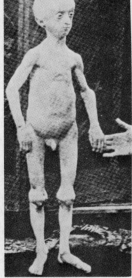

(a)

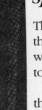

(b)

Figure 8–12
(a) A 17-year-old boy with the Hutchinson–Gilford progeria syndrome. *(b)* A 48-year-old woman with Werner's syndrome.

Premature aging syndromes

There are some genetically caused premature aging syndromes that may eventually shed some light on the aging process. The most spectacular is **Hutchinson–Gilford progeria disease** ("progeria" means premature old age), a dominant disorder. People with this disease have an average life span of about 11 years and take on the appearance of miniature old people (Figure 8–12a). But they do not have *all* the features that characterize a normally aged old person. They are more of a caricature of aging. The genetic mechanism is not well understood. Another premature aging syndrome is **Werner's syndrome**. This is a recessively inherited trait that accelerates some of the processes we associate with aging (Figure 8–12b).

Both these afflictions produce the appearance of premature aging and indeed express many of the features we find in normal aging. But neither is really an acceleration of the total aging process.

There is no single gene that causes a person to undergo accelerated natural aging. This fact fits in with the idea that aging is controlled by a complex series of integrated genetic regulatory mechanisms involving many genes.

Cancer

Few terms trigger such fear as *cancer*. This dread disease occurs when cells lose their ability to regulate their own division. Such cells proliferate unchecked and spread to other parts of the body (metastasize), interfering with normal functions and often leading to death.

We do not yet understand the causes of cancer, but it is becoming increasingly clear that disruption of the control of gene expression is at the root of the disease. Some regulatory change allows a cancerous cell to proliferate unimpeded; if the immune system cannot check this abnormal growth, cancerous cells continue to grow and to displace normal cells over increasing areas of the body.

The primary suspect in cancer is the cell membrane. Substances on the cell surface, controlled by genes, normally provide signals to cells to slow down or stop dividing. The regulatory function of these substances depends on contact with neighboring cells and is called **contact inhibition**. If a cell loses its contact-inhibitory properties, it is free to divide unchecked. There are solid indications that some tumor cells have lost certain cell-surface substances that regulate cell division. In addition, these cells have gained new substances that enable them to escape being attacked by the body's immune system.

The types of genetic change that can lead to cancer fall into two major classes: changes in gene regulation, and structural gene mutation. There are some indications that the genome of a tumor cell is normal but that its regulatory mechanisms are not. An experiment involving the Lucké carcinoma (frog kidney tumor) showed that a regulatory mechanism defect sometimes leads to tumor production. It was found that nuclei from frog tumor cells, when transplanted into enucleated ova, can direct the production of normal larvae and adults (Figure 8–13). This finding indicates that the tumor cell was free of any major genetic alteration. In another experiment, cancer cells were placed into a developing mouse embryo. The new cells became incorporated into the cell population that constituted the embryo—thus creating a mosaic mouse—and produced cell lines that became part of various mouse tissues. These cells functioned normally and did not become malignant (Figure 8–14). This experiment indicated that these cancer cells at least contained the normal set of genes. Somehow these cells must have lost a regulatory mechanism that was reinstated when they were placed in a normal environment.

Though these and some other malignancies may arise from changes in regulatory functions, we now think that most cancers involve structural gene mutations, that is, genes that code for structural proteins. But these proteins may regulate cell functions. Mounds of evidence indicate that mutagenic (mutation-causing) agents such as certain chemicals and ionizing radiation are also carcinogenic (cancer-causing). This is a strong argument in favor of the idea that many cancers are the result of an alteration in the DNA itself.

Viruses and cancer

As we know, cancer is a disease in which the regulation of cell growth and division has broken down. Viruses have been implicated in cancer for over 70 years, since the Rous sarcoma virus (RSV) was shown to be the cause of connective tissue tumors (sarcomas) in chickens in 1912. But how extensive is the viral connection to

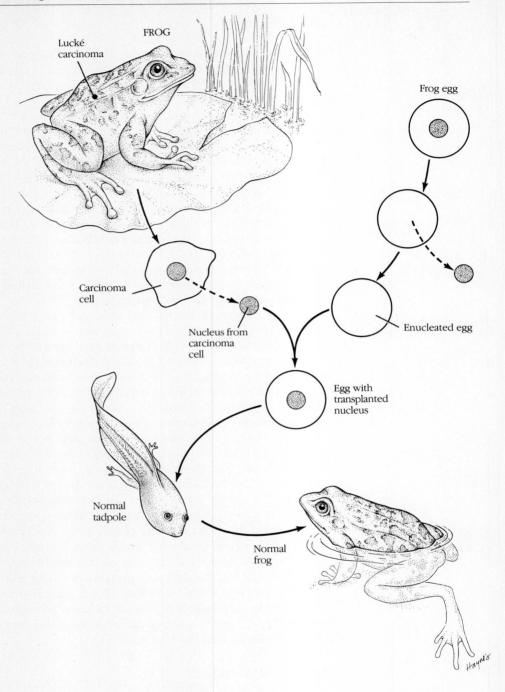

FROG

Lucké
carcinoma

Frog egg

Carcinoma
cell

Enucleated egg

Nucleus from
carcinoma
cell

Egg with
transplanted
nucleus

Normal
tadpole

Normal
frog

Figure 8–13
Nuclei from tumor cells in frogs can direct the production of normal larvae (tadpoles) and adults.

cancer? Recent research suggests that viruses may be the major cause of many, perhaps even all, cancers.

Not all viruses cause cancer, but all invade cells and make use of the cellular DNA-replicating machinery to replicate their own nucleic acids. A **virus** is an entity

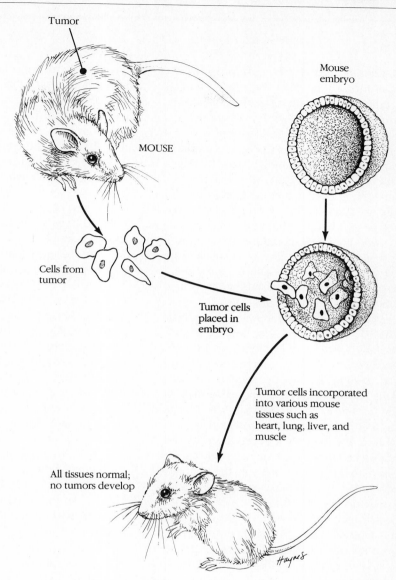

Figure 8–14
When tumor cells are placed in a normally developing mouse embryo, they function normally and do not become malignant tumors.

whose genetic material can function only within the cell of some organism. It lives a parasitic existence, invading living cells of a specific host organism, multiplying within the cells and often killing them. The virus uses the cell's chemical machinery to reproduce. Progeny viruses exit from the host cell and are ready to infect other cells.

Each virus has its own genetic material enclosed within a protein coat known as a *capsid*. The capsid may also contain some lipid molecules. The coat proteins are coded for by the viral genes and assembled from materials in the cellular cytoplasm. The genetic material of viruses can be either DNA or RNA.

Both the DNA and the RNA virus groups include viruses, known as **tumor viruses**, that cause tumors in various organisms. As a result of intensive research

Box 8–1   Have You Heard About the Clone Who . . .

Clones. Cloning. The words confront us everywhere we turn. Clone jokes for a time almost replaced ethnic jokes as the favorite scapegoat of American humor. Everybody knows about clones—*all* about clones. Or do they? When asked to explain the process, most people go mildly blank and mumble something about "biological copy machines." So what on earth *is* a clone?

Generally speaking, a clone is a group of individuals or organisms all of whom are "identical" in genotype. We hedge a little here, because there will be *some* little genetic differences even in the best of conditions, owing to rare changes in genes during the growth of the organism. But these changes are usually insignificant; for most purposes they are safely ignored.

And clones are something new in science, right? Well . . . yes and no. No, because we've been making *some* kinds of clones for probably thousands of years. If your mother ever planted a "slip" or "cutting" from your neighbor's favorite geranium, she was making a "clone," a plant genetically "identical" to the original. It is "identical" because there is no sexual reproduction, no meiosis, involved; the new plant is produced by only *mitotic* divisions from the original. And mitotic divisions faithfully perpetuate only the genotype with which they begin; unlike meiosis and fertilization, they do not scramble it.

Orchardists and nurserymen are very familiar with this type of cloning. It has been used to propagate specific varieties of dozens of different crops—apples, oranges, grapes, and bananas, to name a few. The avocado industry depends heavily on rootstocks that are cloned from types originally selected for their genetic resistance to root-rotting fungi. Cloning, then, is not so new at all.

But some types of cloning *are* rather new, and others are *very* new. As a general statement, the story begins over 30 years ago in the laboratories at Indiana University, where Briggs and King first developed techniques for nuclear transplantation. Nuclei were removed by tiny glass pipettes from the eggs of frogs. A diploid nucleus from the cell of a frog embryo or other larval stage was then transplanted from its original cell into the enucleated egg. Such nuclei often proved themselves "totipotent"; that is, they could undergo mitotic divisions and eventually develop a complete new frog. Thus if 10 such transplanted nuclei from one original embryo were implanted into 10 enucleated eggs, they would give rise to 10 healthy frogs, all of which were genetically identical to the original embryo from which they began. A clone indeed—of 10 hopping, jumping frogs, each of which, in popular terminology, is referred to as a "clone." The term thus suffers a bit from double usage.

But for years following the Briggs and King paper of 1952, we could clone only amphibians. The eggs of other organisms were too small, too fragile, or just too perverse. But researchers kept working, because there were obvious rewards in store. A highly desirable organism could be propagated *without* disturbing the precious and fortuitous combination of genes. Would a racehorse owner be happy to have a clone of Secretariat? Or a dairyman to have clones of highly productive milk cows? Or—heaven forbid—an athletic coach to have clones of . . . name your favorite athletic superstar!

We'll forgo discussion of the obvious ethical problems involved here; let us stay with the technical business of cloning. The term now has become even further clouded in the public mind; it is often used to refer to a wide variety of techniques of genetic and reproductive manipulation. We can, for instance, treat a prize cow with fertility hormones, causing multiple ovulations. We can then flush the ova from her reproductive tract, fertilize them with sperm from prize bulls, and implant each resulting zygote into a surrogate mother—a lesser cow who may lack the best quality gene combinations, but who is very capable of incubating the prize embryo through pregnancy to birth. This means that many more offspring from the desirable cow can be obtained than would be possible otherwise. Since 1977 we have also been performing this trick with humans; a few dozen babies have now been born through this technique, primarily in England, Australia, and the United States.

Are these embryos clones? No. Genetically, they are not necessarily any relation whatever to the surrogate mother, are no more like their genetic parents than any normally born offspring, and certainly are not genetically identical with anyone else.

Other techniques, some already functional, others still under development, come closer to true cloning. Suppose we put *two* of a cow's egg nuclei into one egg, thus making the cell diploid but with only one parent, the mother, as the source of genes. Will this be a clone? Technically no, but we have no other word to designate it. But what if two haploid nuclei from a bull were implanted into one enucleated egg, so that only paternal genes were present? Again, this is not technically a clone; the genotype will not be exactly the same as the parent.

But to get on with our story. True clones have now been obtained not only in amphibians but in fruit flies, carp, zebra fish, and mice. The mice are of particular interest, since they are mammals, and the techniques developed for the mice should be able to produce clones of humans just as well. Do we want to do that? Definitely not at the moment, and probably not at all. The mouse techniques are far from foolproof; in one series of experiments nearly 800 nuclei were transplanted but only 3 developed to become normal mice. One just doesn't tamper around that way with human embryos. So far as can be documented, no human has ever been cloned.

But what of the widely touted published case (Rorvik, 1978) of the American millionaire who supposedly had himself cloned, so that he could leave to the world an identical copy of his highly esteemed genes? Unfortunately—or fortunately—there is not a shred of reliable evidence to indicate that the story is true, and a great deal of evidence to imply that it is not. For one thing, we have never succeeded in cloning an *adult* organism; we must (so far) always take nuclei from very young donors. The allegedly cloned millionaire was said to be in his sixties. Further, apparently no person on earth who could have performed the cloning was involved; none could be found that could confirm the story told. Anyway, considering the very high risk of producing abnormal fetuses, surely no respectable scientist would have attempted the procedure. For these and other reasons, the world's scientific community denounced the book from the start. In a lawsuit on the matter that was recently concluded, the judge declared the book to be "a fraud and a hoax" (Broad, 1981). The publisher has publicly asserted that the book had been published by the company in good faith but that the company now believes it was duped. The author, however, has never recanted and insists that he cannot produce the millionaire or the cloned son because he must preserve their privacy. Perhaps so. But certainly the evidence is so overwhelming against the validity of the story that no one could be blamed for choosing to accept the judge's and the company's verdict: a fraud and a hoax.

So what of the future of cloning? It provides opportunity for such powerful research and application that it will undoubtedly be perfected and more widely applied. But will it ever be applied to humans? One may engage in fantasies and write science fiction about that to one's heart's content, of course, but in the real world the answer to the question *seems* to be: not for a long, long time—if ever.

into the two classes of tumor viruses, we are now beginning to understand how viral genes can transform normal cells into cancerous ones.

DNA tumor viruses   DNA tumor viruses are better understood than their RNA counterparts. Among the DNA viruses are **SV40** (simian virus 40), a monkey tumor virus; **herpes**, a group of viruses that cause numerous ailments, including cold sores, genital sores, mononucleosis, and probably certain cancers; and **adeno-viruses**, a group of viruses that induce symptoms like those of the common cold and may also cause tumors to develop. To get some idea of how a DNA virus may induce cancer, let us focus on the SV40 virus.

The SV40 virus is composed of a protein coat surrounding a double-stranded DNA molecule. We think that this virus may contain only three genes: (1) VP1: major coat protein; (2) VP2: minor coat protein; and (3) T-antigen gene: T-antigen protein.

When the SV40 virus infects a monkey cell, one of two things happens. Either the virus is inactivated and ultimately destroyed by the host's defense mechanisms or it replicates in the cell's nucleus, destroying the cell while producing hundreds of thousands of progeny. The infection is limited to a small number of cells, and though these may be destroyed, the host is not harmed. But when the virus enters a foreign host, such as a mouse or a human, it often *transforms* normal cells to cancerous ones.

When the SV40 virus enters a cell, whether of a monkey or of a foreign host, the first gene to be activated is the T-antigen gene, which produces a protein called the **T antigen** (T for tumor; antigens are discussed in Chapter 11). The T antigen moves to the nucleus, where it appears to be involved in the induction of genes that code for the host cell's DNA-replicating enzymes. These enzymes begin replicating the DNA of both the host and the virus. As replication begins, the coat-protein genes are activated and the coat-protein mRNAs make their appearance.

In a foreign-host cell that is undergoing transformation (e.g., an SV40-infected mouse cell), T antigen, T-antigen mRNA, and coat-protein mRNA are all present. But the coat proteins themselves are not. Apparently, the translation process in these cells is somehow defective with respect to viral coat-protein mRNA, but we do not know the nature of the defect. Transformed cells do not have any visible SV40 particles; that is, the virus never enters its replicating phase and produces progeny viruses. But we can find SV40 DNA inserted into the host cell's DNA. Evidently, all or part of the SV40 DNA, integrated as it is in the host's chromosome, now behaves as a regular Mendelian unit; that is, it is inherited from generation to generation. On the other hand, if an SV40 transformed mouse cell is fused (see Chapter 5) with a normal monkey cell that is free of SV40 infection, the hybrid cell produces SV40 viruses. Clearly, something in the normal host cell—we don't know what—enables the transformed mouse cell to bypass whatever blocked the normal life cycle of the virus.

When the SV40 DNA is integrated into the host's DNA, the T-antigen gene remains active. The T antigen is evidently the culprit in SV40-induced tumors, but precisely how it affects the host cell is not at all clear. We do know that it causes the continuous synthesis of DNA-replicating enzymes, leading to unchecked growth. But it may also activate other host genes, genes that are normally not active at this stage of the host's development; and these genes, acting out of phase, may cause the cancer.

RNA tumor viruses   The RNA tumor viruses are structurally more complex than the DNA tumor viruses. Their RNA chromosome, containing about 10 genes, is surrounded by a protein coat, which is enclosed in a cell-type membrane composed of lipids (fatty substances) and proteins. Glycoproteins (molecules composed of sugar and protein) are imbedded in the surface of the membrane (Figure 8–15).

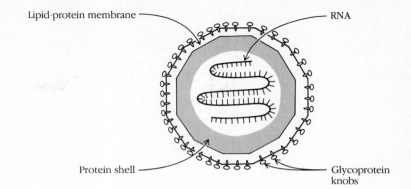

Lipid-protein membrane

RNA

Protein shell

Glycoprotein knobs

Figure 8–15
Schematic diagram of an RNA tumor virus, such as the Rous sarcoma virus (RSV).

Since most of the research on RNA tumor viruses has focused upon the Rous sarcoma virus (RSV), we will discuss it in detail. It is similar to other RNA tumor viruses in its structure and life cycle.

RSV is a normal parasite of chickens. It penetrates the chicken cell, its protein coat is shed, and its RNA is transcribed into a complementary DNA strand by a virally coded enzyme. This RNA–DNA hybrid then is transcribed into double-stranded DNA, which circularizes, enters the nucleus, and integrates into the host cell's DNA. Once integrated, the RSV DNA is transcribed into RSV mRNA, which is translated into RSV proteins. The RSV proteins encapsulate the RSV RNA, and this complex then moves to the cell membrane, which buds out around it and breaks off (Figure 8–16). As a result of this budding, the RSV has a cell membrane around it, but the membrane is covered with glycoproteins produced by and characteristic of the virus. Under normal conditions, this life cycle does not cause the cell to die.

How then does the RSV sometimes cause cancer? Some have suggested that the budding process disrupts the cell membrane, which sometimes causes the cell to lose control of its growth. This idea seems unlikely since RSV strains exist that cannot complete their life cycle but can still cause cancer. A strain also exists that is able to complete its life cycle but is not able to cause cancer.

The latter mutant strain is especially interesting. It reproduces normally, is unable to transform a normal cell into a cancerous one, and carries a deletion of part of its normal genome. The deleted information has been replaced by another gene called the **oncogene**, or "cancer gene," since only those RSVs that have it can cause cancer. Since the oncogene is not essential to the virus life cycle, it may be host gene material that has been inserted into the viral genome and replicated along with the viral RNA. When this gene is in its normal location in the host cell, it is under normal regulatory control. But when the gene is inserted into the RSV genome, it is no longer regulated by the cell and is activated at times when it should be repressed. In RSV, we suspect that the oncogene is actually a gene that normally controls embryonic development in the chicken. If this gene continues to function during the chicken's later development, or is induced to function through RSV infection, cells continue to multiply as if they were embryonic, thus generating tumors.

What is most compelling about this theory is the proposition that *all* cells in the body have an oncogene. In normal cells it is repressed, or active only at the appropriate stages of development. In cancerous cells, however, the oncogene is activated at inopportune times.

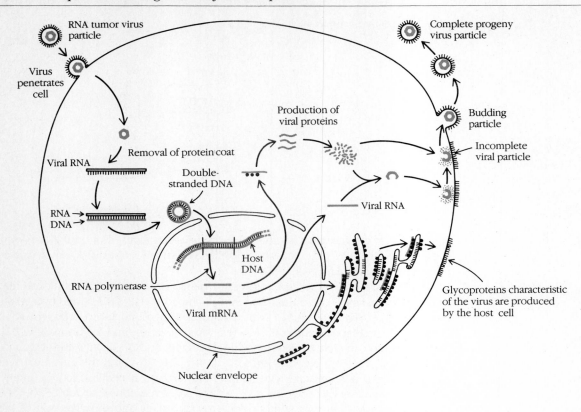

Figure 8–16
Infection of a cell by an
RNA tumor virus.

Chromosomes and cancer

One of the most perplexing problems in cancer biology today is the relationship between cancer and chromosomes that are abnormal in structure and number. There are two competing views on this issue, and no clear resolution is in sight.

One view is that chromosomal changes, perhaps induced by a viral infection or a new mutation, *cause* a malignancy to develop. Supporting this view is the observation that most cancer cells have chromosomal abnormalities. Recall from Chapter 6, for example, the role of the Philadelphia chromosome in leukemia, the $11p^-$ deletion in Wilms' tumor and the $13q^-$ deletion in retinoblastoma. We have also known for a long time that people with Down syndrome (trisomy 21) develop leukemia at a rate 11 times that of the general population.

The second view is that structural and numerical alterations are an *effect* of a cellular change that transforms a normal cell into a malignant one. In support of this view is the observation that some cancers have no perceptible chromosomal aberrations associated with them. Further, in most cancers where there are chromosomal aberrations, the aberrations are not usually specific. That is, the chromosomal changes usually seem to be random.

A few cancers have a clear genetic basis and are associated with specific chromosomal alterations. Ataxia telangiectasis, Bloom's syndrome, Fanconi's anemia, and xeroderma pigmentosa are all recessive disorders (as discussed in Chapter 6) associated with chromosome instability and/or defective DNA repair, and all show a marked predisposition to malignancy. In these cases, at least, a genetic defect apparently leads to a chromosomal change that then leads to the development of cancer.

We have no firm link between chromosomal alterations and cancer, but certain patterns are emerging from the studies. The loss or gain of specific chromosomes seems to greatly increase the likelihood of cancer in many cases. Specific chromosomal rearrangements are associated with certain cancers. And specific chromosomal regions are candidates for causing cancer when they are somehow altered. The alteration may well be the result of disrupting normal regulatory patterns controlled by a certain gene, or oncogene.

Summary

1 Though essentially all of a person's cells have the same genes, different genes are expressed at different times and in different parts of the body. The mechanisms responsible for **differential gene activity** are the subject of much current research.

2 **Nongenetic regulatory mechanisms** control gene expression by controlling enzyme activity. One such mechanism is the conversion of an inactive form of an enzyme to an active form under specific conditions. Another is **feedback inhibition,** in which the end product of a metabolic pathway inhibits the enzyme that catalyzes the first step. Nongenetic regulatory mechanisms, which are very important in maintaining **homeostasis,** do not interfere with the translation of genes into proteins.

3 **Genetic regulatory mechanisms** control gene expression by controlling gene activity; that is, they affect the transcription and translation of genes into proteins. Studies of human **hemoglobin** show dramatically that genes are switched on and off during development. Studies of **lactate dehydrogenase** show that transcription or translation rates can be regulated to produce different amounts of the same substance in different tissues. Gene activity can be regulated at the level of transcription or of translation, or, in eukaryotes, by **posttranscriptional regulation** of RNA.

4 A classic model for the regulation of transcription is the *lac* operon, a system for regulating lactose metabolism in the bacterium *E. coli*. This operon is composed of three **structural genes** (protein-producing genes), two **regulatory genes** called the **promoter** and the **operator,** and a **repressor** gene. All these genes are contiguous on the chromosome. Contiguity of **coordinately regulated genes** is a feature of the operon system of regulation.

5 In humans and other eukaryotes, coordinately regulated genes are usually not contiguous and may be on different chromosomes. Models of human transcriptional regulation are complex and still very tentative.

6 Translation can be regulated through differential ribosome attachment, availability of the molecules necessary for protein synthesis, or modifying mRNA.

7 Many regulatory molecules are involved in the control of transcription and translation, including regulatory proteins and **hormones**. An example of hormonal regulation is the development of the mammary gland in the human female.

8 A good deal of eukaryotic gene regulation is posttranscriptional. Eukaryotic hnRNA is processed as it leaves the nucleus, becoming mRNA. This **RNA processing** involves the removal of many nucleotide sequences. Apparently most of the genes in each cell are transcribed, and the sequences not needed by a particular

cell are eliminated by RNA processing. We do not know how this selective elimination is regulated.

9 We still know very little about how cells differentiate into different types during embryonic development. Nuclear transplant studies have shown that each of an organism's cell nuclei carries a full set of potentially functional genes, implying that **cell differentiation** is the result of differential gene activity. Other studies of this type have shown that differential activity during development is controlled by the cytoplasm.

10 The process of cell differentiation requires that cells first become **genetically determined** or committed to a specific phenotype, then become **differentiated** by expressing that phenotype.

11 Because the female contributes more cytoplasm to the zygote than the male, the offspring's final phenotype may be influenced more by the mother than by the father.

12 The **Lyon hypothesis** interprets the darkly-staining **Barr body** as an inactive X chromosome. The hypothesis states that in every mammalian cell all X chromosomes but one (randomly selected) are inactivated. Thus, normal females have one active and one inactive X chromosome per cell. **Chromosome inactivation** is apparently a regulatory mechanism for achieving **dosage compensation** in females. Understanding how it works may help us understand how genes are selectively inactivated during cell differentiation.

13 Aging is the steady, irreversible loss of adaptability. **Somatic mutations** and repair dysfunctions lead to a general decline in bodily functions. This is clearly a genetic problem, since the general decline is similar in all individuals, and since the life spans of all species are fairly fixed. The **Hayflick limit** is the maximum number of cell divisions attainable by a cell line in a given species, which may tell us something about the maximum attainable life span.

14 The **codon-restriction model of aging** interprets aging as a playing out of the "genetic tape," in which **translational languages** change in an orderly way. The **error–catastrophe model** argues that a single dysfunction has a cascading effect, leading to an ever-increasing number of dysfunctions. The elimination of **genetic redundancy,** through somatic mutations and other mechanisms, could explain species-specific life spans, since different species have different degrees of redundancy.

15 Two genetic diseases, the **Hutchinson–Gilford progeria syndrome** and **Werner's syndrome**, are premature aging syndromes. They share some of the features of natural aging, but not all.

16 Cancer is probably caused by changes in genes that affect the cell-membrane substances regulating cell division. In most cases, the changes probably involve genes that code for proteins, proteins that may function in a regulatory manner.

17 DNA and RNA **tumor viruses** may cause cancer by causing specific cell membrane antigens to be either synthesized or not synthesized.

18. The **oncogene** theory suggests that cancer results from viral genes that perhaps originated from cellular DNA. These oncogenes become integrated into the host cell genome and cause disruptions in gene regulation.

19 Cancer cells commonly have chromosome anomalies. It is not yet clear whether those anomalies are the cause of the cancer or the result of some other change that caused the cell to become cancerous.

Key Terms and Concepts

Differential gene activity

Hemoglobin

Lactate dehydrogenase

Cellular respiration

Isozyme

Hemeostatis

Nongenetic regulatory mechanisms

Metabolic pathway

Trypsinogen

Trypsin

Enterokinase

Feedback inhibition

Positive-control system

Negative-control system

Negative feedback

Positive feedback

Genetic regulatory mechanisms

Lactose metabolism

E. coli

Substrate

Permease

β-Galactosidase

Glucose

Galactose

Transacetylase

Structural genes

Regulatory genes

Operator

Promoter (gene)

Operon

lac operon

Repressor

Inducer

Allolactose

Polycistronic RNA

Coordinately regulated genes

Hormones

Posttranscriptional gene regulation

RNA processing

Enucleation

Barr body

Lyon hypothesis

Dosage compensation

Cell differentiation

Genetic determination

Differentiation

Hayflick limit

Codon-restriction model of aging

Translational languages

Somatic mutations

Error–catastrophe model of aging

Hutchinson–Gilford progeria syndrome

Werner's syndrome

Contact inhibition

Virus

Tumor virus

Oncogene

Problems

8–1 A mutation in the repressor gene results in the *lac* operon being continuously active. How would you explain this?

8–2 Another mutation in the repressor gene causes the operon to be permanently inactive, irrespective of the presence or absence of inducer. Explain this.

8–3 What can you conclude about the repressor protein from Problems 8–1 and 8–2?

8–4 What would be a reasonable prediction of the consequences of deleting part of the *O* region?

8–5 Suppose you had a partially diploid bacterial cell that was normal in all respects except one: one chromosome carried the repressor mutation described in Problem 8–1, and the other chromosome carried the mutation described in Problem 8–2. Would this cell produce *lac* mRNA? Explain.

8–6 A cell does not produce any *lac* mRNA. We find that the I gene is normal, the O region is normal, and the Z, Y, and A genes are normal. How would you explain this?

8–7 With respect to the error-catastrophe model of aging, we offer the following observations:

(a) Abnormal enzymes accumulate with age.

(b) Viruses grown on aged cells are no different from viruses grown on young cells.

How do these observations relate to the theory?

8–8 Actinomycin D, an inhibitor of RNA synthesis, does not seriously affect protein synthesis in newly fertilized eggs. It also does not seriously affect protein synthesis in embryos until the blastula or even gastrula stage. After this, protein synthesis declines, and the embryo's further development ceases. How would you interpret this?

8–9 Some investigators have pointed out that the Hayflick limit does not pertain to all cells. Assuming that this is true, must we reject the pertinence of the Hayflick limit to the aging process?

8–10 How have the nuclear transplantation studies contributed to the concept of the gene?

8–11 We find that on occasion a woman who is identified as a carrier of the X-linked gene for red–green color blindness has some of the red–green color-blind symptoms. Explain this.

8–12 Is it possible for a mutation in the promoter region to result in the permanent activation of the *lac* operon? Explain.

Answers

See Appendix A.

References

Baltimore, D. 1976. Viruses, polymerases, and cancer. *Science* 192:632–636.

Behnke, J. A., C. E. Finch, and G. B. Moment. 1978. *The Biology of Aging.* Plenum, New York.

Broad, W. J. 1981. Saga of boy clone ruled a hoax. *Science* 211:902.

Caplan, A. I., and C. P. Ordahl. 1979. Irreversible gene repression model for control of development. *Science* 201:120–130.

Darnell, J. E., W. R. Jelinek, and G. R. Molloy. 1973. Biogenesis of mRNA: Genetic regulation in mammalian cells. *Science* 181:1215–1221.

Davidson, E. H. 1976. *Gene Activity in Early Development,* 2nd ed. Academic Press, New York.

Davidson, E. H., and R. J. Britten. 1971. Note on the control of gene expression during development. *J. Theor. Biol.* 32:123–130.

Davidson, E. H., and R. J. Britten. 1973. Organization, transcription, and regulation in the animal genome. *Quart. Rev. Biol.* 48:565–613.

Davidson, E. H., and R. J. Britten. 1979. Regulation of gene expression: Possible role of repetitive sequences. *Science* 204:1052–1059.

Dickson, R. C., et al. 1975. Genetic regulation: The *lac* control region. *Science* 187:27–35.

Gartler, S. M., and R. J. Andina. 1976. Mammalian X-chromosome inactivation. *Adv. Hum. Genet.* 8:1–66.

Gurdon, J. B. 1974. *The Control of Gene Expression in Animal Development.* Clarendon Press, Oxford.

Ham, R. G., and M. J. Veomett. 1980. *Mechanisms of Development.* Mosby, St. Louis.

Harnden, D. G., and A. M. R. Taylor. 1979. Chromosomes and neoplasia. *Adv. Hum. Genet.* 9:1–70.

Holliday, R., and J. E. Pugh. 1975. DNA modification mechanisms and gene activity during development. *Science* 187:226–232.

Jenkins, J. B. 1979. *Genetics,* 2nd ed. Houghton Mifflin, Boston.

Kleinsmith, L. J. 1972. Molecular mechanisms for the regulation of cell function. *Bioscience* 22:343–348.

Knudsen, A. G. 1977. Genetics and etiology of human cancer. *Adv. Hum. Genet.* 8:1–66.

Kohn, R. R. 1971. *Principles of Mammalian Aging.* Prentice-Hall, Englewood Cliffs, NJ.

MacLean, N. 1976. *Control of Gene Expression.* Academic Press, New York.

McMahon, D. 1974. Chemical messengers in development: A hypothesis. *Science* 185:1012–1021.

Markert, C. L., and H. Ursprung. 1971. *Developmental Genetics.* Prentice-Hall, Englewood Cliffs, NJ.

Martin, G. R. 1982. X-chromosome inactivation in mammals. *Cell* 29:721–724.

O'Malley, B. W., H. C. Towle, and R. J. Schwartz. 1977. Regulation of gene expression in eukaryotes. *Annu. Rev. Genet.* 11:239–276.

Opitz, J. M., et al. 1979. Terminological, diagnostic, nosological, and anatomical-developmental defects in man. *Adv. Hum. Genet.* 9:71–164.

Rorvik, D. M. 1978. *In His Image: The Cloning of a Man.* Lippincott, Philadelphia.

Smith, D. W. 1982. *Recognizable Patterns of Human Malformation,* 3rd ed. Saunders, Philadelphia.

Stein, G. S., J. S. Stein, and L. J. Kleinsmith. 1975. Chromosomal proteins and gene regulation. *Sci. Am.* 232:46–57 (Feb.).

Temin, H. M. 1976. The RNA provirus hypothesis. *Science* 192:1075–1080.

Wagner, R. P., B. H. Judd, B. G. Sanders, and R. H. Richardson. 1980. *Introduction to Modern Genetics.* Wiley, New York.

Watson, J. D. 1976. *Molecular Biology of the Gene,* 3rd ed. Benjamin/Cummings, Menlo Park, CA.

Weiss, R., N. Teich, and H. Varmus. 1982. *RNA Tumor Viruses: Molecular Biology of Tumor Viruses,* 2nd ed. Cold Spring Harbor Laboratories, Cold Spring Harbor, New York.

Wolf, S. F., and B. R. Migeon. 1982. Implications for X-chromosome regulation from studies of human X DNA. *Cold Spring Harbor Symp. Quant. Biol.* (in press).

9 Gene Mutations

Despite the highly evolved and accurate process of DNA replication, genetic errors can and do occur. Often these mistakes are corrected by cellular repair processes, but sometimes they are not. This is not an altogether bad thing, for the diversity of life on this planet would not be possible unless the genetic material changed over time. Most of the genetic variation we see in human beings and other creatures is the result of different gene combinations, but the differences in the alleles themselves must arise, originally, from mutations. The rate at which mistakes occur is very low—estimated to be about one in every 10^{10} base pairs—but the consequences they have can be far-reaching. The alteration of a single codon can lead to an amino acid change in a polypeptide, which in turn may leave that polypeptide nonfunctional or functioning in a different way. The end result of all this can be a change in the phenotype, a change that may have severe consequences.

In its broadest sense, a **mutation** is any change in the genetic material. This definition includes gross alterations of the genome such as chromosome additions and deletions, and structural alterations such as inversions, translocations, deficiencies, and duplications. We have already discussed these gross changes. In this chapter we shall use the term *mutation* to refer to the subtler alterations that take place within the gene. These intragenic alterations may occur spontaneously during the normal course of replication and development, or they may be induced by external factors that modify the DNA. This chapter will examine the various causes of intragenic mutations, their consequences, and some of the mechanisms by which they are repaired.

Spontaneous Mutations

Some biologists consider the term **spontaneous mutation** a cover-up for our ignorance about the cause of a mutational event. This is not really accurate: by convention we use *spontaneous* to refer to mutations occurring under the natural conditions of DNA replication, repair, and recombination. **Induced mutations**, on the other hand, are caused by specific environmental agents. The types of mutations that are induced are of the same type that occur spontaneously; only the *frequency* of their occurrence is different. There are two major classes of intragenic mutations: **base pair substitutions** and **frameshift mutations**.

Base-pair substitutions

Two main types of base pair substitutions may occur in a gene: transition and transversions. A **transition** is a base-pair change in which the orientation of the bases

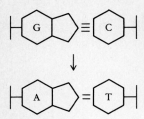

Figure 9–1
A base-pair substitution, transition variety.

remains the same. In other words, the position of the **purine bases** (A and G) and the **pyrimidine bases** (T and C) does not change. From our discussion of DNA's structure in Chapter 7, recall that each base pair consists of a pyrimidine and a purine. In transition mutations, a purine base substitutes for a purine during DNA replication, and a pyrimidine substitutes for a pyrimidine (Figure 9–1). Transition mutations are:

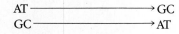

Transition mutations commonly arise during DNA replication because of a phenomenon called **tautomerism** (Figure 9–2), a natural but rare shifting of electrons that gives a molecule a slightly altered form. A tautomeric shift in a DNA base alters its pairing properties so that an A may pair with a C, and a T with a G. One more round of replication converts an AT pair to a GC pair, or vice versa.

Transversions are base-pair substitutions in which the purine–pyrimidine orientation is altered. A purine substitutes for a pyrimidine, and a pyrimidine substitutes for a purine (Figure 9–3). Transversions are:

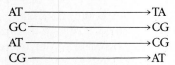

The cause of transversions is usually different from that of transitions. Transversions do not occur during DNA replication. Rather, they appear to be linked to DNA repair by an error-prone repair system. We will discuss DNA repair later in the chapter.

The consequence of a base-pair substitution is either a missense mutation or a nonsense mutation. In a **missense mutation**, one amino acid is substituted for another in a polypeptide chain (Figure 9–4). A well-studied example of missense mutation is sickle-cell anemia, which is the result of an amino acid substitution in the beta globin protein of hemoglobin. A **nonsense mutation** occurs when a codon specifying an amino acid mutates into one that specifies "stop," causing the polypeptide chain to be terminated prematurely. These protein-terminating signals in mRNA are UAG, UAA, and UGA (Figure 9–4). There are some well-known hemoglobin diseases that are the result of nonsense mutations, and we'll discuss them in the next chapter.

Figure 9–2
Base-pair substitution mutations (transitions) that arise when the bases are in their rare enol or imino forms, designated by an asterisk (*).

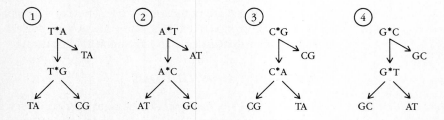

Frameshift mutations

The second major class of mutations is called the frameshift mutation. This type of mutation occurs when one or more base pairs is added or deleted from a DNA molecule. The result of this addition or deletion is a shift in the reading of codons. In Figure 9–4, the added bases cause the reading frame to shift two bases to the right, changing the composition of an entire series of codons. The shift can be corrected if two bases are deleted farther down the line so that the reading frame shifts two bases to the left. But the intervening codons remain altered. The mutant survives only if the protein can sustain these changes without loss of function.

Figure 9–3
A base-pair substitution, transversion variety.

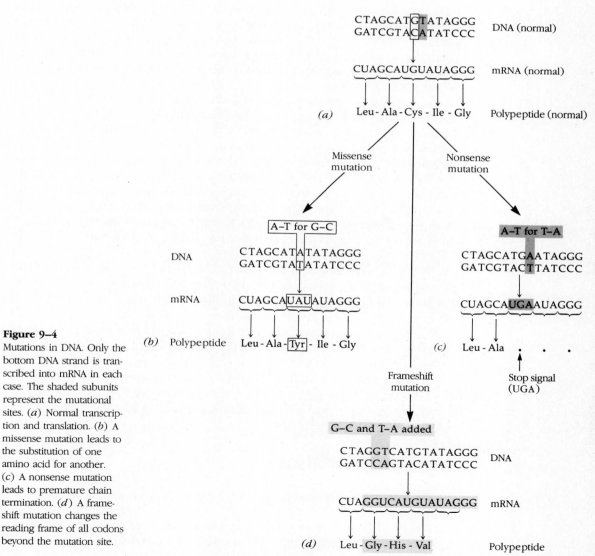

Figure 9–4
Mutations in DNA. Only the bottom DNA strand is transcribed into mRNA in each case. The shaded subunits represent the mutational sites. (*a*) Normal transcription and translation. (*b*) A missense mutation leads to the substitution of one amino acid for another. (*c*) A nonsense mutation leads to premature chain termination. (*d*) A frameshift mutation changes the reading frame of all codons beyond the mutation site.

Frameshift mutations account for most of the spontaneous mutations that occur in higher organisms. They may occur during replication or recombination, or they may be the result of faulty repair activity.

Mutation rates

New mutations are constantly being introduced into the human gene pool. Some of these mutations have very obvious and harmful effects; others have very subtle yet damaging effects over the long run; and still others may have either no effect or very minor effects. Rarely does a mutation have a positive effect on a person's life. Nevertheless, from an evolutionary perspective, natural selection has caused favorable mutations to accumulate over several billion years, and these mutations are the basis of the adaptation of species to their environments.

Given the importance of mutations to our evolutionary history, and to our own lives, we should like to be able to estimate the rate at which new mutations are being introduced into the human gene pool. Out of the several million sperm cells produced by a human male in a single ejaculate, how many of them carry a new mutation? And what are the chances that the egg released at a woman's ovulation carries a new mutation?

It is difficult to estimate the mutation rate for recessive genes, but we are often able to estimate the mutation rate for dominant genes by analyzing pedigrees. Consider the dominant trait achondroplastic dwarfism (Figure 4–26). If this trait is not caused by a new mutation, an affected person should have at least one affected parent. If someone who is expressing a dominant trait has normal parents, we can conclude as one possibility that the person's condition arose as a result of a new mutation in one of the parents. In an analysis of achondroplasia pedigrees, we find that four out of every five achondroplastic dwarfs are produced by normal parents, so they were probably the result of new mutations. By examining all such instances of dwarfs born to normal parents, we estimate that the mutation rate for this gene is 1 in 20,000 gametes.

It may be reasonably suggested that our analysis by this method can be flawed by incomplete penetrance. A parent may in fact be a nonexpressing heterozygote. But in the cases where normal parents produce an achondroplastic dwarf, we find no incidence of the trait in the parents' history. If incomplete penetrance were an issue here, we would expect the grandparents or great-grandparents of the affected person to express the trait. If they do not, we may conclude that new mutations account for the affected children born of normal parents.

In our studies so far, we find that the spontaneous mutation rate for achondroplasia is one of the highest yet discovered. The estimated rate for Huntington's chorea is 1 in 400,000, and for retinoblastoma about 1 in 200,000. These and other rates are summarized in Table 9–1.

Induced Mutations

A wide variety of agents are able to greatly increase the frequency of mutations. Most of these agents are chemicals or forms of radiation. Both radiation and chemicals can induce the types of mutations we have already discussed: base-pair substitutions, and additions or deletions leading to frameshifts. Chromosome breaks may also be induced.

Table 9–1	Estimates of mutation rates for certain genes from nonmutant to mutant

Trait	Mutant Gene per 100,000 Gametes
AUTOSOMAL DOMINANTS	
Huntington's chorea	0.25
Nail-patella syndrome	0.2
Epiloia (type of brain tumors)	0.4–0.8
Aniridia (absence of iris)	0.5
Retinoblastoma (tumor of retina)	0.5
Muscular dystrophy	0.7
Multiple polyposis (formations of polyps) of the large intestine	1–3
Achondroplasia (dwarfness)	4–12
Neurofibromatosis (tumors of nervous tissue)	13–25
X-LINKED RECESSIVES	
Hemophilia A	2–4
Hemophilia B	0.5–1
Duchenne-type muscular dystrophy	4–10

Radiation-induced mutations

One of my most vivid childhood memories is of a machine they used to have in shoe stores, a sort of "pedoscope." You stuck your feet into a slot, then looked down a long tube and had the thrill of watching the bones of your feet move as you wiggled your toes. The effect was achieved by using X rays, and the machine was in use some 20 years after H. J. Muller and L. Stadler discovered that X rays and other forms of high-energy radiation have an enormous capacity to induce mutations. These pioneering investigators used *Drosophila* and barley to prove their point about high-energy radiation, but their findings apply to all living systems.

Before considering the effects of radiation on biologic systems, we will describe the high-energy radiation we are specifically concerned with. There are two main classes of high-energy radiation: **electromagnetic** and **corpuscular**. Both are known as ionizing radiation because they form highly reactive ions when they impact with biologic material. An ion is an atom or molecule (combination of atoms) that is positively or negatively charged when an electrically neutral molecule loses or gains an electron (electrons are negatively charged). Because of their charge, ions can undergo various chemical reactions. They are very reactive, so if a portion of a DNA molecule becomes ionized, it may engage in chemical reactions that change its structure.

Electromagnetic radiation   This type of radiation can be described as waves traveling through space. A scale of electromagnetic radiation is shown in Figure 9–5. The shorter the wavelength, the higher the energy content. Electromagnetic radiation is mutagenic at wavelengths of less than 10^{-4} cm. It causes mutations by imparting high amounts of energy to the DNA molecule, thereby forcing electrons into

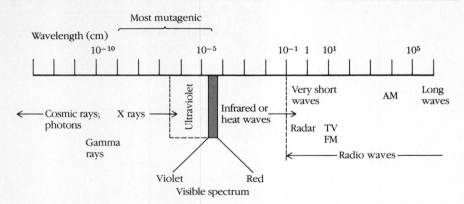

Figure 9–5
A scale of electromagnetic radiation (10^5 cm = 1 km; 10^2 cm = 1 m; 10^{-7} cm = 1 nm).

different, higher energy orbits, disrupting chemical bonds, and promoting chemical reactions that may destroy the integrity of the molecule. Ultraviolet (UV) irradiation has enough energy content to move electrons into new orbits but has little penetrating power. It does not, for example, penetrate through the skin to human gonadal tissue. Rather, it is absorbed in the epidermal layers, and this is where UV has most of its effects, notably skin cancer and eye damage. X rays and gamma rays are high-energy and very penetrating. They easily reach the gonadal tissue and wreak havoc with DNA by knocking out electrons, protons, and neutrons and generating secondary reactions that disrupt DNA's structure.

Corpuscular radiation   This type of radiation consists of particles, such as protons or neutrons, that are propelled at high speeds (often by special pieces of equipment) so that when they collide with biologic material, including genes, they literally shatter the DNA molecule.

Measuring radiation   In older literature, radiation is commonly measured in units called **roentgens (R)**. (One R is the amount of ionizing radiation that will produce one electrostatic unit of electricity in a cubic centimeter of dry air at 0°C and standard atmospheric pressure.) It is more common these days to see radiation measured in **rads**, a unit of the energy *absorbed* from ionizing radiation (equal to 100 ergs per gram of irradiated material). One rad equals 1.8 R. Another unit of measure is the **rem**. This is the amount of radiation it takes to *generate the same effect* on biologic tissue as 1 R of X rays. It takes different amounts of different kinds of radiation to produce similar biologic effects; so if it is these effects we are interested in, it is convenient to discuss radiation in terms of rems. For those of us who remember the frightening events at Three Mile Island in 1979, when the reactor's nuclear core came close to melting down, the terms *rem* and *millirem* ($\frac{1}{1000}$ rem) tend to make us rather uneasy.

Risk from radiation   Radiation does not induce mutations that do not occur normally as spontaneous events, but it can greatly increase their frequency. In addition to causing structural alterations in chromosomes, radiation induces frameshift, transition, and transversion mutations.

In trying to evaluate the genetic risk we face from radiation, we need to know several things:

1 What does the radiation do to the genetic material?

2 What is the degree of exposure to the radiation?

3 How much is the mutation rate increased over the spontaneous rate?

4 What are the consequences of this mutation rate over the long term to the well-being of the population?

5 How do we wish to balance the genetic risks against the benefits that may accrue from the radiation (medical and dental X rays, nuclear power facilities)?

There is an extensive literature on the biologic effects of radiation, and this book can do no more than summarize the primary findings. Most of the research into the biologic effects of ionizing radiation has been carried out with mice, but there is no reason to suppose that the results do not extend to humans. In fact, whenever comparisons have been possible, we have found that human tissue responds to ionizing radiation in the same way that mouse tissue does, with only minor variations. The work with mice has led to the following conclusions:

1 Radiation induces a broad spectrum of mutational phenomena. These include lost chromosomes, chromosomes with structural aberrations, and gene mutations. Though these mutations occur in all cell types, some cell types are more sensitive than others. These are usually cells that occur in tissues with high oxygen levels, such as brain cells.

2 As a general rule, the rate of gene mutation is directly proportional to the radiation *dose* (Figure 9–6). This same proportionality holds for the induction of single breaks in chromosomes. But for two-break events, such as deletions from the middle of a chromosome or inversions, the curve is exponential. The reason for this exponentiality is that the probability of inducing a two-break event is equal to the product of the probability of each separate breakage. As the probability of a single break increases, the product of two such probabilities increases in an exponential fashion.

3 As a general rule, the *intensity* of the dose makes no difference in the mutation rate (Figure 9–7a); it is the total radiation that matters. Thus, a dose of 3500 R produces the same frequency of mutations whether delivered at one time or over a period of several days. In other words, the effects of radiation

Figure 9–6

(*a*) The production of gene mutations is directly proportional to X-ray dose. In this particular case, the induction of sex-linked recessive lethals in *Drosophila* is monitored as a function of increasing radiation. A straight-line curve like this implies that there is a one-to-one relationship between dose and mutation. (*b*) The production of two-break events, such as interstitial deletions, rises exponentially with X-ray dose. Terminal deletions involve only one break, so they increase in direct proportion to the radiation dose. This curve is based on *Drosophila* and mouse studies, and the fact that it is curved implies an exponential relationship between dose and mutation.

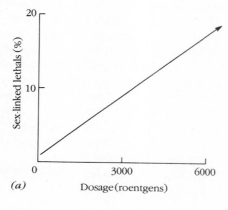

(*a*)

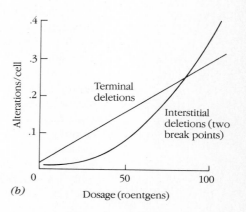

(*b*)

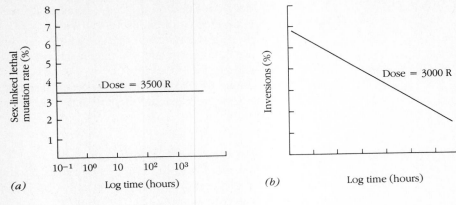

Figure 9–7
(*a*) In general, the intensity of the radiation does not affect the mutation rate. That is, the same effect is produced whether 3500 R is delivered all at once or over a period of 1000 hours. (*b*) The intensity of radiation does make a difference in two-break events, such as inversions. As the time interval between doses increases, there is repair of the first induced break before the second break is induced.

are cumulative. Recent studies with mice, however, indicate that dose intensity may make a difference. These studies show that *under certain conditions* a specific dose given at a low rate produces fewer mutations than the same dose applied at a higher rate. For two-break events, the intensity does make a difference (Figure 9–7b), because the time between doses allows for the repair of the first break before the second one is induced.

4 Structural and numerical anomalies of the chromosome are commonly screened out of developing germ cells by the meiotic process, which tends to eliminate cells with such aberrations. This is as true of the zygote as of other cells: a radiation-induced chromosomal anomaly usually causes the zygote to cease dividing and to perish. But some radiation-induced aberrations do persist and can create serious problems.

5 Radiation-induced gene mutations are transmitted from generation to generation (and from cell to daughter cell in the case of somatic mutations).

To what extent are we exposed to damaging radiation in our daily lives, and what effect, if any, is it having on our genetic material? We are exposed to numerous sources of ionizing radiation on a regular basis. Some of this radiation comes from naturally occurring radioactive materials in the soil and buildings, and some is from the cosmic radiation we all absorb. Over a 30-year period, we may absorb 4 rems of radiation from these sources, though this varies from one locale to another. A person living at the high altitude of Denver is exposed to higher levels of cosmic radiation than someone living at sea level in Atlantic City.

The other prime source of ionizing radiation is our civilization. At the top of this category is radiation absorbed from medical and dental work—about 2 rems over a 30-year period, on the average, though the amount could be much higher for a given person. Other civilization-based sources of radiation, which may account for somewhat less than 1 rem over a 30-year period, include nuclear technology, fallout from nuclear weapons testing, television, and luminous dials on wristwatches. Increased air traffic has disturbed the ozone layer of the upper atmosphere and thus increased our exposure to ionizing radiation from outer space. As you may well imagine, the biologic effects of all this low-level radiation are difficult to assess. For high doses of radiation, there is no question what the effects are: high cancer (and death) rates, high rates of chromosome anomalies, and high gene-mutation rates. Much of our information on the effects of low-level radiation on humans has come from the long-term studies of survivors of the atom bombs dropped on Hiroshima and Nagasaki during World War II.

One important finding of the Hiroshima-Nagasaki studies is that the rate of gene mutations did increase as a consequence of exposure to the radiation. This rate was determined from the inferred increase in X-linked recessive lethal mutations among over 46,000 children born of parents who had been exposed to the blasts at various distances from ground zero. There was a shift in the sex ratio to more females, which is what we would predict if their mothers were carrying radiation-induced X-linked recessive lethal mutations. A woman carrying such a gene will pass it on to half her sons and half her daughters. The sons who receive this gene will die because there is no normal allele to mask its lethal effects; the recipient daughters will live because a normal allele dominates. The net result is a shift toward females. An X-linked dominant mutation induced in mothers is transmitted to and expressed in both males and females. Since these mutations cancel each other out, we cannot screen for them.

If an X-linked dominant lethal gene is induced in a male exposed to the radiation, it will be passed to females only. Because of its dominant effects, the females who receive it will not survive, and the ratio in this instance will shift to males. X-linked recessive lethals induced in males will not be immediately apparent, because they are transmitted to the female children and masked by a normal allele from the mother. The predicted sex-ratio shifts are exactly what was observed in children born to parents of whom only the mother or only the father was exposed to the radiation. When both parents were exposed to the radiation, the effects canceled each other out and no shift in the sex ratio was observed. Among the atom bomb survivors, the closer people were to the blast at ground zero, the higher their frequency of chromosomal aberrations (Figure 9–8). Likewise, the closer people were to the blast, the greater the likelihood that they later developed leukemia, a disease related to structural aberrations of the chromosomes (Figure 9–9).

There is also some good evidence that chromosome anomalies increase upon exposure to even modest levels of radiation. In some parts of Brazil and India, the

Figure 9–8
Frequency of lymphocytes with chromosomal aberrations relative to the distance from the center of the atom bomb blast.

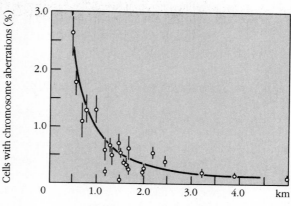

Distance from ground zero of atomic bomb

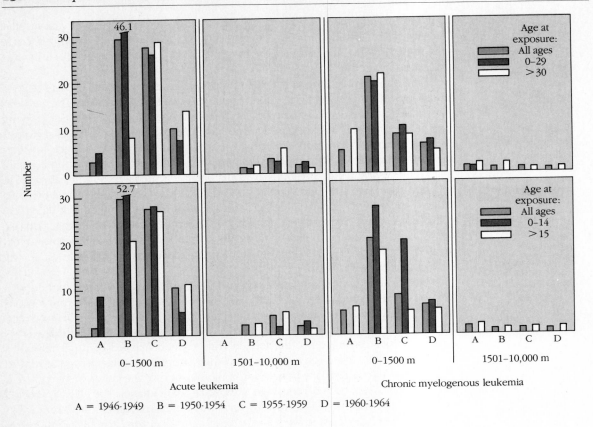

A = 1946-1949   B = 1950-1954   C = 1955-1959   D = 1960-1964

Figure 9–9
Leukemia rates among people exposed to irradiation from the atom bomb blasts in Japan. This study was based on 326 cases of leukemia in people who were from 0 to 10,000 meters (m) away from ground zero. The population was subdivided into two age groups.

soil is exceptionally radioactive because of naturally occurring compounds. People here receive 10–100 times the normal background radiation, and in both regions there is an elevated level of chromosome aberrations.

In summary, there is no doubt that ionizing radiation is capable of doing great harm to the human gene pool. But at the lower end of the radiation exposure scale, it is difficult to assess the genetic risks. One estimate is that 1 rem increases the probability of a single break in a chromosome by about 5%, of aneuploidy (chromosome loss) by about 1%, of a new gene mutation by about 0.2%. All things considered, it is best to avoid any radiation unless the rewards of exposure outweigh the potential damage.

Chemically induced mutations

In 1942, C. Auerbach and J. M. Robson discovered that nitrogen mustard is a **mutagen**—an agent that causes mutation. Their results were not published until 1946 because the compound was being manufactured and stored for possible military use in World War II. Since 1946, hundreds of compounds have been tested and found to be mutagenic. We will examine some of the different classes of mutagenic compounds and the mechanisms by which they induce mutations in DNA. We will also look at some of the problems we face from potential mutagenic agents in the environment.

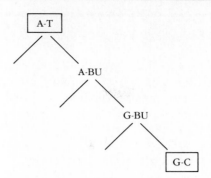

Figure 9–10
The mechanism of bromouracil mutagenesis is shown. BU is incorporated into a replicating DNA molecule as a T; then it behaves as a C in the next round of replication. The result is an AT → GC transition.

Chemical mutagens can be categorized by the effect they have on DNA. Some, called *radiomimetic chemicals*, mimic the effects of ionizing radiation; nitrogen mustard is a good example. Other mutagens cause base-pair substitutions, while still others cause frameshift mutations.

Chemicals that cause base-pair substitutions do so either by mimicking natural DNA bases and substituting for them during DNA synthesis or by modifying the pairing properties of DNA bases. Chemicals that mimic bases are called **base analogs**. For example, bromouracil can be incorporated into DNA in place of a T, then act as a C and pair with a G, causing an AT → GC transition (Figure 9–10). Caffeine, found in such common beverages as coffee, tea, and colas, is a base analog that is mutagenic under certain conditions.

Agents that affect base-pairing properties have no catchy name; they are known as **agents that act on nonreplicating DNA**. These mutagens alter the chemical configurations of the bases so that later, during replication or repair, a mutation may occur. For example, nitrous acid (HNO$_2$) removes an -NH$_2$ from C, converting it to U (normally found only in RNA), which base-pairs like T. This results in a GC → AT transformation. Nitrous acid also affects the pairing properties of A and of G.

Mutagens that cause frameshift mutations, such as the acridine dyes (Figure 9–11), generally bind to the DNA helix or are actually inserted into the helical structure of the molecule. The inserted acridine molecule causes the DNA molecule to stretch and become distorted, leading to the addition or deletion of DNA bases (discussed in Chapter 6).

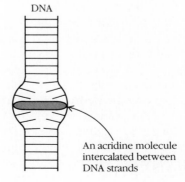

Figure 9–11
The intercalation of the acridine between the DNA strands often results in the addition or deletion of a base pair.

Box 9–1   *Influenza Virus—The Viral World's Quick-Change Artist*

It's like a chess game, in many ways: The viruses make a move, we make a move; then we wait for the viruses again. But by the very nature of the game, we are almost always on the defensive. The game has gone on for thousands of years at least and shows little sign of ending.

In the twentieth century, we humans have developed a battery of tools to try to seize the offensive. And in the game with at least one virus—smallpox—the game seems to be over, and the virus has lost. No natural case of smallpox, once one of the most dreaded killers and maimers of mankind, has been reported anywhere on earth since 1977.

But the key to conquering smallpox virus was its general reliability, its commendable decency in playing the game. With a few exceptions, it always presented the same face, the same game plan. This behavior enabled us to develop vaccines that made humans immune to the virus. With no victims susceptible, the virus has no host within which to reproduce—and that, we hope, has been fatal for the virus. As far as we can currently demonstrate, the virus exists today only in the culture media of selected laboratories, where it is kept under close control.

All we have done, of course, is to assist the human body's natural defense system. When a virus invades our tissues, it injects its nucleic acid core—its genes—into a given cell, where the genes then direct the cell's metabolic machinery to turn out many copies of the virus. Each new virus is loaded with a newly replicated copy of the viral genes, which are carefully packaged into a specifically made protein coat. It is this protein coat that, when the new virus is released from the host cell, is responsible for attacking a new host cell and starting the cycle all over again. But the immune system of the body "learns" to recognize the protein coat and manufactures antibodies that specifically attack it, thus initiating the destruction of the virus. Eventually, then, the immune system will render the body no longer susceptible to the virus—so the virus will have to find another host. In the case of the smallpox, people all over the world received vaccines—a noninfective form of the protein coat. Their bodies then manufactured antibodies against the protein coat, so that when the real virus entered the body, the body was already resistant. Smallpox apparently is now unable to find susceptible hosts anywhere in the world.

This system of building up immunity explains why, once having had a viral disease (smallpox, mumps, chicken pox, measles, etc.), we rarely ever get it again; our bodies "know" the protein coat after the first attack and are thereafter ready with antibody production whenever that same virus appears again, wearing its usual protein garb.

Influenza, however, has not been so decent. Influenza virus specializes in instability. That sounds like a defect, but it is actually an advantage for the virus, because it presents us with an incredibly frustrating game plan. For as quickly as we develop antibodies against one form of the protein coat, the virus develops another. No one really knows how many different viral coats (strains of the virus) currently exist or have existed in the past. Certainly they number in the hundreds. We have developed an elaborate nomenclature system to keep track of them, to unravel their genealogies, and to help us understand the mutative events by which they change. For mutation, pure and simple, is the secret to influenza's "instability." Some strains are the result of merely minor mutations; others come via rather large-scale shifts. They vary in their virulence to man, some causing only minor effects, others being potent enough to kill, particularly if their victims are aged, very young, or otherwise have weakened immunity.

It is hard for us in the 1980s to grasp the fear and helplessness generated by the word *flu* six decades ago, But in 1918–1919 the world passed through an epidemic caused by a killer strain of the virus that afflicted millions and killed hundreds of thousands. It was apparently first seen in a U.S. Army camp, where, on March 11, 1918, a Sergeant Gitchell reported in sick to the post doctor. Before the day was over, more than 100 other soldiers were down with the same symptoms:

fever, muscular pains, headache, sore throat, listlessness. In the next few weeks on the post, more than 1000 soldiers were afflicted; 46 never recovered. The disease rapidly spread to other military posts and then to Europe via military units, ravaging the military and civilians alike. By summer it was into the U.S. civilian population, and eventually it spread around the world. Public meetings were canceled, movie theaters and other places of entertainment were closed, workers refused to go to work for fear of contracting the disease. Factories and even cities closed their operations.

Only against this background can we understand the concern surrounding the U.S. "swine flu" scare of 1976–1977. Once again, the point of origin of a "new" flu was a U.S. military post. But the virus appeared to be the same as that responsible for the 1918–1919 epidemic. Evidence suggested it had been hiding in pigs (birds and horses are also important alternate hosts for flu virus) but had now shifted to again attack humans. The federal government, remembering the previous record of the virus, launched a massive program of vaccine preparation and inoculation of persons considered to be at high risk (e.g., the elderly). As it turned out, the epidemic never materialized, and the government was strongly criticized for overreacting. The government may be loath to react similarly the *next* time a novel and apparently virulent strain appears. But what if the consequences are more serious?

At present, a worldwide surveillance network under the direction of the World Health Organization (WHO) constantly monitors influenza outbreaks. Viral strains isolated from patients all over the world are sent to two identification centers in the United States and Great Britain. If it seems necessary, the pharmaceutical industry is asked to produce new vaccines.

But is it not possible to produce a "super-vaccine"—one that could lead the immune system to recognize *all* forms of the virus? At the moment, no. The virus has been found to be capable of altering its internal materials as well as the surface proteins; indeed, a patient afflicted with two different strains may provide an "incubator" in which components of the the two strains may be recombined to give rise to yet a third strain, and multiple strains may be in circulation at the same time. It has not yet been possible to identify any one common element that can form the basis for a universal vaccine against a "flu virus."

At the moment, then, influenza virus can make "chess moves" faster than we can analyze and react. We are still largely on the defensive. Though a number of recent developments give hope for better control, influenza's quick-change mutating act seems to ensure that it will be with us for a long time to come.

Our environment contains an increasingly large number of chemicals that may induce genetic damage and it is urgent that we be aware of their potential hazards. Should the chemicals we come in contact with induce mutations, we may be condemning future generations to lives of overall lower quality. As for us, these chemicals may be condemning us to higher rates of cancer, lowered vitality, and other afflictions. Nowhere are these words more poignantly illustrated than in the recent tragedy of the Love Canal area near Niagara Falls.

The Love Canal is an uncompleted canal that was used as a dump site for over 20,000 tons of toxic chemicals by the Hooker Chemical Company from 1940 through 1977. Among these chemicals were solvents, pesticides, and dioxin, one of the most toxic substances known. Thousands of people were moved out of the area because of the biologic hazard posed by this chemical waste. From preliminary studies carried out on the population that lived in the Love Canal area, it appears that their rates of birth defects, miscarriages, stillbirths, and cancer are significantly higher than normal.

A study that has been severely criticized states that out of a sample of 36 people who lived in the canal area, 11 had chromosomal anomalies in some of their cells—an astonishingly high frequency. However, a review of the study points out a glaring weakness that throws it into a state of disrepute: the investigation lacked simultaneous controls, that is controls matched by age, sex, medical history, and geographic area. Many people in the Love Canal region think the attempt to discredit the study is a clumsy effort to whitewash the whole affair. The Environmental Protection Agency (EPA) commissioned this study and is responsible for the release of the results; a National Institute of Environmental Health Sciences panel condemned it. At present, the question of chromosome damage in people who lived in the Love Canal area is unresolved. Clearly, the study will have to be done again, and in such a way that the results will be as scientifically sound as possible. The people of Love Canal have experienced, and continue to experience, an inordinate amount of stress, and deserve all the help and information they can get.

Though most of us are not exposed to the intense concentrations of highly toxic chemicals found in the Love Canal area, we are, nevertheless, still exposed to a multitude of chemicals that can inflict damage on our DNA. This damage can result in mutations and/or cancer. Many industrial compounds, such as alkylating agents, vinyl chloride (used in plastics), styrene (used in plastics, resins, and rubber), formaldehyde, and trichloroethylene (used in solvents for dry cleaning and degreasing metal parts) are mutagenic in laboratory animals. Caffeine and food additives such as saccharin (a sweetener) and sodium nitrite (a preservative and color enhancer in bacon, ham, and lunch meats) are all mutagenic in laboratory animals. If a compound is mutagenic in nonhuman animals, it is probably mutagenic in humans, though perhaps not at the same level.

There are some factors which complicate our ability to assess the genetic risks we face from environmental chemicals. Toxic chemicals may be deactivated by metabolic reactions in the body; or seemingly innocuous chemicals may be converted to potent mutagens in the body. As with radiation, it is usually best to avoid substances that have the potential to alter the genetic material, though we must sometimes make hard decisions. For example, is it more dangerous to be 10 pounds overweight or to regularly drink diet sodas that contain saccharin? There is no simple answer to this question other than having a sensible diet that precludes the necessity of diet sodas in the first place.

The Ames Test: A Quick Way to Screen for Mutagens

Each year we are inundated with a variety of new chemicals that need to be tested for their capacity to induce mutations or cause cancer, two related processes. One of the quickest ways to screen these chemicals is the Ames test (developed by Bruce

Ames), a clever system that uses bacteria.

The Ames test employs strains of bacteria (*Salmonella*) that carry either a base-pair substitution (BP) or a frameshift mutation (FS) in a histidine gene. Thus, they are designated his$_{BP}^-$ or his$_{FS}^-$ and will grow only on a medium that has had histidine added to it. The bacteria, which are pathogenic (disease-producing), have been genetically engineered so that they are unable to live outside the laboratory. They are also unable to repair their DNA.

To simulate the conditions a chemical encounters when it enters the human body, rat-liver extracts are used. This is a truly effective procedure, because it is in the liver that toxic chemicals are detoxified, or innocuous chemicals are converted into mutagens and/or **carcinogens** (substances that cause cancer).

The chemical being tested is mixed with the rat-liver extract and with bacteria carrying the his$^-$ mutations (Figure 9–12). The bacteria are placed on an agar medium that is missing histidine, so that the only cells able to grow are his$^+$ cells. If the chemical is able to reverse either the his$_{BP}^-$ or the his$_{FS}^-$ mutation, the his$^+$ revertants

Figure 9–12

Ames test. A suspected mutagen is mixed with rat-liver extract and *Salmonella* carrying either a frameshift or base-pair substitution in the histidine gene. The control mixes both *Salmonella* mutants with the liver extract. The bacteria are then plated on a his$^-$ medium so that only reverse mutations to his$^+$ can grow. In the experiment shown, the suspected mutagen causes frameshift mutations.

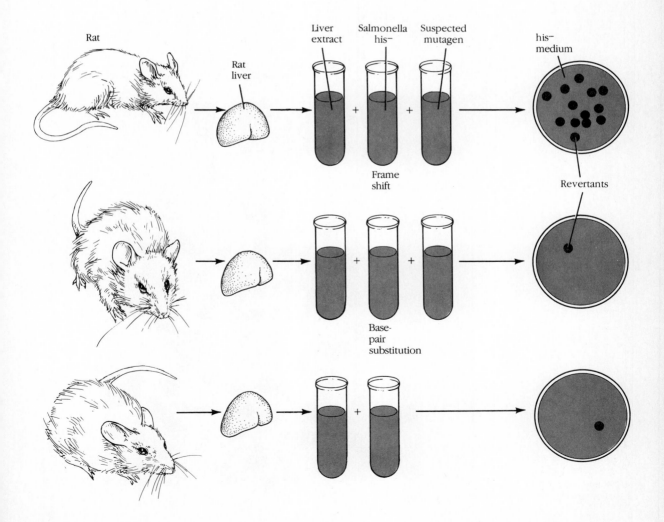

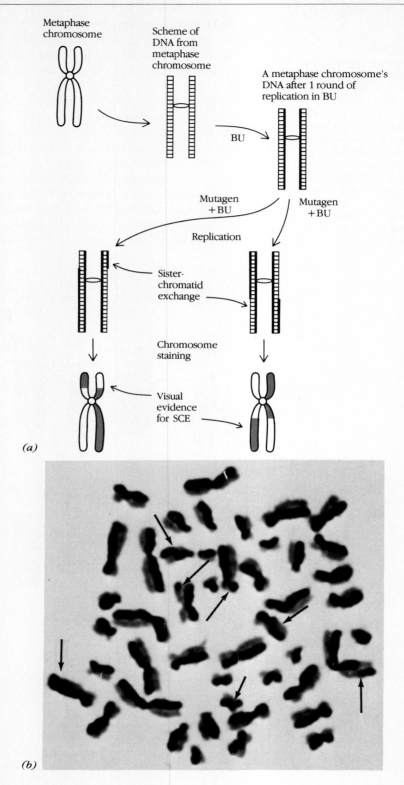

Metaphase chromosome

Scheme of DNA from metaphase chromosome

A metaphase chromosome's DNA after 1 round of replication in BU

BU

Mutagen + BU

Mutagen + BU

Replication

Sister-chromatid exchange

Chromosome staining

Visual evidence for SCE

(a)

(b)

Figure 9–13
(*a*) Formation of sister-chromatid exchanges.
(*b*) Sister-chromatid exchanges (arrows).

will grow into colonies. Controls are run in which the cells are exposed to just the liver extract.

Because this test is so easy and quick, it is used in scores of laboratories all over the world.

Sister Chromatid Exchange: A Means to Detect Mutagens

As we noted in Chapter 3, a duplicated chromosome consists of two strands known as sister chromatids. Sometimes these chromatids trade segments with each other, an exchange called sister-chromatid exchange (SCE). The formation of SCEs is a normal event in cells, though it is infrequent. But the frequency of SCE formation is a useful indicator of mutagenic activity. The recent development of the sister-chromatid exchange test has revolutionized the cytogenetic approach to the identification of chemicals that may be mutagenic or in other ways hazardous.

The technique we employ for an SCE analysis is outlined in Figure 9–13a. A replicating chromosome is in a medium with the base analog bromouracil (BU), which was discussed earlier. After one round of semiconservative replication, each chromatid has one DNA strand with BU and one without. (Remember that BU acts like T and is paired with A during replication. We will ignore its mutagenic capacity for now.) After two rounds of semiconservative replication, one chromatid has two DNA strands with BU, and the sister chromatid has only one DNA strand with BU. Between the first and second replication cycles, a chemical with suspected mutagenic properties is added to the medium. If it is mutagenic, it induces an increase in the formation of SCEs. The SCEs are detected using a special dye (Hoechst 33258) that acts differentially on the two chromatids. Those chromatids with two BU-labeled DNA strands stain differently from chromatids with only one BU-labeled DNA strand (Figure 9–13b). The SCEs appear as either dark or light segments in a chromatid. This is an extremely sensitive technique, and though we do not fully understand the mechanism of SCE formation, we do know that it differs from classic recombination mechanisms. Chemicals that induce SCE formation do not induce chromosome breaks but do lead to gene mutations. To determine the mutagenic potential of a compound, we count the SCEs in a set of treated chromosomes and compare the number with the number of SCEs in a set of untreated (control) chromosomes. The SCE-inducing capacity (and thus potential hazard) of some chemicals is recorded in Table 9–2.

SCE analysis may be of value in some cancer therapies. Many chemotherapeutic drugs are potent cytostatics—agents that stop cancer cells from dividing. They will also stop normal cells from dividing, though normal cells are less sensitive to the cytostatic effects. Unfortunately, these same chemotherapeutic drugs are often potent mutagens. Three patients recently treated for breast tumors with a particular drug all developed leukemia while their breast cancers went into remission. It seems reasonable to suggest that the drug successfully combated the one form of cancer while inducing another.

It may be that these three patients were more sensitive than others to the drug. If so, we may be able to screen a person for this sensitivity by using the SCE analysis. Someone who is exceptionally sensitive to the drug may show a higher frequency of

Table 9–2   Chemically induced SCEs in mammalian cells in vitro*

Chemical	Dose (molarity)	SCE/Cell Control	SCE/Cell Experimental
Acetaldehyde	1.8×10^{-4}	4.7	28.4
Acridine orange	2.3×10^{-6}	10.8	16.9
Barbital	8×10^{-3}	7.7	9.7
Butanol	10^{-2}	5.5	5.7
Caffeine	10^{-3}	4.6	5.5
Chlorambucil	9.6×10^{-6}	3.3	56.0
Diepoxybutane	3×10^{-6}	12.0	90.0
Diethylstilbestrol	10^{-4}	7.7	7.8
Ethanol	1.7×10^{-2}	4.5	4.8
Ethyl methanesulfonate	2.3×10^{-3}	15.1	35.0
Hoechst	10^{-5}	12.0	60.0
Lead acetate	10^{-5}	4.1	4.8
Methanol	2.5×10^{-2}	4.5	4.2
Mitomycin C	10^{-7}	12.0	120.0
Nitrogen mustard	10^{-6}	12.0	107.0
Proflavine	1.2×10^{-5}	10.8	13.1
Propanol	1.3×10^{-2}	5.5	4.7
Saccharin sodium	4.9×10^{-2}	8.8	12.5

*These data represent the average of several experiments using the chemicals indicated.

SCEs than a person who is less sensitive. If sensitivity is detected, an alternative therapy can be used.

Gene Repair

If DNA replication proceeded flawlessly, mutations would not occur unless they were induced by some kind of external agent. But replication is not flawless. We find that about one out of every 10 billion base pairs is in error after a round of replication, a rate of about 10^{-10}. This mutation rate is much lower than expected on the basis of DNA polymerase's error rate during replication. Indeed, we have isolated mutations in the DNA polymerase gene itself that cause the error level to increase substantially. But one particular property of DNA polymerase makes the high fidelity of DNA replication understandable. In addition to adding DNA subunits in a $5' \longrightarrow 3'$ direction, DNA polymerase can reverse direction and *remove* them in a $3' \longrightarrow 5'$ direction. If an incorrect base is inserted in a growing DNA strand, the enzyme stops, backs up, removes the incorrect base, and then moves forward again to continue replication. DNA polymerase thus has a **proofreading function**, which helps to minimize the errors (Figure 9–14). Proofreading is a **repair function** directly connected to the replication function of the enzyme. As you will see, polymerase and other enzymes involved in replication also function in other kinds of repair.

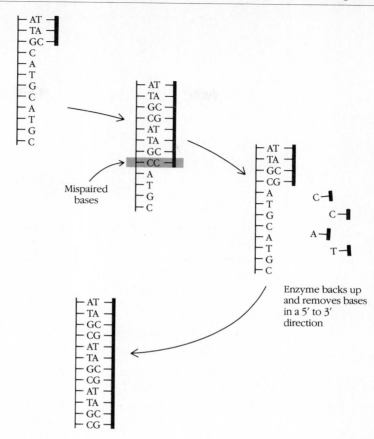

Figure 9–14
Proofreading function of DNA polymerase. The incorrect base pair is corrected.

In addition to errors of replication, DNA is subject to the damaging effects of environmental agents, such as ultraviolet (UV) light from the sun, ionizing radiation, and chemical agents, as we just discussed. Many of these environmental insults that damage DNA also induce cancers. Fortunately, we are able to repair a great deal of this damage; unfortunately, we are not always able to do so.

Consider the seemingly innocuous and often pleasurable activity of sunbathing. Sunlight, in addition to its heating rays, is composed of tanning rays that include UV radiation. A person lying in the sun is absorbing the full spectrum of the sun's radiation from infrared through UV, and will burn or tan according to the intensity of the sunlight, the duration of exposure, and the amount of pigment in the skin. At the same time, the person is accumulating genetic damage to the exposed cells that, if left unrepaired, will eventually lead to cell death or cell transformation into malignancy.

UV radiation interacts strongly with DNA and causes the formation of **thymine dimers**. A dimer is a molecule composed of two subunits—in this case, two single molecules of thymine fused together (Figure 9–15). Thymine dimers distort the DNA molecule and interfere with replication; they can cause mutations and cell death. Normally, however, the DNA is repaired by enzymes that recognize the distortion, bind to the damaged region, and either remove the dimer or split it (Figure 9–16). If the dimer is removed, DNA polymerase fills in the gap, with the help of

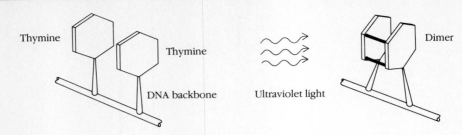

Figure 9–15
Thymine dimer, caused by
UV light.

DNA ligase. If the dimer is not removed, an error-prone repair system enables replication to bypass the distortion; but, as a result of this bypass, incorrect bases are inserted opposite the dimer. This process causes a high mutation rate (Figure 9–17).

Chemicals may also damage DNA. For example, certain industrial by-products can alter a DNA base, changing its pairing properties and causing a mutation. In these cases, an enzyme often removes the altered base from the DNA strand. Another enzyme removes the sugar and phosphate, and then DNA polymerase working with DNA ligase fills in the gap.

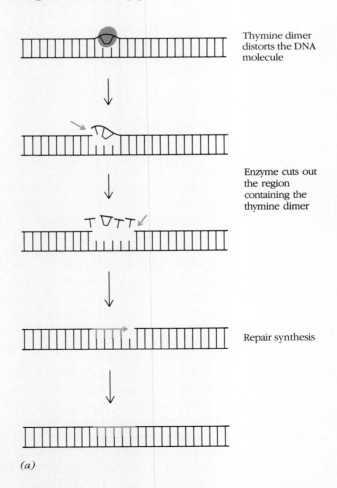

Thymine dimer distorts the DNA molecule

Enzyme cuts out the region containing the thymine dimer

Repair synthesis

Figure 9–16a
DNA repair. Removal of
thymine dimer.

(a)

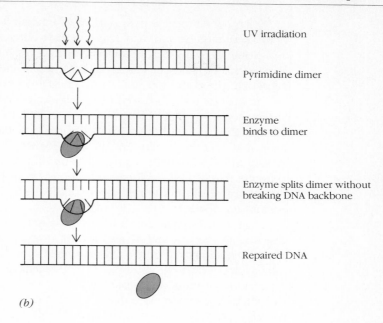

Figure 9–16b
DNA repair that does not involve removal of DNA bases.

UV irradiation

Pyrimidine dimer

Enzyme binds to dimer

Enzyme splits dimer without breaking DNA backbone

Repaired DNA

(b)

We are finding that humans, and indeed all organisms, have an enormous variety of mechanisms for repairing genetic damage. Natural selection has favored mechanisms that ensure that the genetic material will be faithfully replicated and that it can properly direct the development of a new organism. On the other hand, errors of replication and repair still occur, and this persistent imperfection allows enough aberrations to accumulate to guarantee variety in the living world.

Malfunctions in repair and replication

Since the repair of damaged DNA is controlled by enzymes, and since enzymes are coded for by genes, it follows that mutations can affect the repair function. The recessive disease xeroderma pigmentosum (XP) is an example of a malfunctioning repair process. Cells taken from a patient with XP and grown in culture have a very poor survival rate when exposed to UV radiation. These cells are unable to excise the thymine dimers that are induced by the UV, a defect that prevents DNA repair. People with XP are extremely sensitive to sunlight and have a high incidence of skin cancer.

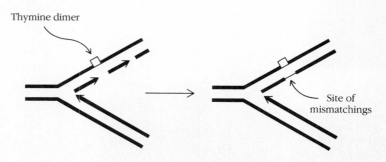

Thymine dimer

Site of mismatchings

Figure 9–17
Error-prone repair. The gap opposite the dimer is filled, but because base pairing is obstructed, incorrect bases are added.

XP is not uniform in its symptoms. The repair function can be disrupted at any one of several steps, and each defect can have different consequences. For example, any of the following may be defective: the enzyme that recognizes the dimer; the enzyme that removes the damaged DNA; the polymerase that fills in the gap; or the ligase that closes the gap. XP is a genetically heterogenous condition.

In Chapter 6 we discussed three autosomal recessive diseases characterized by chromosome instability: Bloom's syndrome (BS), Fanconi's anemia (FA), and ataxia telangiectasia (AT). All three exhibit a high frequency of chromosomal structural anomalies (Figure 9–18a). And in all three there is a greatly enhanced risk of cancer. In FA cells, there is a defect in excising and repairing DNA damage; in BS cells, the rate of DNA replication is seriously retarded. These findings support the theory that repair and/or replication defects lie at the root of the **chromosome-instability diseases**.

Can our understanding of XP shed any light on the chromosome-instability diseases? Do BS, FA, and AT also involve thymine-dimer formation? If thymine dimers are not removed from a DNA molecule, they do not serve as templates for new DNA synthesis, and chromosome breaks sometimes occur (Figure 9–18b). XP cells have more chromosome breaks after exposure to UV irradiation than normal cells do, which supports the idea that a defective repair system can lead to an increased frequency of breaks. But unirradiated XP cells do not produce the kind of breaks that occur spontaneously in BS, FA, and AT cells. So perhaps the molecular defects in XP are different from those in the chromosome instability diseases, even though all may involve replication and/or repair.

(a)

Figure 9–18
(*a*) A child with Fanconi's anemia. (*b*) Formation of chromatid breaks. Thymine dimers, if not removed, can cause chromatid breaks.

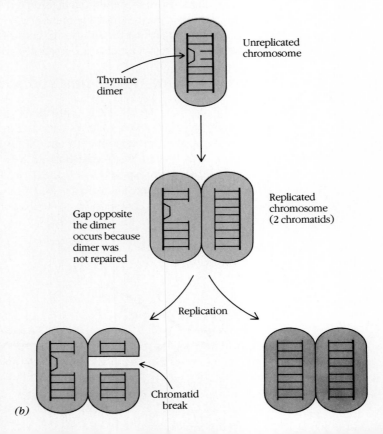

Thymine dimer

Unreplicated chromosome

Gap opposite the dimer occurs because dimer was not repaired

Replicated chromosome (2 chromatids)

Replication

Chromatid break

(b)

Summary

1 **Mutation** refers broadly to any change in the genetic material, including a gross chromosomal change. More narrowly, mutation means an alteration in an individual gene.

2 All gene variation arises from mutation. Mutations may have serious or minor consequences for the individual; only rarely are their effects desirable.

3 **Spontaneous mutations** occur during normal replication or development. **Induced mutations** are caused by external agents. Spontaneous mutations include **base-pair substitutions** and **frameshift mutations**. There are two types of base-pair substitutions: **transitions** and **transversions**.

4 In transitions, a purine base substitutes for another purine, and a pyrimidine substitutes for a pyrimidine. Transitions occur during DNA replication and are often caused by **tautomerism**, a natural shifting in the configuration of a base that changes its pairing properties.

5 In transversions, a purine substitutes for a pyrimidine, and a pyrimidine for a purine. Transversions are associated with faulty DNA repair.

6 The consequence of a base-pair substitution of either type is either a **missense mutation**, in which a wrong amino acid appears in a polypeptide, or a **nonsense mutation**, in which the polypeptide is prematurely terminated.

7 Frameshift mutations occur when one or more base pairs are added to or deleted from a DNA molecule, shifting the reading of codons. Most spontaneous mutations in higher organisms are frameshift mutations.

8 It is difficult to estimate mutation rates for recessive genes. Mutation rates for dominant genes can be estimated by examining pedigrees for new occurrences of a trait. Spontaneous-mutation rates vary greatly; one of the highest is 1 per 20,000 gametes for achondroplastic dwarfism.

9 External agents can greatly increase the rate of particular mutations, but they do not induce any kind of mutation that does not occur spontaneously. Most mutagenic agents can be classified as radiation or chemicals.

10 Some **electromagnetic radiation,** including X rays and ultraviolet rays, is mutagenic. **Corpuscular radiation**, consisting of high-speed atomic particles, is also mutagenic; both kinds include types known as **ionizing radiation**. Ionized portions of DNA molecules undergo chemical reactions that change their bonding properties and cause mispairing.

11 Radiation may be measured in **roentgens**, **rads**, or **rems**. A rem is a measure of the biologic effects of radiation.

12 In general, the mutation rate is directly proportional to the total radiation received, regardless of the intensity of the individual doses. For two-break chromosomal events, such as inversions, the mutation rate increases exponentially with the total radiation, and higher individual doses produce higher mutation rates.

13 High doses of radiation cause high cancer rates, high rates of chromosomal anomalies, and high gene-mutation rates. Much of our information on high exposure comes from survivors of the atomic bombings of Hiroshima and Nagasaki.

14 The effects of low-level radiation are hard to assess. Sources include natural background radiation from cosmic rays and naturally radioactive materials, and manufactured sources such as X rays and nuclear fallout.

15 Chemicals that cause base-pair substitution may be **base analogs**, which substitute for DNA bases during replication, or **agents that act on nonreplicating DNA**, which alter the configuration of bases so that mispairing occurs during replication or repair. Some chemicals cause frameshift mutations; these attach to or are inserted into the DNA molecule.

16 Hundreds of chemical compounds are known to be mutagenic, and we are increasingly exposed to a multitude of possibly mutagenic substances. The problem is illustrated by the experience of the residents of the Love Canal area, but there are conflicting reports on the results of their exposure.

17 The Ames test and sister-chromatid exchange are valuable tools used in testing the mutagenicity of compounds.

18 The mutation rate is much lower than we would expect from our knowledge of mutational mechanisms, because there are mechanisms for the repair of genetic damage. DNA polymerase has a **proofreading function:** when an error in replication occurs, it removes bases in reverse order until the error is eliminated.

19 Other enzymes involved in replication also have a repair function. Ultraviolet light from the sun causes formation of **thymine dimers**, which distort DNA, but most of these are removed or split by enzymes that function in DNA repair.

20 Genetically caused malfunctions of genetic repair systems can lead to specific diseases. In xeroderma pigmentosum, the ability to repair UV-damaged DNA is impaired and skin cancer occurs frequently. The **chromosome-instability diseases**, which exhibit a high frequency of chromosomal anomalies, may also be caused by defective repair systems.

Key Terms and Concepts

Mutation	Missense mutation	Mutagen
Spontaneous mutation	Nonsense mutation	Agents that act on nonreplicating DNA
Induced mutation	Electromagnetic radiation	Carcinogen
Base-pair substitution	Corpuscular radiation	Proofreading function
Frameshift mutation	Roentgen	Repair function
Transition	Rad	Thymine dimer
Transversion	Rem	Chromosome-instability diseases
Purine base	Mutagen	
Pyrimidine base	Base analog	
Tautomerism		

Problems

9–1 How does a transition differ from a transversion mutation?

9–2 How does a frameshift mutation differ from a base pair substitution?

9–3 If one DNA base participated in more than one codon, what prediction would you make about base-pair substitutions?

9–4 Can spontaneous mutations have a positive effect on the survival of a species, such as our own? Explain.

9–5 What are the two main classes of high-energy radiation? How does a rem differ from a roentgen?

9–6 What accounts for the exponential curve we find when we follow the induction of two-break events as a function of radiation dose?

9–7 What will the sex ratio be among the offspring of a mother who carries an X-linked recessive lethal mutation?

9–8 What do we call chemical mutagens that mimic the effects of ionizing radiation?

9–9 What is a base-analog mutagen?

9–10 Compare the effects that an acridine dye has on DNA with the effects of bromouracil.

9–11 What is the significance of DNA polymerase's ability to remove DNA subunits in a $3' \rightarrow 5'$ direction?

9–12 Why is excessive suntanning dangerous?

9–13 What accounts for the different types of xeroderma pigmentosum we find in humans?

9–14 How is DNA replication related to recombination?

9–15 How many sister chromatid exchanges are evident in these chromosomes?

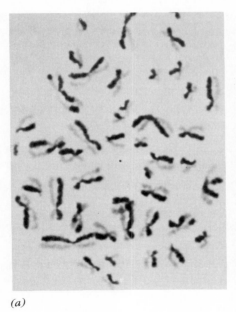 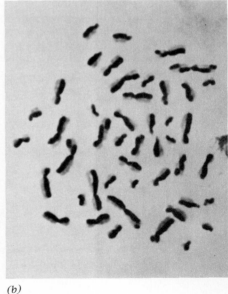

(a)   *(b)*

Answers

See Appendix A.

References

Auerbach, C. 1976. *Mutation Research.* Chapman and Hall, London.

Auerbach, C., and B. J. Kilbey. 1971. Mutation in eukaryotes. *Annu. Rev. Genet.* 5:163–218.

Ayala, F. J., and J. A. Kiger. 1980. *Modern Genetics.* Benjamin/Cummings, Menlo Park, CA.

BEIR (Biological Effects of Ionizing Radiation). 1972. *The Effects on Populations of Exposure to Low Levels of Ionizing Radiation.* National Academy of Sciences, National Research Council, Washington, DC.

Bloom, A. D. 1972. Induced chromosomal aberrations in man. *Adv. Hum. Genet.* 3:99–172.

deSerres, F. J., and A. Hollaender, eds. 1971–1982. *Chemical Mutagens: Principles and Methods for Their Detection.* 7 vols. Plenum, New York.

Devoret, R. 1979. Bacterial tests for potential carcinogens. *Sci. Am.* 241:40–49 (Aug.).

Drake, J. W. 1970. *The Molecular Bases of Mutation.* Holden-Day, San Francisco.

Fishbein, L., W. G. Flamm, and H. L. Falk. 1970. *Chemical Mutagens: Environmental Effects on Biological Systems.* Academic Press, New York.

Hanawalt, P. C., and R. B. Setlow, eds. 1975. *Molecular Mechanisms for Repair of DNA.* Plenum, New York.

Jenkins, J. B. 1979. *Genetics,* 2nd ed. Houghton Mifflin, Boston.

Latt, S. A., et al. 1980. Sister chromatid exchanges. *Adv. Hum. Genet.* 10:267–332.

McElheny, V. K., and S. Abrahamson, eds. 1979. *Banbury Report One; Assessing Chemical Mutagens: The Risk to Humans.* Cold Spring Harbor Laboratories, Cold Spring Harbor, New York.

Neel, J. V., H. Kato, and W. J. Schull. 1974. Mortality in the children of atomic bomb survivors and controls. *Genetics* 76:311–326.

Sankaranarayanan, K. 1974. Recent advances in the assessment of genetic hazards of ionizing radiation. *Atomic Energy Rev.* 12:47–74.

Scholte, P. J. L., and F. H. Sobels. 1964. Sex ratio shifts among progeny from patients having received therapeutic X-radiation. *Am. J. Hum. Genet.* 16:26–37.

Singer, B., and J. T. Kusmierek. 1982. Chemical mutagenesis. *Annu. Rev. Biochem.* 51:655–694.

Vogel, F., and R. Rathenberg. 1975. Spontaneous mutation in man. *Adv. Hum. Genet.* 5:223–318.

Watson, J. 1984. *Molecular Biology of the Gene,* 4th ed. Benjamin/Cummings, Menlo Park, CA.

10 Human Biochemical Genetics

The pioneering study of human gene activity was published by Archibald Garrod, an English physician, in 1902. In this study, Garrod analyzed the metabolic defect known as alkaptonuria and made two extremely important points: (1) Mendel's laws apply to metabolic disorders, and (2) Mendelism is a concept that can be successfully applied to humans. Garrod possessed that rare ability to make a connection between a new discovery and seemingly unrelated concepts. He tied in his discovery of a human biochemical disorder with the laws of inheritance.

The link between the gene and the phenotype was further clarified during the early 1940s when the one gene–one enzyme theory was proposed. This theory served its useful purpose and then gave way to new ideas when it was discovered that not all genes code for enzymes. Some genes, it will be recalled, are transcribed into RNA that is not translated (tRNA and rRNA); some code for hormones, antibodies, or structural proteins; some are regulatory; and some code for only part of an enzyme—a polypeptide that joins with other polypeptides to form the functional enzyme.

In this chapter, we want to explore the connection between the gene and the phenotype by looking at some of the consequences of both normal and altered gene activity. In doing so, we will tie together many of the ideas of earlier chapters—the basic principles of inheritance, the regulation of gene activity, and mutation. Specifically, we will consider a few of the metabolic disorders caused by mutant genes and look at the genetic basis for hemoglobin synthesis and variations.

Metabolic Disorders

Metabolic disorders, also known as **inborn errors of metabolism**, are diseases caused by defective or missing enzymes. Enzymes control a highly integrated network of metabolic reactions concerned with the synthesis and breakdown of amino acids, carbohydrates, lipids, and nucleotides. A defective enzyme often results in a metabolic block, which can cause a serious disease. For example, consider the following sequence of metabolic reactions:

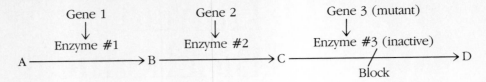

A mutation in the gene that codes for enzyme #3 causes a block in the $C \rightarrow D$ reaction. Substance C increases in concentration as a result of this block, and D is not produced. A deficiency in D may have pathologic consequences, and so may the buildup of C in the cell.

The metabolic block just diagrammed may have much more complicated consequences. Any increased quantity of C may be shunted into another metabolic pathway, where it may cause serious problems. D may normally feed into other metabolic sequences, so its absence may cause a wide range of disorders to develop. Because of the complex interrelated nature of metabolic processes there is no easy way to evaluate metabolic disorders. Nevertheless, let us take a look at some examples of interruptions in the network of human metabolism and their major effects.

Disorders of carbohydrate metabolism

The substances known as carbohydrates include sugars, starches, cellulose, and glycogen, all composed of one or more sugar units, or **saccharides**. The simple hexose (six-carbon) sugars, such as glucose, galactose, and fructose, have molecules consisting of a single sugar unit and are called **monosaccharides** (Figure 10–1a). The **disaccharides** are double sugars, with molecules consisting of two units each; lactose, for example, consists of one molecule of glucose and one of galactose, while sucrose, or table sugar, consists of glucose and fructose (Figure 10–1b). Longer-chain carbohydrates are called complex carbohydrates, or **polysaccharides** (Figure 10–1c). Starch, cellulose, and glycogen all consist of long chains of glucose units, but the chains of each substance are branched in a distinctive manner.

When humans digest starch, intestinal enzymes break down the long polysaccharide chains into monosaccharides. (We have no enzyme that can break down cellulose—popularly known as "fiber"—which passes through the human intestine undigested.) Monosaccharides are absorbed into the blood through the intestinal wall and are transported to the cells, where they enter into metabolic reactions. The monosaccharides fructose and galactose are converted into glucose in the cells, and glucose enters into energy-producing metabolic reactions.

Humans and other mammals store excess carbohydrate in the form of **glycogen**, which is synthesized from glucose units in the liver and, to a lesser extent, in other tissues. When a cell needs extra glucose for energy production, it breaks down its stored glycogen. If still more glucose is needed, glycogen from the liver is converted to glucose and transported in the blood to the tissues that need it.

You can see from all this that carbohydrate metabolism includes many reactions, all of them, of course, mediated by enzymes. It is not surprising, then, that there are many disorders of carbohydrate metabolism—the lack of a single essential enzyme may have any of the various consequences we described earlier. Let us consider just a few inborn errors of carbohydrate metabolism.

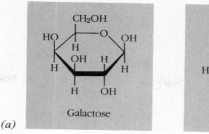
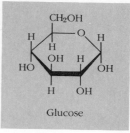

(a)

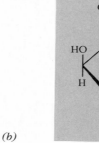

(b)

Figure 10–1
Structure of some common carbohydrates. (*a*) Two monosaccharide molecules. Each corner of the ring structure represents a carbon atom. (*b*) A disaccharide. Lactose is a combination of galactose and glucose; it is found only in milk. (*c*) A polysaccharide. Each circle represents one glucose unit in this section of a glycogen molecule. Glycogen is "animal starch," found primarily in the liver and muscles of vertebrates.

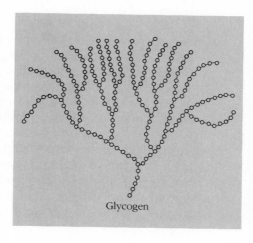

(c)

Galactosemia   A newborn girl snuggles in her mother's arms and eagerly accepts the milk offered her. Day after day she drinks her milk, but she fails to thrive. Weight gain is slow, the liver becomes enlarged, eye cataracts develop, and she begins to show signs of neurological damage. The child is dead before her first birthday. The disease she had is called galactosemia (which was briefly discussed in Chapter 4), and it is caused by the inability to metabolize galactose. The baby took in the lactose and broke it down into glucose and galactose, which were carried to her cells. Here the galactose should have been converted to glucose, but this process was blocked by the absence of the enzyme **G1PUT** (galactose-1-phosphate uridyl transferase), as shown in Figure 10–2. G1PUT deficiency is a recessively inherited condition, as are most enzyme defects. Most of the pathological effects we described in the baby are

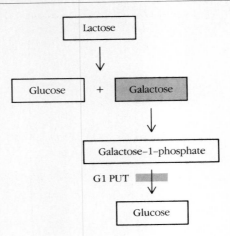

Figure 10–2
Galactose metabolism. Ga-
lactosemia occurs when the
enzyme G1PUT is absent or
defective.

the result of increased concentrations of the first intermediate product (galactose-1-phosphate), or to high concentration of galactose, whose breakdown is slowed. Usually the effects of galactosemia can be prevented by restricting galactose intake, that is, by using a milk substitute in the child's diet, but this measure requires an early recognition of the disorder.

Glycogen storage disease   A young boy lived next door to us for about a year. He had a disproportionately large abdomen, was quite short, and, although only 13 years old, had the face of a much older person. This boy was suffering from a genetic disorder known as **glycogen storage disease**, or **glycogenosis**. The pathways for the synthesis and breakdown of glycogen are complex, involving many enzymes. Numerous disorders of glycogen synthesis are known. What was unusual about the boy next door was that he was able to synthesize glycogen but unable to break it down—a very rare occurrence. His distended abdomen was the result of a greatly enlarged liver, packed with accumulated glycogen; the excess glycogen in liver and muscle tissue had retarded his growth and led to altered facial features. Special treatment is required to get rid of the glycogen stores in such a case. The boy is receiving this special treatment and is doing quite well so far.

Several different enzymes are involved in the synthesis and breakdown of glycogen. Defects in any of these enzymes disrupt glycogen metabolism and may lead to the accumulation of glycogen in various tissues. The boy we described was missing a functioning phosphorylase enzyme.

G6PD deficiency diseases   During the Korean War, an interesting disorder came to light in a most unusual way. To protect soldiers against the ravages of malaria, a parasitic infection of the red blood cells, doses of a drug called **primaquine** were routinely given out. Dark-pigmented people seemed to be especially sensitive to primaquine. About 1 out of every 10 black soldiers who were given this drug suffered severe **hemolysis**—destruction of the red blood cells. Yet only about 1 in every 1000 white soldiers had this reaction. Similar reactions had been recorded in the medical literature earlier, but they did not attract attention until the incidence increased during the Korean War. In the Mediterranean area, a similar hemolytic

Table 10–1	Some variants of G6PD	
Variant Allele	Red Blood Cell Activity (% of Normal)	Incidence
Normal	100	Commonest type in all populations
A	90	Common in blacks
Athens	25	Common in Greece
A^-	8–20	Common in blacks
Canton	4–24	Common in Southeast Asia
Mediterranean	0–7	Common in Mediterranean and Middle East area

disease, called **favism**, occurs in some people who eat the fava bean (broad bean) *Vicia faba*.

In 1954, it was discovered that the common denominator in these hemolytic attacks is an inherited enzyme defect in the red blood cell. The missing enzyme, coded for by an X-linked gene, is **G6PD** (glucose-6-phosphate dehydrogenase), which is involved in one of the pathways of glucose metabolism. This pathway is a relatively minor one in terms of energy production, and blocking it has no noticeable effects most of the time. However, it is the absence of a by-product of the blocked pathway that ultimately causes red cells to react to a substance in fava beans as well as to primaquine and (as was later discovered) numerous other drugs.

The G6PD deficiency diseases vary in their symptoms and severity because G6PD production is controlled by a multiple allele system, with the different mutant alleles producing enzyme variants that have slightly different properties. As in some other genetic diseases, the allele frequencies vary among different ethnic groups (Table 10–1).

Disorders of amino acid metabolism

A single amino acid may be involved in hundreds of metabolic pathways, each involving many enzymes. Therefore, the same amino acid may be implicated in a variety of metabolic diseases. Let us consider three disorders that can arise from enzyme defects in the pathways of phenylalanine metabolism (Figure 10–3).

Alkaptonuria   Garrod's analysis of alkaptonuria was the first study that identified a genetically caused disruption of the metabolism of an amino acid. Alkaptonurics excrete a compound called **homogentisic acid** (or "alkapton") in the urine; normal people do not. In the presence of oxygen, the homogentisic acid turns black, a potentially embarrassing situation that is not otherwise very serious. In later life, alkaptonurics may acquire deposits of dark pigment in their cartilage and connective tissue and therefore become arthritic; but they are usually basically healthy otherwise.

Figure 10–3
The metabolic blocks in three genetic disorders involving the metabolism of phenylalanine and tyrosine. (*a*) In alkaptonuria, homogentisic acid cannot be broken down further and appears in the urine. The missing enzyme is homogentisic acid oxidase. (*b*) In phenylketonuria (PKU), phenylalanine cannot be converted to tyrosine; consequently, phenylalanine and its breakdown products accumulate, resulting in abnormal neurological development. The missing enzyme is phenylalanine hydroxylase. (*c*) In classic albinism, tyrosine cannot be converted to DOPA, and so the formation of melanin pigments is prevented. The missing enzyme is tyrosinase.

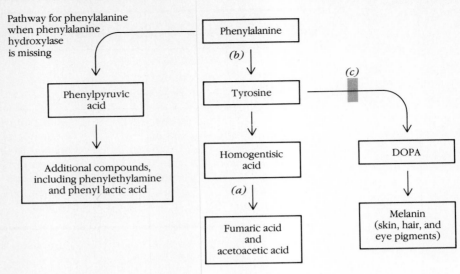

Garrod found that when alkaptonurics are fed a dose of homogentisic acid, they excrete the entire amount, whereas normal people excrete little or none. Since this substance is not a normal part of the diet, Garrod suspected that the homogentisic acid regularly secreted by alkaptonurics is an intermediate product of some metabolic pathway. He was able to demonstrate that homogentisic acid is a breakdown product of the amino acid phenylalanine. Most people, he concluded, are able to break homogentisic acid down as a next step, but alkaptonurics have a metabolic block that prevents them from doing so.

It was not until 1958 that the nature of the metabolic block was discovered. In normal people, phenylalanine is enzymatically degraded to fumaric acid and aceto-acetic acid via several intermediate products, including homogentisic acid (Figure 10–3 *middle*). With the help of the enzyme **homogentisic acid oxidase**, homogentisic acid is then converted to maleylacetoacetic acid. But alkaptonurics lack this enzyme, so the conversion is blocked, and the excess homogentisic acid is passed out in the urine. Homozygous recessive genes account for the lack of homogentisic acid oxidase.

PKU Other defects in phenylalanine metabolism have much more serious consequences. If the first enzyme in the pathway of phenylalanine breakdown is missing, the debilitating condition known as **phenylketonuria**, or **PKU**, results. The enzyme in question, phenylalanine hydroxylase, or PH, normally converts phenylalanine to tyrosine (Figure 10–3 *middle*). If the enzyme is missing, phenylalanine cannot, of course, be metabolized in the usual manner. So it accumulates in the tissues and, via diversion to other pathways, is converted to phenylethylamine, phenylpyruvic acid, and phenyl lactic acid (Figure 10–3 *left*). The excess of phenylalanine and of its by-products then causes a wide range of secondary metabolic disturbances, chief among them being abnormal neurological development. The mental development of PKU victims is so retarded that their IQs are usually around 20, and so they are generally classified as mentally defective. In addition, since the melanin pigments are generated by the phenylalanine metabolic pathway (Figure 10–3 *right*), people with PKU usually have less pigment than normal.

Most states now require that each newborn child be tested for PKU. If PKU is detected, the child is put on a phenylalanine-restricted diet, which prevents the neurological deterioration. The human body does not synthesize its own phenylalanine; it derives this amino acid from external sources and the internal breakdown of proteins. Since phenylalanine is an essential amino acid, the restricted diet must supply enough of it for normal growth, but no excess. Once neurological growth is completed (usually at about age six), the person with PKU can usually go on a regular diet.

Albinism   Albinism—the partial or total absence of normal pigmentation of the skin, hair, and eyes—is also related to phenylalanine metabolism. There are several types, but we will mention only the most common, classic, type. The albino is unable to convert tyrosine to DOPA (3,4-dihydroxyphenylalanine), an intermediate in melanin biosynthesis (Figure 10–3 *right*). Melanin-producing cells are present in normal numbers, but they are unable to make melanin. Coloration in the albino consists of pinks, blues, and yellows, all the result of the red blood color seen through a colorless layer of epithelium, and the carotenoid pigments (yellows and oranges) that are produced in normal amounts by most albinos.

Drug sensitivities

Some enzyme defects can cause extreme sensitivity to certain drugs, which can lead to serious clinical problems. We have already discussed the hemolytic anemia that develops when people with G6PD deficiency are given the drug primaquine. Investigation of unusual inborn drug sensitivities is a rapidly developing area of human genetics called **pharmacogenetics**.

Succinylcholine sensitivity   The discovery of genetically based drug sensitivity is usually totally accidental. David Jones was scheduled for routine surgery in a Philadelphia hospital and was given the drug **succinylcholine** as a muscle relaxant. Upon administration of the drug, David entered a period of prolonged muscular and respiratory paralysis, and only emergency procedures kept him alive. After careful study, it was found that David was missing a normal form of the enzyme called **serum cholinesterase**. We are unsure of this enzyme's normal function, but it does not appear to be crucial to normal development, because people who lack it seem to suffer no consequences except when they are given succinylcholine.

Succinylcholine is used as a relaxant because it interferes with the transmission of impulses from one nerve ending to the next and thus decreases muscle activity. In normal people, serum cholinesterase quickly breaks down the drug into inactive components. But if the normal form of this enzyme is not present, succinylcholine persists around the nerve endings much longer than it is supposed to. Under these conditions, the effects of the drug increase and relaxation turns into total paralysis.

As with G6PD, several abnormal forms of serum cholinesterase are known, and each causes a different degree of drug sensitivity. Here again we see the workings of a multiple allelic system.

Isoniazid sensitivity   The drug **isoniazid**, used widely to treat tuberculosis, can present a similar problem. When the drug is administered, it undergoes an inactivation reaction that is catalyzed by a specific enzyme. Some people inactivate this drug rapidly, whereas others inactivate it slowly. The slow inactivators may develop neurological problems because of the length of time the drug remains in their system in an active state.

The rapid and slow types of inactivation are genetically determined traits. The slow inactivator is homozygous for a pair of "slow" alleles; the rapid inactivator is either homozygous or heterozygous for the "rapid" allele. The "rapid" allele seems to act as a dominant, but there are indications that the heterozygote may be somewhat intermediate.

Other metabolic disorders

We have described but a few of the hundreds of known metabolic defects. There are, of course, inborn errors in the metabolism of substances other than carbohydrates and amino acids—nucleotides, for example. A defect in one of the enzymes involved in the degradation of purine (double-ring) nucleotides such as guanine and adenine can lead to the Lesch–Nyhan syndrome, whose bizarre manifestations were described in Chapter 4. A different defect in the same enzyme can lead to the very different disease of **gouty arthritis**. The Lesch-Nyhan syndrome and gouty arthritis provide another example of a multiple allelic system. Both involve increased production of uric acid. The different alleles of the gene known to produce the enzyme in question (hypoxanthine guanine phosphoribosyltransferase, or simply HGPRT) code for variant forms and thus lead to variant metabolic symptoms.

Some metabolic disorders are known as storage diseases, in which breakdown products of metabolism accumulate in the **lysosomes**. Lysosomes are the cell organelles responsible for storage and breakdown of diverse and complex large molecules. They represent a kind of intracellular digestive system. If a particular enzyme involved in the breakdown of some substance is missing, the substance itself and the intermediate products of its breakdown may accumulate in the lysosome and hinder its functioning. Tay–Sachs disease, described in Chapter 4, is a lysosome storage disease. In this condition, a chemical constituent of nerve cells called a **ganglioside** fails to break down and accumulates in the lysosome, ultimately interfering with the nervous system.

Metabolic defects may involve proteins other than enzymes—**transport proteins**, for instance. The function of these proteins is to attach to small molecules, such as amino acids, fatty acids, and monosaccharides, and transport them into the cell. In **cystinuria**, the oldest known **transport defect**, defective transport proteins in the kidney tubules prevent the reabsorption of **cystine** and other amino acids into the kidney tissue, so that they become concentrated in the urine. Cystine may eventually become so concentrated that it crystallizes as a kidney stone.

When you consider that we have thousands of metabolic pathways involving tens of thousands of enzymes, and that these pathways influence one another in elaborate and intricate ways, you may find it surprising that metabolic disorders are as rare as they are. Each of us has tens of thousands of genes that could be defective in some way, yet most of us function amazingly well. Our typical freedom from inborn errors of metabolism is testimony to the remarkable powers of genes to replicate faithfully, to repair themselves, and to be transcribed and translated almost entirely without error.

Detecting metabolic disorders

From our discussions so far, you can see how important it is to detect a metabolic disorder as quickly as possible. The devastating effects of galactosemia or PKU can be circumvented by restricting the intake of galactose or phenylalanine, respectively; drug sensitivity reactions can be avoided by changing drugs. When an infant is born, the attending physician should be acutely aware of any kind of abnormality that may indicate a metabolic disturbance. Checking the odor and color of the urine is one diagnostic avenue to explore in the newborn child, since many metabolic disorders cause intermediate products of metabolism to be excreted in the urine in high concentration. Abnormal neurological reflexes may also indicate a metabolic disorder. The sooner the anomaly is recognized, the sooner the treatment can begin and the greater the chances for success. In some metabolic disorders, it is possible to detect the carrier of the defective gene and prevent a potentially disastrous genotype from occurring. Many enzyme defects can also be spotted by means of amniocentesis, allowing the parents to decide whether or not to continue the pregnancy.

Human Hemoglobin Variation

The study of hemoglobin occupies a central position in genetics, particularly in the burgeoning field of molecular genetics. The association of abnormal hemoglobins with mutant alleles was an early link in the chain of events showing that genes code for polypeptides. When it was discovered that sickle-cell anemia is caused by a hemoglobin molecule that differs from normal hemoglobin by only one amino acid, it became obvious that a gene mutation accounts for the alteration of a single amino acid. The study of hemoglobins has been and continues to be intense because it has established a clear, unquestionable link between gene and gene product, and because hemoglobin anomalies cause many serious diseases.

Hemoglobin structure

Hemoglobin is found in the red blood cells, or **erythrocytes.** A hemoglobin molecule (Figure 10–4) consists of four polypeptide chains, each one bound to an iron-containing **heme** molecule, also called a **heme group**. Heme, the pigment that gives blood its red color, combines with oxygen in the capillaries of the lung and transports it to the body's tissues. The four polypeptide chains together, without the heme groups, are called the **globin** part of the molecule.

Hemoglobin variation does not usually involve the heme group. There are exceptions—such as **porphyria**, which we described in Chapter 1. Porphyria results from a defective enzyme in the pathway of heme synthesis and causes intermediate products to accumulate in the body. We do not know exactly how these products produce the characteristic symptoms: intestinal pain and neurological disturbances that can lead to mental confusion, partial temporary paralysis, and fits of uncontrolled rage. Interestingly, porphyria is considered a dominant trait, unlike most enzyme defects. It is probably similar to the "dominant" traits we discussed in Chapter 4, traits in which the heterozygote expresses a visible intermediate phenotype (achondroplastic dwarfism, for example) and homozygosity is lethal. A homozygous fetus unable to synthesize heme at all would not survive to be born.

Box 10–1   *Of Straitjackets—and the Genes of Kings*

The stern and commanding features of Queen Victoria have long graced the pages of genetics texts. The story of how she apparently obtained a mutated and defective gene for blood clotting from one of her parents, and faithfully passed it to certain of her progeny, makes a fascinating tale. And there are often gentle speculations of how Russian and world history might have been somewhat different had the mutant gene not contributed to the rise to power in Russia of the debauched monk, Rasputin. This "mad" monk had great influence on the czar and czarina because of his uncanny ability to ease the miseries of their hemophilic son, who inherited the defective gene from his mother, a granddaughter of Victoria.

But we wish here to consider not Victoria, but her grandfather, King George III. Indeed, the story runs back a full seven generations further, to another very famous queen—Mary, Queen of Scots.

Medical records from Mary's day are poor, of course, but one of her sons (King James) was treated by his physician for "colics," which he said he had inherited from his mother. Colics. An innocent-enough term to hide a variety of respiratory problems, but hardly the stuff from which history is made. Ah, but there is more. Mary died when only 45; had she lived longer one wonders what other symptoms she might have developed. For James suffered also from another characteristic, one that is a clear indicator of a now-recognized genetic disease. His urine, he reported, was "dark," reddish; indeed, it matched the color of his favorite Alicante wine. This condition characterizes an autosomal dominant genetic disease now called porphyria, and colics and red urine are only the beginnings of the miseries it causes. The basic problem involves porphyrin, a pigment contained in hemoglobin. But as with other body chemicals, the cell destroys the porphyrin, as a part of the normal process of aging and death of red blood cells. But, if a person lacks the ability to actively drive reactions that break down the porphyrins, they build up to abnormal levels. Some are excreted in the urine and account for the

"wine" color. But other more serious characteristics may develop: a racing pulse, hoarseness, weakness of muscles (particularly in the arms and legs), and delirium. In a severe attack, the patient suffers excruciating pain, paralysis, and mental derangement. And this is where Queen Victoria's grandfather, King George III, comes into our account.

George has been popularly caricatured as "the Mad King." He was active in the passage of the notorious Stamp Act of 1765, which led to the Boston Tea Party, the American Revolution, and England's loss of the American colonies. No doubt some of the subsequent portrayals of him as "mad" are motivated by a desire to explain this British loss, but George's behavior was unquestionably sufficient to justify much of the uncomplimentary comment. At the height of his porphyria attacks, he was completely out of control mentally. He was too weak to walk or stand, could swallow only with the greatest of difficulty, could not sleep, and suffered from excessive excitement, endless babbling, headaches, tremors, dizziness, visual disturbances, and convulsions. He was repeatedly placed in straitjackets, while Parliament debated for weeks how to resolve the crisis to government posed by an "insane" sovereign. The best doctors available attended him, but there is little evidence that they did any real good. Nevertheless, his apparently spontaneous recovery from the attacks was hailed as evidence of medical skill—and the citizens rejoiced. After recovery from his first and worst attack, medallions were struck proclaiming, "Britons rejoice—Your King's Restored." Josiah Wedgwood, the famous potter, produced a porcelain plaque that sported a likeness of the king and the inscription "Health Restored."

But recovery was not permanent. In 1811, 23 years after his first attack, he was removed from the throne and replaced by his son. The "Mad King"—though widely versed in arts and sciences and considered in earlier accounts to be one of the most learned of England's monarchs—was shamed and strapped in straitjackets.

He died at age 81, blind and senile. It is not clear what role the porphyria played in his final decline and demise.

According to the most readable account of this tragedy (Macalpine and Hunter, 1969), George's illness greatly stimulated the development of psychiatry and focused attention on the care of the mentally disturbed. And his gene for porphyria apparently did *not* play any really critical role in the loss of the American colonies. George ruled for 51 years (1760–1811). Although one may question the wisdom of the English policies that led to the Revolution during George's

rule, the policies were certainly not all of his doing; Parliament must also share the blame. There is no evidence that the disease impaired George's competence during the two decades critical to the Revolution. His first real attack came when he was 50, in fall 1788, well after the war was over. But the gene certainly *did* play a role in his deposition as monarch and his name in history. It came to him through a lengthy line of royalty and continues today among his descendants—a perverse sort of legacy from Mary, Queen of Scots.

Most hemoglobin variation involves the amino acid composition of the globin molecules. Globin variation can inhibit the binding of oxygen to heme or distort the shape of the entire red blood cell, causing problems ranging from minor to lethal. Such variants can involve amino acid substitutions in any of the globin chains.

The four polypeptide chains in the adult human are of two types, alpha and beta. The molecular formula for the common form of adult hemoglobin, symbolized as HbA, is $\alpha_2\beta_2$. The α and β chains are coded for by α and β genes, respectively, and recent studies have shown that there are two separate α gene loci, each producing the α polypeptide. The α and β polypeptides are similar in many respects. The α chain has 141 amino acids, the β chain 146.

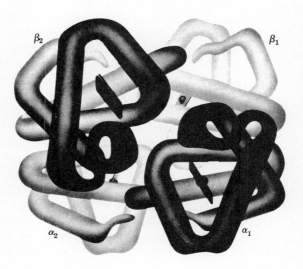

Figure 10–4

Diagram of a hemoglobin molecule, showing the two α- and two β-polypeptide chains. The rectangle in the center of each chain represents an iron-containing heme group.

There are other forms of normal hemoglobin. In addition to the common adult variety HbA, there is a minor adult hemoglobin, HbA$_2$, which accounts for about 2% of adult hemoglobin. It contains two α chains and two chains of a type called delta (δ). Its molecular formula is thus α$_2$δ$_2$. The δ chain, coded for by its own locus, has 146 amino acids and differs from the β chain by only 10 amino acids. There are also four kinds of fetal hemoglobin, with their own characteristic chains, γ, ζ, and ε. The common fetal hemoglobin, HbF, has the formula α$_2$γ$_2$.

Hemoglobin disorders fall into two categories: the **hemoglobinopathies**, in which qualitative changes occur in the amino acid composition of the globin chains, and the **thalassemia syndromes**, in which the quantity of normal α or β globin is reduced.

Globin variation: The hemoglobinopathies

We are aware of hundreds of variants in the various globin chains. The β chain, for example, has about 160 known variants, all of them with substitutions for a single amino acid. Some globin variants produce clinical symptoms, others have no noticeable effects.

Sickle-cell anemia: A β-chain variant   The most famous of the β-chain variants is sickle-cell anemia. This is a severe and often fatal hemolytic disease in which the red blood cells become sickle-shaped when the oxygen concentration in the blood is low (Figure 10–5)—as during ordinary physical exertion. All of the disease's symptoms are related to the abnormal shape of the erythrocyte. The affected person has severe anemia and jaundice caused by the destruction of many of the red cells. During a "sickling crisis," when the cells assume the sickle shape, they block the smaller blood vessels, causing painful necrosis (tissue death) in the areas where the blood supply is disrupted. Other severe symptoms also occur during such a crisis. Figure 10–6 outlines the many clinical problems that may develop.

Sickle-cell anemia is caused by a homozygous recessive gene, but the heterozygote can exhibit some of the sickling properties when exposed to very low oxygen levels—at very high altitudes, for example. We say the homozygote has **sickle cell disease**, or sickle-cell anemia, and the heterozygote has the **sickle-cell trait**. The gene is especially frequent in equatorial Africans and their descendants, and it is present in about 1 out of every 10 American blacks. The gene is also found in nonblacks, but not nearly as often.

When, in 1949, Linus Pauling compared the hemoglobin of normal people and sickle-cell patients, he discovered an important difference. When the normal and abnormal hemoglobins are broken down into small pieces and placed in a gel in an electric field, the pieces migrate to the positive electrode at different rates according to size and net charge. This technique, called **electrophoresis,** is widely used in biologic research for separating or distinguishing among different substances. Pauling found that one of the fragments of sickle-cell hemoglobin had no counterpart in normal hemoglobin. In 1957, Vernon M. Ingram showed that the sickle-cell fragment differed from its normal counterpart by a single amino acid (Figure 10–7). Normal hemoglobin has glutamic acid at position 6 of the β chain, whereas

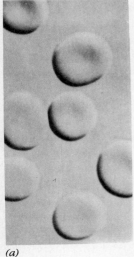

(a)

(b)

Figure 10–5
Sickle-cell anemia, a molecular disease. Photomicrographs show normal (*a*) and sickled (*b*) red blood cells (X1500). Both samples were treated with a reducing compound that lowers oxygen retention.

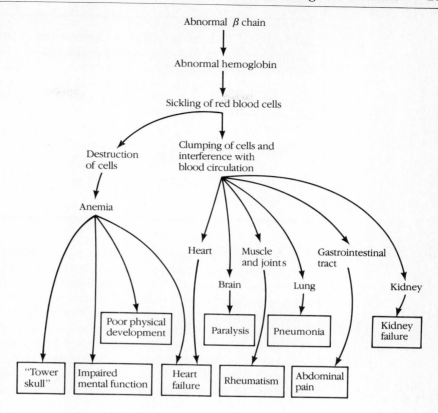

Figure 10–6

Summary of the consequences of sickle cell homozygosity.

Figure 10–7

The first seven amino acids of the β chain of human hemoglobin; the β chain consists of 146 amino acids. A substitution of valine for glutamic acid at position 6 is responsible for the severe disease known as sickle-cell anemia.

sickle cell hemoglobin has a valine at that position. This difference of one amino acid distorts the three-dimensional configuration of the hemoglobin molecule. This distortion causes the red cell to assume its abnormal shape, which in turn leads to the tragic "sickling crises" we have described.

Hemoglobin from sickle-cell anemia is referred to as HbS. (The gene for this abnormal hemoglobin is designated Hb^S; the gene for normal hemoglobin is designated Hb^A.) The homozygote produces all HbS, and the heterozygote with the normal allele produces both types: $Hb^S Hb^A$. The formula for HbS is $\alpha_2^A \beta_2^S$, which means that the α chains are normal and the β chains are sickle-cell type. This information is summarized in Table 10–2.

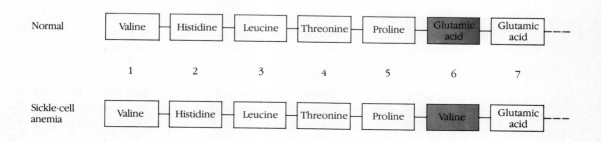

Table 10–2   A comparison of normal hemoglobin composition with sickle-cell hemoglobin

Phenotype	Hemoglobin	Hemoglobin Composition	Genotype
Normal	HbA	$\alpha_2^A\beta_2^A$	$\alpha\alpha\beta\beta$
Sickle-cell trait	HbA/HbS	$\alpha_2^A\beta_2^A$, $\alpha_2^A\beta^A\beta^S$, and $\alpha_2^A\beta_2^S$	$\alpha\alpha\beta\beta^S$
Sickle-cell disease	HbS	$\alpha_2^A\beta_2^S$	$\alpha\alpha\beta^S\beta^S$

Oddly enough, although the *disease* sickle-cell anemia is commonly fatal, the *trait* (heterozygous condition) causes only minor problems with respect to the anemia and actually confers a great advantage with respect to malaria. The malarial parasite is unable to infect red cells that carry HbS, so the sickle-cell trait is a boon to people who live in regions of the world where malaria occurs. Two heterozygotes living in a malaria-infested area tend to produce children in the ratio of 1 Hb^A to 2 Hb^A/Hb^S to 1 Hb^S. The Hb^S child usually dies from anemia; the Hb^A child gets malaria; but the two Hb^A/Hb^S children do quite well. Thus, half the children produced by heterozygous couples are well adapted to the environment where malaria is prevalent. The children of homozygous Hb^A parents are all Hb^A and invariably come down with malaria. So even though two heterozygotes may produce some children who are fatally stricken with sickle-cell anemia, such parents are reproductively better off than a homozygous Hb^A couple. The reproductive advantage of heterozygotes explains the persistence and high frequency of the Hb^S allele in Africa and other malarial areas of the world.

Termination mutants, deletions, and duplications   At the end of a gene coding for a polypeptide, there is normally a stop signal, which causes elongation of the polypeptide chain to terminate. There are three codons that function as termination signals: UAA, UGA, and UAG. It stands to reason that if mutations can occur in amino acid–specifying codons, they can also occur in termination codons. And indeed they do.

The hemoglobin variant known as Constant Spring (HbCS) is an example of a termination mutant. (The name of an aberrant hemoglobin commonly includes the name of the locality in which it was found.) Instead of having an α chain with 141 amino acids, HbCS has one with 172. The sequence of the additional 31 amino acids bears no resemblance to the sequence in any of the other chains. HbCS apparently arose when a DNA mutation changed the UAA codon at position 142 on the mRNA into an amino acid–specifying codon, CAA. A glutamine was inserted at this position and translation continued for 30 more coding units until another stop signal was encountered (Figure 10–8b).

Interestingly, people who are homozygous for the Hb^{CS} allele, have less than 10% of their hemoglobin in this elongated form. The rest is normal HbA. This situation reinforces what we said earlier: that there are normally two α gene loci

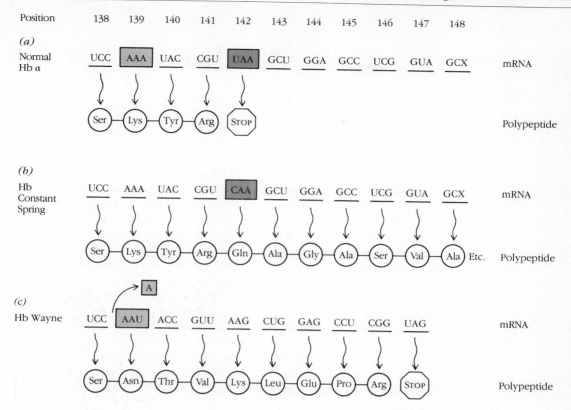

| Position | 138 | 139 | 140 | 141 | 142 | 143 | 144 | 145 | 146 | 147 | 148 |

(a)

Normal Hb α   UCC AAA UAC CGU UAA GCU GGA GCC UCG GUA GCX   mRNA

Ser—Lys—Tyr—Arg—STOP   Polypeptide

(b)

Hb Constant Spring   UCC AAA UAC CGU CAA GCU GGA GCC UCG GUA GCX   mRNA

Ser—Lys—Tyr—Arg—Gln—Ala—Gly—Ala—Ser—Val—Ala Etc.   Polypeptide

(c)

Hb Wayne   UCC AAU ACC GUU AAG CUG GAG CCU CGG UAG   mRNA

Ser—Asn—Thr—Val—Lys—Leu—Glu—Pro—Arg—STOP   Polypeptide

Figure 10–8

(a) Portion of the normal hemoglobin α chain. *(b)* Hemoglobin Constant Spring. A base-pair substitution at codon 142 changes a stop signal to an amino acid–specifying signal. The result is a long α polypeptide. *(c)* Hemoglobin Wayne. A deletion of a base in codon 139 causes a reading frameshift. The result is an abnormally long α polypeptide.

per chromosome and that only one or two of the four alleles are mutant in HbCS. It also suggests that the rate of synthesis of the HbCS α chains is much lower than that of normal α chains.

An unusual example of a deletion that alters the termination signal of the normal chain is seen in the variant known as **Hb Wayne**. Here the α chain is 146 amino acids long instead of 141. But what makes this variant especially interesting is the abnormality of the amino acids from position 139 on. The best way to interpret this abnormality is as a frameshift mutation in which a base in codon 139 has been lost (Figure 10–8c). This loss shifts the reading frame over by one base and makes the UAA stop signal in position 142 read AAG, which codes for lysine. At position 147, a stop signal occurs.

In both HbCS and Hb Wayne we see an important confirmation of colinearity (i.e., the gene and the protein it determines are in the same linear order) and of the genetic code. In both instances, we are able to account for the amino acid alterations by specific changes in the coding units. This is especially evident in the Hb Wayne anomaly.

The deletion of one or more bases from the DNA molecule can occur when homologous chromosomes mispair during meiosis (discussed in Chapter 3). **Hybrid genes** may be another consequence of unequal crossing over when homologs mispair. **Hb Lepore** (named after the family in which this rare hemoglobin was originally found) is an example of a series of hybrid genes. Here the α chains are normal; but instead of β chains, there is a hybrid chain with first β and then δ

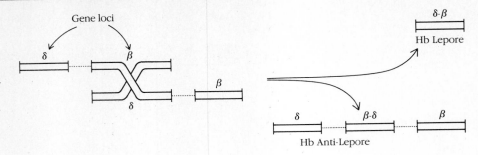

Figure 10–9
Formation of Hb Lepore through unequal crossing over. The Hb Lepore chromosome has a δ-β hybrid gene; the Hb anti-Lepore chromosome has a complete δ and β gene along with a hybrid β-δ gene.

sequences. An interpretation of this variant combines two established facts: the β and δ genes lie very close to one another; and unequal crossing over is a well-documented phenomenon. By putting these two facts together, we can develop a model for the Hb Lepore variants (Figure 10–9). Unequal crossing over occurs in the β and δ genes, producing one chromosome with just a hybrid δ-β gene (Lepore) and another chromosome with a δ, a β, and a hybrid δ-β gene (**anti-Lepore**). In all the Hb Lepore conditions, which show moderate to severe symptoms of hemolytic anemia, both the δ and β genes are missing. The hybrid gene that takes their place varies according to the position of the exchange; consequently, the hybrid hemoglobin chain varies in its amino acid composition. For this reason we speak of Hb Lepore as a series rather than a single variant. The anti-Lepore hemoglobins are classic examples of gene duplications.

Reduced globin synthesis: The thalassemia syndromes

In the thalassemia syndromes, the amino acid sequence of the globin chains is normal, but a particular chain is synthesized in reduced amounts. The thalassemias are divided into those involving the β chain (β thalassemia) and those involving the α chain (α thalassemia).

β thalassemia   β thalassemia is relatively common in certain populations around the Mediterranean, such as Southern Italy and Greece. We also find it fairly often in India and the Far East (10–15% frequency of heterozygotes). It is a serious health problem in all these areas. The β-thalassemia defect, when homozygous (**thalassemia major**), is usually fatal by early adulthood. In this homozygous state, the disease is commonly called **Cooley's anemia**, and without treatment it is fatal in early childhood. Even if the disease is treated, though, the patient will probably die. Ironically, death is usually a consequence of the treatment: repeated transfusions of blood and the administration of iron-binding agents tend to cause a hopelessly toxic iron buildup in the body. But at present there is little else that can be done. The heterozygous condition (**thalassemia minor**) is less serious.

What genetic defect causes reduced β-chain synthesis? We have found that there are actually two types of β thalassemia: β^+ and β^0. In β^+ thalassemia, the amount of β globin mRNA, and therefore of β globin, is greatly reduced. The cause of this reduced mRNA concentration is either a decrease in transcription rate or a defect in the processing of the RNA after it has been transcribed. We think that the latter may be the case, because the *Hbβ* gene contains intervening sequences of nontranslated

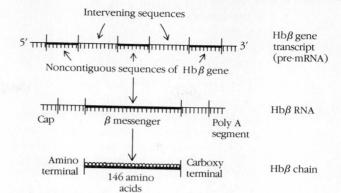

Figure 10–10

A scheme of the *Hb*β gene transcript, its mRNA, and the Hbβ chain.

nucleotides, and the pre-mRNA it produces requires considerable tailoring before it becomes a functional mRNA molecule (Figure 10–10). If something interfered with this processing, it would account for the scarcity of functional β-globin mRNA.

In β⁰ thalassemia, β-globin production is completely absent in the homozygote, and in some homozygous patients no β-globin mRNA is present. The lack of mRNA may be the result of either a deletion of all or part of the *Hb*β gene or a mutation that prevents RNA transcription. Other homozygous β⁰ patients do have the β-globin gene and β globin mRNA, but the mRNA is not translated. Perhaps there is no translation because the mRNA has termination codons in the middle of the gene sequence; perhaps the initiation sequence is deleted; or perhaps the RNA processing is defective. Figure 10–11 summarizes these ideas about β thalassemia.

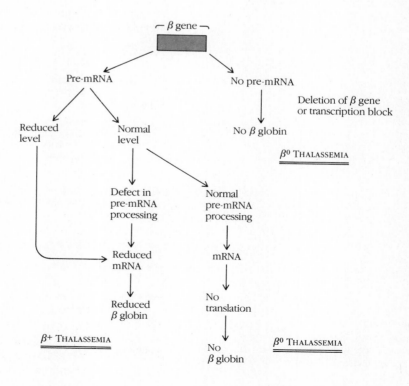

Figure 10–11
Summary of β thalassemia.

*α **thalassemia*** Each member of chromosome 16 has two *Hbα* genes. Consequently, there are four types of α thalassemia, a disease characterized by a decrease in, or total absence of, α-globin synthesis. Each of the α thalassemia diseases is due to the loss of a different number of *Hbα* genes:

1 Loss of one *Hbα* gene: decreased α-globin synthesis; no clinical symptoms; individual referred to as a "silent carrier."

2 Loss of two *Hbα* genes: red cell abnormalities; very slight anemia problems; condition referred to as the α-thalassemia trait.

3 Loss of three *Hbα* genes: *HbH disease*, characterized by hemoglobin composed of four β chains. HbH precipitates in cells and causes a mild hemolytic anemia.

4 Loss of four *Hbα* genes: *Hydrops fetalis—α* thalassemia, a fatal disease. The fetus dies in utero (within the uterus) from heart failure and severe anemia.

As in so many other genetic diseases, there are some clear ethnic differences in the frequency of the various α thalassemias. In East Asians, for example, both *Hbα* genes are commonly missing from the same chromosome, and hydrops fetalis will tend to occur in one quarter of the offspring if each parent carries such a chromosome. In Mediterranean and black populations, on the other hand, hydrops fetalis is very rare, because when two Hbα genes are missing, one is usually missing from the same locus of each chromosome. If each parent has two such chromosomes, the couple cannot produce a child who is missing all four *Hbα* loci (Figure 10–12). Though people in different ethnic groups may show the same trait, we know from the different kinds of offspring they produce that the location of the two deleted genes is different in the two groups described.

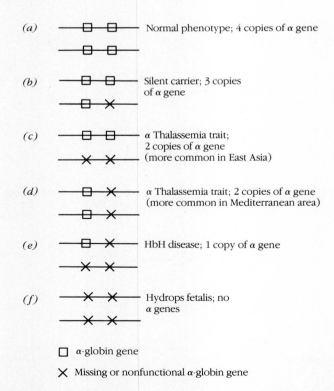

Figure 10–12
Summary of α thalassemia. The diploid state is shown with various numbers of α gene loci.

Location and evolution of the hemoglobin genes

There is now a substantial body of genetic and biochemical evidence confirming the location of the hemoglobin genes within the human genome. Two γ genes, called γ^G and γ^A, lie close to each other on chromosome 11. Each produces its own polypeptide, and the two polypeptides differ by a single amino acid. Also on chromosome 11 is a δ gene and a β gene. We do not yet know where the ζ and ε genes are located. The two α genes are located on chromosome 16. Each of the genes is apparently transcribed into a large molecule of heterogenous nuclear RNA that contains nontranslatable, intervening sequences. The removal of these sequences allows the nuclear RNA to transform into a functional mRNA.

All the hemoglobin genes are remarkably similar. The three-dimensional structures of their by products are virtually identical, and their amino acid sequences are the same in many parts of the molecule. All this similarity argues that the different globin genes have a common evolutionary origin. The most likely mechanism for the evolution of these molecules is gene duplication followed by nucleotide-sequence diversification. As a general rule, the more similar the amino acid sequence of two globin chains, the more recent is their presumed evolutionary split. Since the γ^G and γ^A gene products differ by only a single amino acid, the split must have occurred relatively recently, in evolutionary terms. The α and β genes have more differences between them, which indicates that they diverged from each other longer ago.

A scheme for the evolutionary origin of the various *Hb* genes is presented in Figure 10–13. It is based on the supposition that there was a successive series of gene duplications through unequal crossing over (see Chapter 5), followed by the independent accumulation of nucleotide alterations, which then led to different amino-acid substitutions. In this scheme, the γ^A and γ^G genes, which code for polypeptides that differ by only one amino acid and by only one base pair, diverged very recently from each other. The two α genes presumably diverged even more recently, since they do not produce different globin chains; the original gene has duplicated, but neither of the new genes has yet had a nucleotide alteration in the coding regions. The δ and β genes have 10 amino acid differences and thus must have diverged longer ago. All of these genes—γ, δ, and β—code for polypeptides of 146 amino acids each, and all are located on the same chromosome.

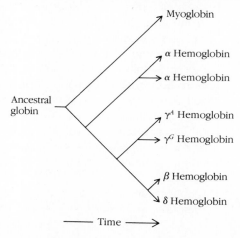

Figure 10–13
Evolutionary relationship among the globin genes.

The α gene product is only 141 amino acids long, and the amino acid differences between it and the non-α hemoglobin chains are greater than the differences among the non-α chains. We can suggest that the α and the non-α split occurred a long time ago, and that the small deletions that subsequently accumulated account for the difference in amino acid number. The duplication that split α from the non-α genes was followed by a translocation that placed the α gene on a different chromosome.

The primary evolutionary divergence occurred when an ancestral gene duplicated and gave rise to two lines: the **myoglobins** and the hemoglobins. **Myoglobin** is a heme-containing protein that acts as a storehouse for oxygen in muscle tissue. In many ways, it is structurally and functionally similar to the globin molecules of the red blood cells.

Summary

1   This chapter examines the connection between the gene and the phenotype by considering metabolic diseases and hemoglobin variation.

2   Metabolic diseases, or **inborn errors of metabolism,** are caused by defective enzymes and other proteins. Missing or inactive enzymes cause metabolic blocks that disrupt the highly integrated network of metabolic reactions. Disorders may result from the lack of the end product; from a by-product of the reaction; from the accumulation of intermediate products; or from the diversion of larger than normal quantities of intermediate products into other pathways.

3   **Galactose**, a component of lactose (milk sugar), is normally converted to glucose in the cells. In **galactosemia**, a disorder of carbohydrate metabolism, an enzyme required for this conversion—**G1PUT**—is missing. Unless put on a lactose-free diet, a baby with galactosemia dies from the effects of accumulated intermediate products.

4   Excess carbohydrate is stored as **glycogen** in the liver and other tissues; glycogen is synthesized from glucose units and is broken down when glucose is needed for energy-producing reactions. One of the many disorders of glycogen metabolism is **glycogen storage disease**, in which glycogen accumulates in the liver in enormous quantities because an enzyme necessary for its breakdown is missing.

5   In the **G6PD deficiency diseases**, including **primaquine** sensitivity and **favism**, ingestion of certain substances causes **hemolysis** (rupture of the red blood cells) owing to a block in a minor pathway of glucose metabolism. The enzyme involved is controlled by multiple alleles, which give rise to different symptoms and vary in frequency in different ethnic groups.

6   A single amino acid may take part in dozens of metabolic pathways. Disruption of the metabolism of the amino acid **phenylalanine** can lead to alkaptonuria, **phenylketonuria (PKU)**, or **albinism**, depending on the enzyme involved. Phenylketonuria, the most serious of these disorders, can be controlled by a low-phenylalanine diet.

7   Both **succinylcholine** and **isoniazid** sensitivity are caused by variations of an enzyme that breaks down the drug in question. **Pharmacogenetics** is a rapidly developing branch of human genetics that studies such inherited drug sensitivities.

8   Metabolic diseases result from enzyme defects in many other pathways—for example those of nucleotide metabolism (Lesch–Nyhan syndrome, gouty arthri-

tis) or of lysosomal storage and breakdown (Tay–Sachs disease). They can also result from defects in the transport proteins, as in **cystinuria**.

9 Adult hemoglobin (HbA) is constructed of two α chains (polypeptides) and two β chains; it has the formula $\alpha_2\beta_2$. Each chain is bound to an iron-containing **heme group**. A minor form of adult hemoglobin (HbA$_2$) contains two δ chains and has the formula $\alpha_2\delta_2$. Fetal hemoglobins have three unique chains; the common fetal hemoglobin is HbF ($\alpha_2\gamma_2$). Each type of chain is coded for by at least one separate gene locus.

10 Variation in the amino acid composition of the polypeptide chains produces **hemoglobinopathies**. The best-known of these is sickle-cell anemia, caused by a variation of the β chain. Sickle-cell hemoglobin, HbS, differs from HbA by the substitution of a single amino acid. The homozygous condition (the disease) is often fatal, but the heterozygous condition (the trait) creates few problems and provides protection against malaria. Many other amino acid substitutions are known.

11 Hemoglobinopathies can be caused by mutations in the termination signals of the hemoglobin genes. This results in polypeptides that are longer than normal, such as **hemoglobin Constant Spring (HbCS)**, and that may also have abnormal sequences before the usual termination point, such as **Hb Wayne**.

12 Hemoglobinopathies can be caused by unequal crossing over between hemoglobin genes. Since the *Hb*β gene and the *Hb*δ gene lie next to each other, unequal crossing over can produce a **hybrid gene** such as **Hb Lepore**.

13 The **thalassemia** syndromes involve reduced synthesis of normal hemoglobin chains. β **thalassemia**, when homozygous, is called **Cooley's anemia** or **thalassemia major** and is commonly fatal. In the β^+ form, both β-globin chains and β globin mRNA occur in reduced amounts; this form may be caused by faulty mRNA processing. In the β^0 form, essentially no β chains occur; this form may be caused by deletion of all or part of the Hbβ gene. The heterozygous condition, **thalassemia minor**, is less serious.

14 The *Hb*α gene loci are duplicated, so there are four kinds of α **thalassemia**, determined by how many genes are nonfunctional. The nonfunctional genes may be on the same or different chromosomes. Four nonfunctional genes produce the lethal condition **hydrops fetalis**. As with β thalassemia, the different kinds of α thalassemia have different ethnic distributions.

15 The hemoglobin genes have a common evolutionary origin and evolved through duplication, translocation, and mutation. They are thought to share their evolutionary origin with the gene for **myoglobin**, a respiratory pigment of muscle tissue.

Key Terms and Concepts

Metabolic disorders (inborn errors of metabolism)

Saccharide

Monosaccharide

Disaccharide

Polysaccharide

Glycogen

G1PUT

Glyogen storage disease (glycogenosis)

Primaquine

Hemolysis

Favism

G6PD

Homogentisic acid

Homogentisic acid oxidase

Phenylketonuria (PRU)

Albinism

Pharmacogenetics

Succinylcholine

Serium cholinesterase

Isoniazid

Gouty arthritis

Lysosome

Ganglioside	Porphyria	Hybrid gene
Transport proteins	Hemoglobinopathies	Hemoglobin Lepore
Cystinuria	Thalassemia syndromes	Hemoglobin anti-Lepore
Transport defect	Sickle-cell disease	β Thalassemia (thalassemia major, Cooley's anemia)
Cystine	Sickle-cell trait	
Erythrocytes	Electrophoresis	α thalassemia
Heme	Hemoglobin Constant Spring (HbCS)	Thalassemia minor
Heme group		Myoglobin
Globin	Hemoglobin Wayne	

Problems

10–1   In the metabolic sequence *A B C*, there is an enzyme defect that disrupts the *B C* step. This disruption should lead to an accumulation of the "*B*" metabolite, but when the body tissues and fluids are examined, the "*B*" level is normal. How would you account for this?

10–2   Two albinos marry and have children. Normally, one would expect such a mating to produce all albino children, but this couple produced all normally pigmented children. How would you explain this?

10–3   A female who has the sickle cell trait marries a male who is homozygous for the HbC trait, which is a β-chain variant carrying a lysine as a substitute for the glutamic acid that normally is at position 6. What kind of children can this couple produce?

10–4   What are the two main classes of hemoglobin disorders, and how do they differ?

10–5   How do the defects that produce β^0 thalassemia differ from those that produce β^+ thalassemia?

10–6   How can mutations such as the one that causes sickle-cell anemia become established in a population?

10–7   Why might mutations in α genes go unnoticed, whereas β-gene mutations would be more apparent?

10–8   How can certain metabolic disorders, such as the G6PD deficiency diseases, go unnoticed? How are such disorders eventually discovered?

Answers

See Appendix A.

References

Bank, A., J. G. Mears, and F. Ramirez. 1980. Disorders of human hemoglobin. *Science* 207:486–493.

Bondy, P. K., and L. E. Rosenberg, eds. 1974. *Duncan's Diseases of Metabolism,* 8th ed. Saunders, Philadelphia.

Brady, R. O. 1976. Inherited metabolic diseases of the nervous system. *Science* 193:733–739.

Brock, D. J. H., and O. Mayo, eds. 1978. *Biochemical Genetics of Man,* 2nd ed. Academic Press, New York.

Desnick, R. J., and G. A. Grabowski. 1981. Advances in the treatment of inherited metabolic diseases. *Adv. Hum. Genet.* 11:281–369.

Gardner, L. I., ed. 1975. *Endocrine and Genetic Diseases of Childhood and Adolescence,* 2nd ed. Saunders, Philadelphia.

Harris, H. 1981. *The Principles of Human Biochemical Genetics,* 3rd ed. Elsevier North-Holland, New York.

Kabat, D., and R. D. Koler. 1975. The thalassemias: Models for analysis of quantitative gene control. Harris, H., and K. Hirschhorn, eds. *Adv. Hum. Genet.* 5:157–222.

Macalpine, I., and Hunter, R. 1969. Porphyria and King George III. *Sci. Am.* 221:38 (July).

McKusick, V. A. 1972. *Heritable Disorders of Connective Tissue,* 4th ed. Mosby, St. Louis.

McKusick, V. A. 1982. *Mendelian Inheritance in Man: Catalog of Autosomal Dominant, Autosomal Recessive, and X-Linked Phenotypes,* 6th ed. Johns Hopkins University Press, Baltimore.

Nyan, W. L., ed. 1974. *Heritable Disorders of Amino Acid Metabolism: Patterns of Clinical Expression and Genetic Variation.* Wiley, New York.

Orkin, S. H., and D. G. Nathan. 1981. Molecular genetics of thalassemia. *Adv. Hum. Genet.* 11:233–280.

Piomelli, S., and L. Carash. 1976. Hereditary hemolytic anemia due to enzyme defects of glycolysis. Harris, H., and K. Hirschhorn, eds. *Adv. Hum. Genet.* 6:165–240.

Rosenberg, L. E. 1976. Vitamin-responsive inherited metabolic disorders. Harris, H., and K. Hirschhorn, eds. *Adv. Hum. Genet.* 6:1–74.

Seegmiller, J. E. 1976. Inherited deficiency of HGPRT in X-linked uric aciduria (Lesch-Nyhan syndrome and its variants). Harris, H., and K. Hirschhorn, eds. *Adv. Hum. Genet.* 6:75–164.

Stanbury, J. B., J. B. Wyngaarden, and D. S. Fredrickson, eds. 1983. *The Metabolic Basis of Inherited Diseases,* 5th ed. McGraw-Hill, New York.

Weatherall, D. J., and J. B. Clegg. 1976. Molecular genetics of human hemoglobin. *Annu. Rev. Genet.* 10:157–178.

Weatherall, D. J., and J. B. Clegg. 1982. Thalassemia revisited. *Cell* 29:7–9.

Winter, W. P., S. M. Hanash, and D. L. Rucknagel. 1979. Genetic mechanisms contributing to the expression of the human hemoglobin loci. Harris, H., and K. Hirschhorn, eds. *Adv. Hum. Genet.* 9:229–292.

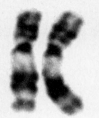

11 Genetics of the Immune System

The human body has a vast defense system whose legions seek out, recognize, and destroy foreign invaders: bacteria, viruses, fungi, other people's tissues, even the body's own altered tissues. We usually take for granted this remarkable protective function, which keeps us alive and well amid a host of besieging pathogens (disease-producing agents). Though we encounter microscopic predators daily, we are seldom sick, and when we do get sick we usually recover without medical intervention. The army that defends us against these alien substances is called the **immune system**.

The immune system provides an especially dramatic example of cellular differentiation. When a foreign agent invades the body, it causes certain white blood cells, the **lymphocytes**, to proliferate. Some of the lymphocytes react directly with the invading substance; others produce circulating **antibodies**, or **immunoglobulins**, that attack the foreign material. Any substance that can evoke this **immune response** is called an **antigen**. Each lymphocyte clone (i.e., duplicated copy) usually reacts to only one antigenic specificity, a remarkable example of the extent to which cells can become specialized.

Animals can become immune to well over a million different antigens, for this is the approximate number of different antibodies that can be produced by a single individual. Still more remarkably, some of these specific antibodies are produced in response to synthetic antigens that an animal would not and could not encounter in its normal habitat. How can the human body generate such a diverse population of antibody-producing cells? Does every lymphocyte have the capacity to produce any one of over a million different antibodies? Is each different antibody coded for by a different gene, so that the code for the immune system consists of over a million distinct genes? How does a lymphocyte recognize and respond to a specific antigen? These are some of the many questions we can ask about the immune system. So far, despite intensive research, we have only partial answers.

The Immune System

Humans, like all vertebrates, have a two-part defense system. The **cellular immune system** consists of circulating lymphocytes called **T lymphocytes** (or **T cells**), which constitute the body's primary line of defense against viral, fungal,

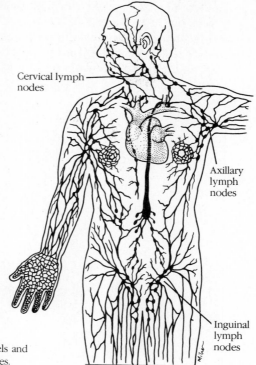

Cervical lymph nodes

Axillary lymph nodes

Inguinal lymph nodes

Figure 11–1
Major lymphatic vessels and groups of lymph nodes.

and parasitic infections, cancer, tissue transplants, and plant toxins such as poison ivy. The T lymphocytes recognize, bind to, and eliminate the foreign material. The **humoral immune system** consists of antibodies that circulate in the blood and lymph (*humor* is an archaic term for body fluid). The antibodies are produced by noncirculating **plasma cells**, which are in turn derived from lymphocytes called **B lymphocytes** (or **B cells**). Plasma cells and B lymphocytes are found chiefly in the spleen, lymph nodes, and other parts of the **lymph system** (Figure 11–1). Antibodies are the body's major means of defense against bacteria, and they are also good at fighting off many viruses and other foreign substances.

The T and B lymphocytes are nearly indistinguishable in appearance, and both are derived from precursor (ancestral) cells called **stem cells**, found in the bone marrow. The T lymphocytes develop their specialized function after passage through the thymus gland (hence the letter T for these lymphocytes). It is not clear how the B lymphocytes acquire their special properties in mammals. In birds, these lymphocytes are processed in an organ called the **bursa** (hence the designation B), and they probably undergo a similar processing in mammals, perhaps in the bone marrow.

When microorganisms or other foreign materials enter the body, they generally stimulate the division of both kinds of lymphocyte. Newly produced T lymphocytes rush to the site of the infection, creating the characteristic pus and inactivating the invading antigens by binding to them. If the antigens are tissue cells (rather than, say, toxins or bacteria), their cell membranes are disrupted and they undergo lysis (rupture). The stimulated B cells give rise to plasma cells, each of which produces about 2000 antibody molecules per second. The antibodies bind to the antigens, causing them to clump together. The clumps are then destroyed by scavenger cells called **macrophages**, large white cells that **phagocytize** (engulf and destroy) foreign bodies. Normally, both the T cells and the antibodies

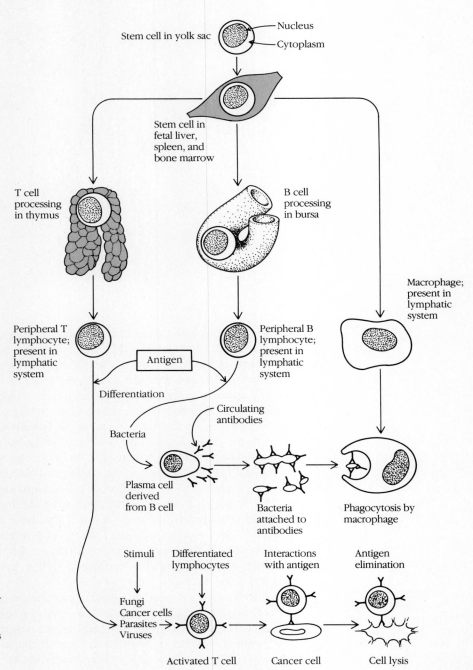

Figure 11–2
The development and processing of lymphocytes, and the differentiation, interaction, and elimination events that occur when they are activated.

bind only to the specific antigen that stimulated their production. The origin and basic function of the two cell types are summarized in Figure 11–2.

Clonal selection

The highly specific nature of the immune response is best explained by the **clonal selection theory**. We know that lymphocytes have specific receptor molecules on their membranes; each kind of cell has uniquely shaped receptors, which fit with a specific antigen in lock-and-key fashion (Figure 11–3). In the case of B cells, these receptors are antibodies. In the case of T cells, they are fragments of antibodies. According to the clonal selection theory, each one of a person's B and T lymphocytes is committed, *before being exposed to any antigen*, to the synthesis of one particular antibody. This antibody is the one that appears on the cell surface as a receptor. A small clone of each type of lymphocyte is present initially, and there are over a million types, so that we have a tremendous variety of receptors.

When the cell encounters an antigen that fits its receptors, it is stimulated to divide, forming a large clone of lymphocytes of the same type. After several rounds of division, the B cell descendants (plasma cells) begin to produce large quantities of circulating antibodies, and the T cells with antibody fragments in their surface membrane are ready to complex with the antigen. *How* the interac-

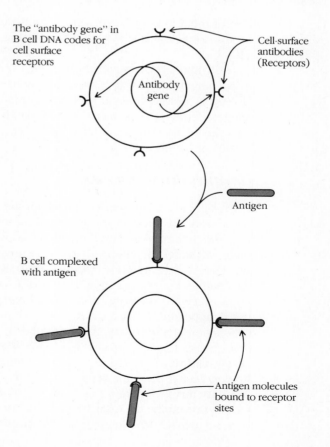

The "antibody gene" in B cell DNA codes for cell surface receptors

Antibody gene

Cell-surface antibodies (Receptors)

Antigen

B cell complexed with antigen

Antigen molecules bound to receptor sites

Figure 11–3
Interaction of a B cell with an antigen. Each B cell produces a specific antibody that interacts with a specific type of antigen.

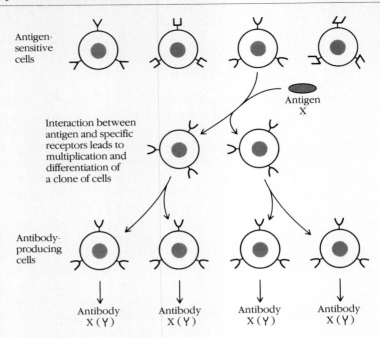

Antigen-sensitive cells

Interaction between antigen and specific receptors leads to multiplication and differentiation of a clone of cells

Antigen X

Antibody-producing cells

Antibody X (Y)   Antibody X (Y)   Antibody X (Y)   Antibody X (Y)

Figure 11–4
Clonal selection theory. Each antigen-sensitive cell expresses the genes that code for a single antibody species. The antigen triggers that specific cell type to multiply and differentiate, causing a clone of specific antibody-producing cells to appear.

tion of the B cell or T cell with the antigen triggers cell division and antibody production remains a mystery. The clonal selection theory is illustrated in Figure 11–4.

A lymphocytic receptor molecule has only one specificity; so it is able to bind to only one species of antigen, or occasionally to several closely related antigens. However, a large antigen, such as a virus or a bacterium, has many different antigenic sites—that is, many different kinds of molecules on its surface that can provoke the immune response—and it is therefore susceptible to attack by many classes of lymphocytes.

Immunologic memory

When we encounter a specific antigen for the first time, our immune response may on occasion be too slow to prevent sickness from developing. In most cases, the immune system eventually overcomes the invading pathogen; but while the battle rages, we suffer the effects of, say, a bacterial toxin, as well as some undesirable side effects of the immune response itself. (Rashes, for example, can be caused by the complexing of antigens and antibodies.) But in the case of many antigens, our bodies retain a "memory" of the initial immune response long after the antigen has disappeared. In such cases, the initial immune response has resulted in the production of a **memory cell**—an especially stable line of B and T lymphocytes that persists for decades (Figure 11–5). We refer to this persisting response as **sensitization** to the specific antigen. If the circulating lymphocytes of the stable line encounter the same antigen again, they "remember" it, and they undergo a response that is both more rapid and of greater magnitude than the original one, usually preventing the occurrence of disease symptoms altogether. Immunologic memory explains the long-term, or even life-long, immunity conferred by many

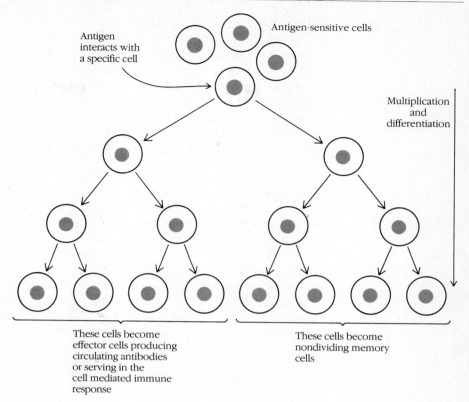

Figure 11–5
The memory cells, unlike the nondifferentiated lymphocyte, require a very little time to respond to an antigen. Therefore, a powerful and rapid immune response occurs when the antigen is encountered at some future time.

illnesses (such as chicken pox) and immunizations, and it also explains the powerful immunizing effect of a "booster shot"—a second immunization given after the initial immune response has occurred.

Antigen, antibody, and complement

There are three key components of the immune response, of which we have so far mentioned two—the antigen and the antibody. Antigens, the substances that induce the immune response, are usually large molecules (e.g., proteins or long-chain carbohydrates) or complexes of large molecules (e.g., a virus). Bacteria and other cells have antigenic molecules on their surfaces—usually more than one kind. Small molecules are not usually antigenic. Antibodies, also termed immunoglobulins, are proteins of a specific structure (to be discussed in the next section) that bind to and inactivate antigens.

The third component of the immune response is **complement**, a collection of about nine protein factors in the blood serum. When an antibody has complexed with an antigen, the complement system is activated and complexes with the antigen–antibody complex. The antigen–antibody–complement complex is then either destroyed by enzymes or engulfed by macrophages. Complement prompts certain cells to release **histamine**, the substance that produces the inflammatory reaction we see when cuts become infected or when the skin blisters after contact with poison ivy.

Antibody Structure

Early in the development of immunology, it was thought that all antibody molecules had the same amino acid sequence as well as the same basic structure. When confronted with different antigens, the antibodies supposedly "molded" around them to form complementary structures that then "recognized" the same antigen if it appeared again. The capacity of the antibody to form different molds was considered to be limitless. But this view had to be abandoned when it was discovered that antibodies produced in response to different antigens differ in their amino acid sequence.

The fact that each antibody appears to be unique raises two interrelated questions: Does each lymphocyte carry the requisite genetic information to code for the million or so antibodies we estimate a person can produce? How does an antigen activate a cell to divide and to begin producing antibodies? We have recently gained insight into these problems, in part as a consequence of our understanding of the structure of the immunoglobulin molecule.

Immunoglobulins belong to a class of proteins called **gamma globulins**; all antibodies are gamma globulins, though not all gamma globulins are antibodies. Restricting our attention to the circulating antibodies produced by the B cells, we find that there are five major kinds of immunoglobulins (Ig) in the blood serum: IgG, IgA, IgM, IgD, and IgE. By far the most abundant and best understood of these is IgG.

The IgG molecule (Figure 11–6) is composed of four polypeptide chains: two identical **heavy (H) chains** and two identical **light (L) chains**. The H chains have about 430 amino acids each, the L chains about 210. The four chains are held together in a flexible Y-shaped structure by **disulfide bonds** consisting of two linked sulfur atoms. A small carbohydrate is linked to each of the H chains.

Each of the five immunoglobulin classes is characterized by a different kind of H chain, that is, one with a different amino acid sequence. There are also two kinds of L chains, called κ and λ, each of which may combine with any of the five kinds of H chains. Thus, the H chains are unique to each class, and the L chains are

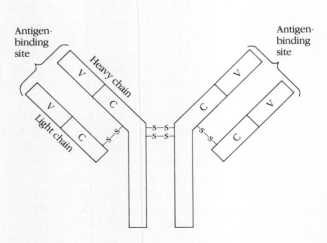

Figure 11–6
The heavy and light chains of the IgG molecule are linked by disulfide bridges (S-S). The antigen-binding sites are located between the variable regions of the heavy and light chains.

Table 11–1   Classes of immunoglobulin molecules

Immunoglobulin	Heavy Chain	Light Chain	Molecular Formula
IgM	μ	κ or λ	$(\mu_2\kappa_2)_5$ $(\mu_2\lambda_2)_5$
IgG	γ	κ or λ	$\gamma_2\kappa_2$ $\gamma_2\lambda_2$
IgA	α	κ or λ	$\alpha_2\kappa_2;\ (\alpha_2\kappa_2)_2$ $\alpha_2\lambda_2;\ (\alpha_2\lambda_2)_2$
IgD	δ	κ or λ	$\delta_2\kappa_2$ $\delta_2\lambda_2$
IgE	ϵ	κ or λ	$\epsilon_2\kappa_2$ $\epsilon_2\lambda_2$

Note: The heavy chains indicate the immunoglobulin class. A particular heavy chain can bind with either a κ or a λ light chain. Each chain, whether heavy or light, represents different gene products.

common to all classes. However, a given Ig molecule has only one kind of L chain. Table 11–1 shows the composition of the different immunoglobulins.

Within each H chain and each L chain of the IgG molecule are two regions: in one the amino acid sequence is constant, and in the other the sequence is extremely variable. We refer to these as the **constant (C)** and **variable (V) regions**, respectively. Thus, an Ig molecule has four basic regions: V_L, C_L, V_H, and C_H. There are two **antigen-binding sites** located on the ends of the Y in the region. The existence of two binding sites permits the antibody molecule to link to two antigens and form a complex of cross-linked antibodies and antigens that can be phagocytized by circulating macrophages or degraded by special enzymes (Figure 11–7).

Our knowledge of immunoglobulin structure emerged slowly because of the tremendous variety of Ig molecules circulating in the bloodstream. Until recently,

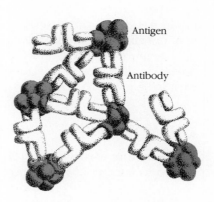

Figure 11–7

Model of how antigens and antibodies combine to form large aggregates.

it was hard to isolate a single species of Ig for study. Fortunately, this difficulty was overcome with the discovery of a certain protein in the urine of patients suffering from multiple myeloma. In this form of cancer, a lymphocyte stem cell in the bone marrow is transformed to a cancerous state and divides unchecked. The tumor that develops is actually a large clone of a single lymphocyte, and this clone produces an excessive amount of a single species of protein, which turns out to be part of an Ig molecule. This excess protein, excreted in the urine of patients with multiple myeloma, is called the **Bence-Jones protein** (named after H. Bence-Jones, who discovered it in 1847). No two patients have the same species of this protein in their urine. We now know that the Bence-Jones protein is composed of two Ig light chains. The individual variation is due, of course, to differences in the V_L region. By being able to isolate a single species of Ig molecule, or at least a part of one, in large quantities, we gained much insight into the structure of the Ig molecule. However, recent developments in cell fusion (Box 11–1) have now made it possible to isolate many Ig species in large quantities.

Genetics of Antibody Formation

The genetics of antibody formation poses problems that take on new and exciting dimensions with each fresh discovery. At one level there is the problem of cell differentiation: Do all lymphocytes have the same information? And if so, how does each clone of cells produce only one antibody? At another level there is the problem of unconventional gene activity: Why is it that the production of a poly-peptide chain with a C and a V region does not conform to what we understand about the synthesis of other proteins?

One of the first discoveries about antibody gene arrangement challenged some established ideas about genes. It turned out that there are separate genes coding for the V and C regions of the L and H chains, genes located some distance apart on the chromosome. The four genes are designated V_L, C_L, V_H, and C_H. This discovery prompts us to question the universality of the one gene–one polypep-tide concept, for here we have a case of "two genes, one polypeptide." For two genes to produce a single polypeptide, either they or the mRNA molecules tran-scribed from them must somehow become joined. We now know that the base sequences intervening between the *V* and *C* genes are excised, perhaps by means of DNA processing, to produce a continuous sequence of coding DNA.

There are three major theories that try to explain how antibody genes can produce such a wealth of diversity. The two older theories, which we will present first, are fairly straightforward, but they do not take into account some recent discoveries about the *V* and *C* genes. The most recent theory, which we will present in simplified form, is more complex, but it seems to account for much of what we know.

The germ-line and somatic-mutation theories

The **germ-line theory** suggests that immunoglobulin variability is the result of an enormous number of *V* genes, one for every variable region. All these genes are

Box 11–1   *Hybridoma*

For many years, the field of immunology languished as a science, making slow and hardly sensational advances. All of this is changed now. Immunology has burst into the 1980s with spectacular discoveries, and even more spectacular developments loom on the horizon. One cause of the euphoric state currently infecting immunologists is the development of a revolutionary new cell-fusion technique. This technique has produced a type of cell line called the **hybridoma**, the product of fusion between a B lymphocyte and a myeloma tumor cell. The reason for the hybridoma's exalted position will soon become clear.

A B cell produces a single species of immunoglobulin, as we have already noted. However, in response to a complex antigenic stimulus such as a leukemia cell, several lines of B lymphocytes

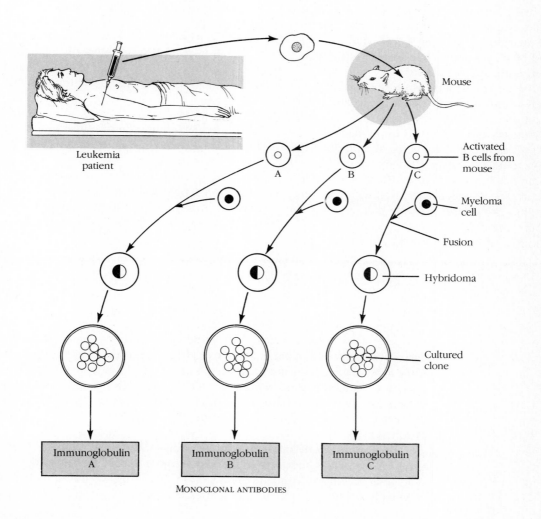

MONOCLONAL ANTIBODIES

may be activated to divide and produce antibodies, because the leukemia cell has many antigenic sites. If we want to study antibody synthesis in a single cell line or to isolate a single species of immunoglobulin in large quantity for study, we need first to isolate individual lymphocytes and then to grow them in culture. We are usually able to isolate these cells, but we cannot get them to grow. Cancer cells, on the other hand, grow magnificently in culture. By fusing a B cell to a cancerous B cell (one that has mutated so that it does not make its own Ig) to produce a hybridoma, we can circumvent the B-cell culturing problem. The hybridoma has the tremendous growth capacity of the myeloma cell and the antibody-synthesizing ability of the B cell. Therefore, we are now able to grow large clones of hybrid cells that produce a single species of immunoglobulin in large quantities. The single-species antibody produced by a hybridoma clone is called a **monoclonal antibody** (see figure).

The potential value of monoclonal antibodies is enormous. Consider, for example, a human cancer cell isolated from a cancer patient and injected into a mouse. The human cell, because it has multiple antigenic sites, induces the mouse to activate several B lymphocyte lines to combat this foreign invader. These activated B cells are then isolated and fused with myeloma cells, and clones are grown in culture. Next, the antibody against the human cancer cell is collected from a clone of hybridomas. It may then be used to attack the malignant cells in the cancer patient.

Monoclonal antibodies may someday also be used to detect the precise location of a tumor and to determine whether it is metastasizing (spreading). To achieve this, we might attach a radioactive label to the antibody and follow its journey through the body. The antibody should go to the site of the tumor, complex with the antigenic sites, and be detected as a concentrated area of radioactivity.

We may even be able to employ the monoclonal antibodies in chemotherapy regimes. One of the shortcomings of chemotherapy is that the therapeutic agent kills normal cells almost as well as it kills cancer cells. Biomedical researchers have long been seeking drugs that are more reactive with cancer cells than with normal ones, but so far in vain. With monoclonal antibodies, we can entertain the possibility of linking a chemotherapeutic agent to the antibody so that the antibody delivers the drug specifically to the tumor cell.

Hybridomas offer us the promise of a vast range of practical and basic research applications. They have a potential role to play in the fight against cancer; they may soon serve as an aid in organ transplants; and they may soon be used to isolate gene products that have so far remained elusive. Hybridoma technology has barely begun to exploit its full range of applications.

present in each of our cells, being derived from the germ cells of our parents, but some form of gene regulation determines that each lymphocyte will produce only one kind of antibody. Since we can produce over a million kinds of antibody, this theory requires that we have over a million genes—about 50 times the number estimated from other data—and that nearly all of them be used for antibody production.

The **somatic-mutation theory** holds that, on the contrary, there are relatively few *V* genes in the germ cells and the zygote. According to this idea, embryonic cells accumulate somatic mutations in their *V* genes during cell division, eventually producing a vast array of lymphocytic stem cells with different *V* alleles.

The gene-rearrangement theory

There is evidence that the V region is repeated on the chromosome—perhaps several hundred times. This means that there could be several hundred *V*-gene variants on a single chromosome. A variable *V* gene would account for some antibody variation, since a single *V* gene could be excised and combined with the C region in each lymphocyte, but a few hundred gene variants would not explain the existence of a million or so different antibodies. However, a remarkable discovery about the *V* gene has shown us a likely basis for this tremendous variability.

When the V_L gene was first analyzed, it was found to code for only 98 of the 112 amino acids in the V_L region. This was most perplexing: Where were the other 14 codons? They were eventually found in a short segment of DNA called the **joiner (J) region** which lies between the *C* and *V* genes (Figure 11–8). In cells that are actively producing antibody, we find that the J region and the *V* gene are contiguous, probably joined by the same process that brings the V and C regions together. This means that the variable region of the L chain is coded for by two originally separate chromosome segments! We must now either develop a three gene–one polypeptide model of antibody formation, with *C*, *V*, and *J* genes coding for the L chain, or designate the two separate DNA regions, V and J, as parts of a single gene. We have recently found a J region in the H-chain gene of mice, so it seems probable that humans, too, have J regions in both the *L* and the *H* genes. If so, we would have at this point six genes (or gene segments) coding for an antibody: V_L, V_H, C_L, C_H, J_L, and J_H.

The discovery of the J region allowed the development of the **gene-rearrangement theory** of antibody variability. The J region, like the *V* gene, appears to be repeated on the chromosome, so there can be J variants as well as V variants. According to the gene-rearrangement model, any J variant can end up being combined with any V variant during the excision processes that bring J, V, and C regions together. This would allow for enormous variability in the V region of the polypeptide, for we think that there are at least 200 V_L variants and 4 J_L variants. A new segment has been recently discovered for H-chain genes. This is the *D* segment and it is located between the V and J regions. We estimate 10 different *D* segments, which when coupled with 200 V_H segments and 4 J_H segments, gives us 8000 V_H combinations. If so, there could be 800 (4 × 200) L chains and 8000 H chains (200 × 10 × 4), and these could combine to provide 6.4 million (800 ×

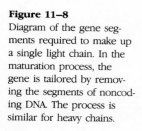

Figure 11–8
Diagram of the gene segments required to make up a single light chain. In the maturation process, the gene is tailored by removing the segments of noncoding DNA. The process is similar for heavy chains.

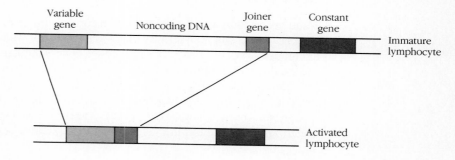

Variable gene — Noncoding DNA — Joiner gene — Constant gene — Immature lymphocyte

Activated lymphocyte

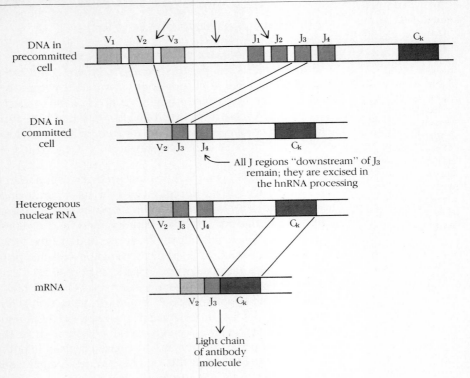

Figure 11–9
Model for how a cell produces a single species of immunoglobulin molecule.

8000) kinds of antibodies. Figure 11–9 shows how progressive excision of V, J, and C regions in the L chain could leave a cell with a single set of one V, one J, and one C for each type of chain.

The gene-rearrangement theory is exciting and is developing rapidly. Even though it is now confirmed, it does not completely invalidate the other two theories. The germ-line theory can be adapted to fit the rearrangement theory: If gene rearrangement is correct, it would mean that the germ cells do carry much of the information necessary for antibody variation, as the germ-line theory maintains. On the other hand, somatic mutation undoubtedly does occur in the V, C, and J regions of the chromosome, and it too may significantly increase antibody diversity.

Gene recombination for antibody diversity

Some recent discoveries about the structure of the DNA on either side of the *V* genes suggests still another way that *V*-gene variability may arise. The sequence of nucleotides on either side of the *V* genes is very similar. These short sequences may facilitate recombination between related genes carried on different members of a pair of chromosomes. One scheme in which this recombination generates diversity is through unequal crossing over (Figure 11–10).

In summary, several factors may contribute to V-gene variability. Somatic mutations probably account for some of it. Recombination between *V* genes gen-

Figure 11–10
Diagram of gene expression and loss through an unequal crossover between two homologous, but nonallelic, hypothetical region sequences. The reciprocal products formed by this event are an expanded set of genes (from three genes to five) and a diminished set of genes (from three genes to one). Such unequal crossovers between homologous, nonallelic genes may account for the diversity in cell types.

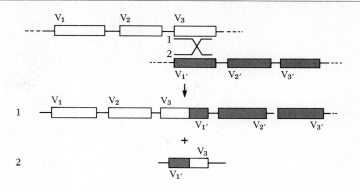

erates additional variability. And the rearrangement of *V* and *J* genes leads to variability in the V region of the polypeptide chain.

Histocompatibility and Transplantation

The body's main line of defense against "foreign" tissues—tissue transplants or its own tumors—is a special type of T lymphocyte called a cytotoxic **killer T-cell** (**T_K cell**). The T_K lymphocytes, when activated, can destroy the cells that caused their activation. The destruction involves contact between the T_K cells and the foreign tissue, but its details are poorly understood.

Killer lymphocytes also provide a primary defense against viral infections. Once the infecting virus has entered a cell, it is hidden from direct attack by the T_K cells. But the viral genes code for proteins that come to be part of the cell surface, and these serve as antigens to activate and attract specific T_K cells, which then multiply and destroy the virally infected cells.

Studies of the defensive reaction against virally infected cells led to a surprising discovery. T_K lymphocytes derived from strain "A" mice that were infected with virus "Q" were able to destroy the virally infected "A" cells, but not cells from other mouse strains that were also infected with virus "Q." T_K lymphocytes from strain "B" mice infected with virus "Q" were able to destroy the virally infected "B" cells, but not "A" or any other strain of cells infected with "Q." It turns out that there are two types of receptor sites on the T_K cells: One site is specific for the foreign antigen (in the case just cited, the viral antigen); the other site is specific for proteins coded for by the host cell's own genome, proteins unique to the individual mouse (Figure 11–11). Thus, members of a T_K cell line that has proliferated in response to a viral infection can bind only to virally infected cells of the same individual. The "individualized" cell-surface proteins are called **histocompatibility antigens** (the Greek prefix *histo-* means tissue). They resemble the immunoglobulins in structure and are evolutionarily related to them. These proteins make every individual histologically unique; that is, they distinguish the tissues of every individual from those of every other, except in the

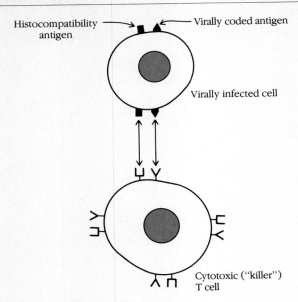

Figure 11–11
Receptor sites on cytotoxic (killer) T cells. One site on T_K cells is specific for the viral antigen; the other site is for an antigen encoded by the cell genome.

case of identical twins or members of the same inbred strain. It is the histocompatibility antigens that cause acceptance or rejection of tissue grafts and organ transplants.

The HLA gene complex

In humans, many of the genes that code for histocompatibility antigens are clustered in a complex called the **human lymphocyte A complex**, or **HLA** (historically, HLA refers to "human leukocyte antigen"). This is the major human histocompatibility region, though not the only one. The HLA complex is on chromosome 6 and carries four HLA loci, A, B, C, and D, along with some other loci with related functions (Figure 11–12). These other loci include some that code for complement factors and some that code for other substances that enhance the immune response. The HLA genes on a single chromosome constitute a **haplotype**, or "half genotype," which is inherited as a Mendelian unit except when crossing over occurs within it. Every person has two haplotypes, with eight main HLA loci.

Each of the four main HLA loci consists of a series of multiple alleles—up to 30 or more per locus. Because of the large number of alleles that can occur, most

Figure 11–12
Map of the HLA region on the short arm of chromosome 6. Other genetic markers are also shown. The HLA region is expanded. The numbers are map distances.

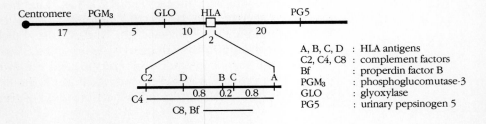

people are heterozygous for every locus. The enormous variety of combinations possible for each haplotype ensures that no two people have the same HLA genotype (except for monozygotic twins).

Transplantation

In organ transplantation, acceptance or rejection depends primarily on how similar the HLA systems are in donor and recipient, and to a lesser extent on the similarity of other histocompatibility loci. The more similar the HLA systems are, the more similar the major histocompatibility antigens of the two individuals and the greater the chances that the transplanted tissue will not be rejected. Transplants between identical twins are usually very successful because all their histocompatibility genes are the same. The case of twins aside, a transplant is still most likely to work if the donor and recipient are siblings, since siblings stand a good chance of being identical for half of their HLA loci. Consider the following haplotypes:

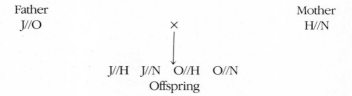

The more distantly related the donor and recipient, the less likely the success of the transplant. Even when the HLA systems are closely matched, other histocompatibility loci can lead to tissue rejection. These are not as powerful, however, in their ability to evoke a rejection response.

In any transplantation operation, the probability of rejection is minimized by using drugs to suppress the immune response and by matching donor and recipient histocompatibility genotypes as closely as possible. Today, international organizations maintain lists of people needing transplants, classifying them according to their HLA and blood group antigens. (We will discuss blood groups shortly.) A potential organ donor is characterized in the same way, and the best matches are made whenever possible. When matches are poor, the donor tissue is rejected, as shown in Figure 11–13.

HLA and disease

An increasingly large number of diseases are now associated with the HLA antigens, though the cause-and-effect connections remain obscure in many cases. Table 11–2 summarizes the associations between HLA antigens and diseases. To date, the strongest association is between the HLA-B27 antigen (coded by allele 27 at the B locus) and a disorder called *ankylosing spondylitis (AS)*. This is a disease of the joints in which ligaments become inflamed and ossified (develop into bone); eventually the joints fuse. In 90% of the cases of AS, the HLA-B27 antigen is present. Only 5% of the general population carries this antigen, and about 3% of these people develop AS, but this is 100 times greater than the risk faced by people who do not have the HLA-B27 antigen.

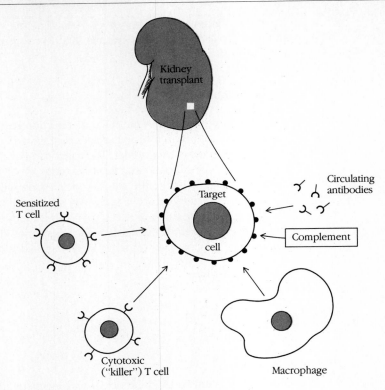

Figure 11–13
Some of the factors involved
in transplant rejection.

There are a number of possible reasons for the association between an HLA antigen and a specific disease:

1 The HLA antigen in the cell membrane may function as a target site for viral infection or attachment of some other pathogenic agent.
2 The HLA antigen may interact with a foreign antigen and either reduce or exaggerate an immune response.
3 Certain HLA alleles may actually be mutant.
4 There may be a very close linkage relationship between the HLA antigen gene locus and another HLA antigen locus that is so far unidentified. This second locus may be the real cause of the disease association. For example, we are able to show that though juvenile diabetes mellitus is associated with HLA-B8, it is much more strongly associated with HLA-Dw3. Since HLA-B8 tends to stay together with HLA-Dw3, perhaps because of a close linkage on the same chromosome, there is a connection between HLA-B8 and juvenile diabetes. But the disease is really associated with HLA-Dw3; HLA-B8 is carried along because of its proximity to HLA-Dw3.
5 The locus for the HLA antigen may be closely linked, but *unrelated*, to another locus that causes the disease in question. In keeping with this idea, which has been getting some support lately, is the linkage of the HLA locus to a locus that intensifies the immune response to the antigen in question.

Table 11–2   The association between diseases and HLA

Type of Disease	Disease	Racial Group	HLA Type	Antigen Frequency Patients	Controls
Gastrointestinal tract	Chronic active hepatitis	White	A1	41.6	28.4
			B8	44.2	20.3
			Dw3	60.0	21.7
	Celiac disease	White	A1	63.7	29.5
			B8	71.2	23.1
			Dw3	98.0	15.0
	Iron metabolism dysfunction	White	A3	78.4	27.0
			B14	25.5	3.4
	Ulcerative colitis	Japanese	B5	80.0	30.8
	Pernicious anemia	White	B7	35.8	22.1
Skin	Psoriasis	White	B13	19.7	4.5
			Bw17	26.2	7.8
			Bw37	7.7	1.4
	Herpes-type dermatitis	White	A1	69.0	30.1
			B8	77.0	24.7
	Pemphigus vulgaris	White	A10	39.3	12.7
	Herpes labialis	White	A1	55.6	25.1
			B8	33.3	16.8
Eye	Anterior uveitis	White	B27	56.8	7.7
	Vogt–Koyanagi–Harada disease	Japanese	B22J	42.9	13.2
Malignancy	Hodgkin disease	White	A1	31.1	39.0
			B5	10.6	16.0
			B8	23.7	29.0
			Bw18	7.1	13.0
	Acute lymphatic leukemia	White	A2	60.0	53.6
			B8	29.0	23.7
			Bw18	29.0	25.2
	Nasopharyngeal cancer	Chinese	Sia2	44.0	21.0
Joints	Ankylosing spondylitis	White	B27	89.8	8.0
		Japanese	B27	66.7	0.0
		Haida Indians	B27	100.0	50.0
		Bella Coola Indians	B27	100.0	20.2
		Pima Indians	B27	36.0	18.0
	Reiter disease	White	B27	78.2	8.4
	Yersinia arthritis	White	B27	79.4	9.4
	Salmonella arthritis	White	B27	66.7	8.6
	Psoriatic arthritis	White	B13	19.8	5.5
	Central joints		B27	40.2	8.7
			Bw17	11.6	5.5
			Bw38	22.7	2.9

(Continued)

Type of Disease	Disease	Racial Group	HLA type	Antigen Frequency Patients	Controls
	Peripheral joints	White	B13	9.9	5.5
			B27	15.5	8.7
			Bw17	24.8	5.5
			Bw38	12.6	2.9
	Juvenile rheumatoid arthritis	White	B27	26.4	8.5
	Rheumatoid arthritis	White	Dw4	42.2	15.7
		White	Cw3	30.0	17.0
Endocrine glands Thyroid	Graves' disease	White	B8	36.7	21.7
			Dw3	50.0	21.7
		Japanese	Bw35	56.8	20.5
	De Quervain thyroiditis	White	Bw35	76.9	12.5
Adrenal	Addison disease	White	B8	50.0	22.7
			Dw3	70.0	21.7
Pancreas	Diabetes	White	B8	36.7	21.8
	Juvenile-onset diabetes		Bw15	22.8	14.9

Blood Group Antigens

No other antigens have been as important in medical genetics as those of the erythrocyte membrane. The erythrocyte has dozens of systems of membrane-bound antigens, each system controlled by a different gene locus, or in some cases, perhaps by several interacting loci.

The familiar ABO and Rh antigen systems play a crucial role in obstetrics and blood transfusion. A transfusion is a tissue transplant, and the red cell antigens are histocompatibility antigens, though they are not coded for by the major histocompatibility locus. But it is vastly easier to match blood donors and recipients for their ABO or Rh types than to match organ donors and recipients for HLA types, because the number of alleles involved is much smaller. Mismatched blood can elicit a strong antibody response in the recipient, causing serious medical problems. And a pregnant woman can develop a strong antibody response to the Rh antigens produced by the fetus.

Other red cell antigens, those of the MNO system, for example, elicit a very weak antibody response. Since they are of little medical importance (although interesting to the geneticist) we will not discuss them in this text.

The ABO, H, and secretor systems

The ABO multiple-allele system, which we discussed in Chapter 4, was the first multiple-allele system to be demonstrated in humans. People can be classified

Table 11–3   The ABO blood groups

Genotype*	Blood Group	Antigenic Specificities on Red Cells	Antibodies in Serum
ii	O	—	Anti-A and anti-B
$I^A i$ or $I^A I^A$	A	A	Anti-B
$I^B i$ or $I^B I^B$	B	B	Anti-A
$I^A I^B$	AB	A and B	—

*i = a non-A or B allele; I^A = allele that codes for A antigen; I^B = allele that codes for B antigen

into four basic groups according to their ABO blood type. Those who carry both the **A** and the **B antigens** on their red cell membranes are type AB; those with just the A antigen, type A; and those with just the B antigen, type B. One who has neither antigen is blood type O. Individuals have blood serum antibodies that complement whatever antigens they carry (Table 11–3); thus, a type A person has anti-B antibodies, but not anti-A, and so on.

Transfusion   If a type A person receives a transfusion of type B blood, his or her antibodies will react with the B antigens on the donor's blood cells. The donor cells will then clump together and block the recipient's capillaries. This reaction is extremely serious, sometimes fatal. A type B recipient, of course, will react similarly to type A blood; and a type O recipient, who has both anti-A and anti-B antibodies, will have an adverse reaction to A, B, or AB blood. Type AB was formerly called "the universal recipient," on the grounds that an AB person could receive any kind of blood; and on similar grounds, type O was called the "universal donor." But it was eventually found that when large amounts of blood are transfused, the *donor's* antibodies can react with the *recipient's* red cells; thus, if a type AB person is given type O blood, the anti-A and anti-B antibodies in the donor's blood can cause clumping. For this reason, type ABO recipient and donor are always matched in blood transfusions.

The distribution of the A, B, O, and AB blood types around the world is not even. We find that certain blood groups predominate in certain parts of the world. We will discuss this distribution later.

ABO antigens and the H substance   The A and B antigens are carbohydrates, and they are bound to fatty acid molecules that protrude from the erythrocyte membrane. The A and B antigenic specificity is determined by the terminal sugars of the carbohydrate (Figure 11–14). The B antigen has a galactose molecule added to the carbohydrate terminus by a specific enzyme, coded for by the I^B allele. The A antigen has a sugar called *N*-acetyl-D-galactosamine in the same position; this sugar is added by a different enzyme, the product of the I^A allele. The *i* allele codes for neither enzyme, so the genotypes $I^A I^A$ and $I^A i$ produce the A antigen, genotypes $I^B I^B$ and $I^B i$ produce the B antigen, $I^A I^B$ produces both, and *ii* produces neither. The carbohydrate molecule without the A or B antigenic terminus is called the **H substance**. Type O persons have this nonantigenic H substance, with no further modification to A or B antigen, on their red cells. Since H

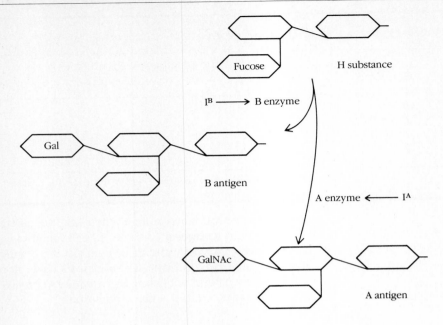

Figure 11–14
The conversion of H substance into A and B antigens. Gal = galactose; GalNAc = *N*-acetyl-D-galactosamine.

is a precursor for A and B antigens, people with blood types A, B, and AB normally produce the H substance.

The Bombay phenotype   The H substance is under control of a separate gene locus with two variants, *H* and *h*. The *H* allele produces an enzyme essential to the manufacture of the H substance, but the *h* allele, which is quite rare, does not. An *hh* person cannot form the H substance, and this can have some strange consequences. For example, an apparently type O woman gave birth to a child who was blood type AB. This should not be possible, because an AB child must get an *A* allele from one parent and a *B* allele from the other, and the mother presumably had neither. But it was soon discovered that the mother was genetically blood type B; she was unable to form the B antigen because she was homozygous for the *h* allele and therefore could not form the H molecule. The *B*-gene product was present, but there was no *H* molecule for it to attach to. The combination of *hh* and the blood type A, B, or AB is called the **Bombay phenotype**, because it was discovered in Bombay, India. As the Bombay phenotype lacks the A, B, and H antigens, it appears to be type O even though the *A* and/or *B* allele may be present.

The secretor locus   We mentioned earlier that the carbohydrates that constitute the A and B antigens and the H substance are attached to fatty acids on the cell membrane. The carbohydrate–fatty acid complex, called a **glycolipid**, can be extracted from the red cell only with alcohol, so the membrane-bound A and B antigens are called **alcohol-soluble antigens**. But the A and B antigens and the H substance are not restricted to the red cell—in most people they are also found in body secretions such as saliva and the mucous secretions of the intestine. But in these secretions the carbohydrate is attached not to a fatty acid but to a protein, forming a water-soluble complex called a **glycoprotein**. The A and B antigens of this type are called **water-soluble antigens**.

The ability to secrete the water-soluble antigens and the H substance in the body fluids is controlled by a **secretor locus**. Secretors—people who secrete A, B, and H—have at least one dominant allele (*Se*) and make up about 80% of the population. Nonsecretors (genotype *se se*) account for the other 20%. The *Se* allele produces an enzyme that forms the H substance on the protein portion of the water-soluble glycoprotein; without it the water-soluble type of antigen cannot be assembled. A different enzyme links the H substance to fatty acids, so a nonsecretor may still have alcohol-soluble A and/or B antigens on the red cells.

The Rh system

The **Rh antigen** (often called the "Rh factor") was first identified in the Rhesus monkey, hence the Rh designation. About 85% of the people in North America have the Rh antigen (or, more precisely, one of a group of Rh$^+$ antigens) and are called Rh-positive, or Rh$^+$. About 15% lack the antigen and are called Rh-negative, or Rh$^-$. Many people are aware of the antigen's existence because they have heard of "Rh babies," though they may have only a vague idea of what the "Rh factor" is. The antigen's structure is not well understood, but we do know it is an integral part of the erythrocyte membrane.

Genetics of the Rh system The genetics of the Rh system is poorly understood. There are two primary and competing systems of genetic notation for the Rh genes: the Wiener system of multiple alleles, and the Fisher–Race system of three closely linked gene loci. The two systems, which are compared in Table 11–4, try to take into account the variant Rh antigens, each of which elicits a different antibody response. According to the Wiener model, there is a single Rh locus with a series of eight alleles. Each of the four dominant alleles produces a specific antigen; the four recessive alleles produce none. In the Fisher–Race model, the

	Fisher–Race Notation	Wiener Notation	Frequency	
	CDe	R^1	41%	
	cDE	R^2	14%	Rh$^+$
	cDe	R^0	3%	
	CDE	R^Z	Rare	
Alleles	*cde*	*r*	39%	
	Cde	*r'*	1%	Rh$^-$
	cdE	*r''*	1%	
	cdE	r^y	Very rare	

Table 11–4   Two different systems of nomenclature for the Rh multiple alleles

Rh system consists of three closely linked gene loci with two alleles each—*Cc*, *Dd*, and *Ee*—and these alleles interact to produce the variant forms of the Rh antigen. However, only when the *D* allele is present is there an Rh antigen of any kind, so this locus might be viewed as the principal Rh locus.

Medically, the presence or absence of the Rh antigen is of more importance than the existence of different kinds of the antigen, so for convenience we may regard the Rh system as consisting only of the *D* locus. Thus, a *DD* or *Dd* person is Rh$^+$ regardless of the other alleles involved, and a *dd* person is Rh$^-$. One of the most important aspects of having or lacking Rh antigen is the role the antigen plays in **hemolytic disease of the newborn**, or **HDN**, the condition that produces the so-called Rh babies.

Hemolytic disease of the newborn   The Rh system differs from the ABO system in that an Rh$^-$ person does not have anti-Rh$^+$ antibodies unless he or she has been exposed to the Rh$^+$ antigen. Thus, an Rh$^-$ person can receive one transfusion of Rh$^+$ blood without ill effects, but this first transfusion will stimulate the production of antibodies, which will react with the Rh$^+$ antigens in any later transfusion.

In HDN, an Rh$^-$ mother is stimulated to produce anti-Rh$^+$ antibodies by the antigens of her Rh$^+$ fetus. Since few red blood cells cross the placenta, little synthesis of anti-Rh$^+$ antibody occurs during the first pregnancy with an Rh$^+$ child. However, fetal blood cells may enter the maternal circulation in larger numbers during birth, when the placenta detaches from the uterus, and this is when the Rh$^-$ woman starts producing antibodies against the fetal Rh$^+$ antigens. These antibodies are synthesized too late to affect the newborn. But if the same woman becomes pregnant again, her anti-Rh$^+$ antibodies will enter the circulation of the fetus and, if the fetus is Rh$^+$, attack and destroy its red blood cells (Figure 11–15). The maternal antibodies remain in the baby's system for some time after birth, causing additional erythrocyte destruction.

Hemolytic disease due to Rh incompatibility is very unlikely in a first child, unless the mother has been previously sensitized by a terminated pregnancy or a blood transfusion. Even in the case of a second Rh$^+$ child born to an Rh$^-$ mother, the damage is usually slight, because the mother's level of Rh$^+$ antibodies is still fairly low. But with each birth of an Rh$^+$ child, the mother is stimulated to produce more antibodies, and in a third or fourth pregnancy the effects may be so severe that the fetus dies in utero. If the infant survives to be born, he or she is usually jaundiced, because red cell destruction releases large amounts of hemoglobin into the body tissues and hemoglobin breaks down into the yellow pigment **bilirubin**. If the high concentration of bilirubin is not lowered immediately after birth by replacing the baby's blood through extensive transfusions, severe and often fatal brain damage occurs.

Owing to Rh incompatibility, HDN can occur only when the father is Rh$^+$ and the mother Rh$^-$. Two Rh$^-$ parents cannot produce an Rh$^+$ child, and an Rh$^+$ woman will not produce anti-Rh antibodies no matter what the Rh type of her fetus. Even if the mother is Rh$^-$ and the father Rh$^+$, the father may be heterozygous and thus have a 50% chance of passing on an Rh$^-$ allele to the baby. If he does, of course, no sensitizing of the mother will occur, nor will an Rh$^-$ fetus carried by a previously sensitized mother be in any danger.

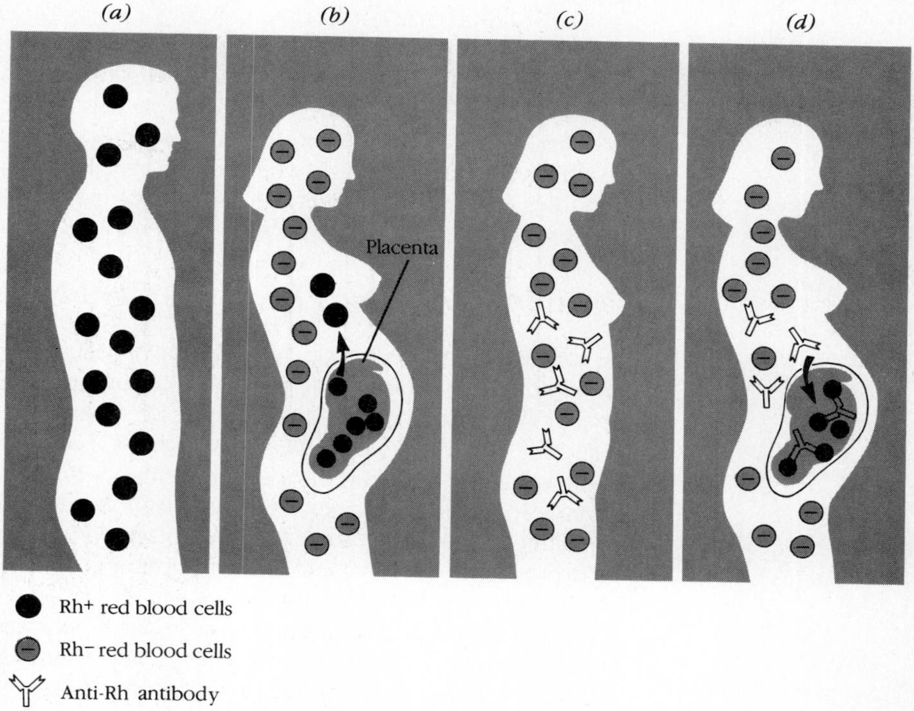

Figure 11–15
The development of HDN.
(*a*) Homozygous Rh⁺ father. (*b*) Rh⁻ mother carrying an Rh⁺ fetus. Fetal Rh⁺ antigens may enter the mother's bloodstream, particularly during birth.
(*c*) The mother produces anti-Rh antibodies against the fetal Rh⁺ antigens.
(*d*) During a subsequent pregnancy with an Rh⁺ fetus, the mother's anti-Rh antibodies cross the placenta and enter the fetal circulatory system, destroying its red blood cells.

● Rh⁺ red blood cells
⊖ Rh⁻ red blood cells
Ⴤ Anti-Rh antibody

Not all combinations of an Rh⁻ mother and an Rh⁺ fetus produce HDN. For example, if the mother has blood type O, she has a built-in defense against sensitization. Recall that people with type O blood carry anti-A and anti-B antibodies. If an Rh⁻ woman who is type O gives birth to an Rh⁺ child who is type A, B, or AB, any fetal red cells that enter the maternal bloodstream are rapidly destroyed by the mother'a anti-A and/or anti-B antibodies. This destruction occurs before the fetal cells can stimulate production of anti-Rh antibodies in the mother. About 13% of all matings in North America are Rh-incompatible, but even before the modern technique of preventing sensitization was developed, only 1–2% of these matings produced HDN. This low figure is due to male heterozygosity, the small number of Rh-incompatible families with more than two children, and biological defense mechanisms like the one just described.

Rh-incompatible pregnancies no longer present the threat they once did, because it is now possible to prevent an Rh⁻ woman from being sensitized by her child. Immediately after the birth of her first Rh⁺ baby, an Rh⁻ woman is given an injection of anti-Rh⁺ antibodies. These antibodies destroy any fetal red cells that are circulating in her blood, before the fetal cells can stimulate production of her own anti-Rh⁺ antibodies. The injected antibodies do not persist very long in the woman's circulation, so they cannot affect the fetus in a later pregnancy. If the same precaution is taken after every birth of an Rh⁺ child, the Rh⁻ woman will never develop anti-Rh⁺ antibodies, unless she has a transfusion of Rh⁺ blood. The wide use of RhoGAM, as the antibody injection is called, began in 1968, and it has made HDN a rare occurrence today.

HDN caused by incompatible blood types is extremely rare in all but the Rh system. For example, about 1% of newborn children from potentially incompatible matings have HDN because of Rh incompatibility, but only about 0.1% of all newborns have HDN because of ABO incompatibility. The reason for this difference may lie in the size of the antibody molecules produced by the mother. The anti-Rh$^+$ antibodies are IgG, whereas the naturally occurring anti-A and anti-B antibodies are IgM. The larger IgM molecules may not be able to cross the placenta and enter the blood system of the fetus.

Resolving dilemmas by using blood groups

The erythrocyte antigens have been extremely valuable in various legal situations involving the identification of persons. For example, Mrs. Jones claims that the father of her child is Mr. Smith. He denies it. Blood typing shows the child to be O, and Mr. Smith to be AB. We know that it is not normally possible for an AB person to have an O child. So we can exclude Mr. Smith as the father. But if Mr. Smith were B, and Mrs. Jones were A (or B or O), then we could not rule out the possibility of Mr. Smith as the father. This type of testing in a paternity suit does not prove that a certain man, who has the appropriate genetic makeup, *is* the father, but it can exclude some other man.

Blood typing has been used in criminal investigations too. For example, a young man is viciously attacked by a man while walking alone late at night. In the struggle, the young man manages to draw blood from his attacker. This blood can be matched by type with the rounded-up suspects either to eliminate them from further suspicion or to strengthen the case against them.

Today, we know of over 80 different human erythrocyte antigens, some of which are listed in Table 11–5. Armed with this knowledge, plus our ability to identify a wide range of HLA antigens, we can now be very accurate in resolving many paternity disputes and cases of criminal identification.

Defects in the Immune System

Several inherited conditions of varying severity can be traced to defects in the immune system. Some of these involve abnormal production of antibodies; others involve a failure to produce antibodies.

Autoimmune diseases

Normally, the immune system does not respond to a person's own antigens or **self-antigens**, but occasionally it does. When this happens, pathological conditions called **autoimmune diseases** may result.

As fetal lymphocytes mature, they build up a tolerance to the self-antigens they are exposed to. If a tissue is removed from an embryo, lymphocytes develop that have not "seen" that tissue before. If the tissue is replaced in the body, descendants of the lymphocytes that developed in its absence will respond to it as to a foreign tissue. An antigen that is sequestered (i.e., secluded) in the body and not normally in contact with lymphocytes may stimulate an autoimmune reaction if it does come in contact with them. Sperm, for example, does not usually reach a

Table 11–5	Some of the human red blood cell antigens, grouped into several systems			

System	Year of Discovery	Number of Antigens Known	Estimate of Average Heterozy-gosity*
ABO	1900	6	0.51
MNS	1927	18	0.70
P	1927	3	0.50
Rhesus	1940	17	0.66
Lutheran	1945	2	0.08
Kell-Cellano	1946	5	0.12
Lewis	1946	2	0.30
Duffy	1950	2	0.52
Kidd	1951	2	0.50
Diego	1955	1	0.00
Auberger	1961	1	0.49
X_g	1962	1	0.46
Dombrock	1965	1	0.46
Stoltzfus	1969	1	?

*Based on the assumption of random mating.

man's lymphoid tissue. Thus, if his sperm is injected into his bloodstream, an immune response occurs.

Autoimmune diseases may develop in two main ways. First, a cell may synthesize a "new" antigen because of a viral infection or a somatic mutation. This new antigen activates an autoimmune response, since it was not present when the immune system was developing a tolerance for self-antigens. Second, a mutation may occur in a T cell or a B cell, which then gives rise to a "forbidden clone." Cells of this clone produce a mutant Ig species that reacts against the normal self-antigens.

An example of the first situation is a condition known as **sympathetic ophthalmia**. It arises when one eye is badly injured and releases antigens that have heretofore been sequestered in the eye. These new antigens prompt an immune reaction against the uninjured eye, which is then damaged, or even destroyed, by the antibody attack. An example of the second situation is **rheumatoid arthritis**. This is a condition in which a mutation occurs in the B cells, resulting in a mutant form of Ig that functions as an autoantibody. In this case it is an IgM antibody that is altered, and in the altered state this molecule is called the **rheumatoid factor**. The rheumatoid factor complexes with normal IgG, causing joint inflammation and swelling of the lymph nodes. We can describe the mutant IgM as an antibody against one's own antibodies.

Systemic lupus erythematosus, or "lupus," is a complex autoimmune disease in which antibodies produced by mutant B clones attack self-antigens such as nucleic acids, heart cells, kidney cells, and epithelial cells, leading to a

---

Table 11–6   Examples of autoimmune diseases

Systemic Diseases

Dermatomyositis, polymyositis
Polyarteritis nodosa
Rheumatic fever
Rheumatoid arthritis
Scleroderma
Sjögren syndrome
Systemic lupus erythematosus

Organ-Specific Diseases

Disease	*Organ Involved*
Addison disease, idiopathic	Adrenal gland
Diabetes mellitus, juvenile*	Pancreas
Encephalomyelitis, acute disseminated	Central nervous system
Goodpasture glomerulonephritis	Kidney
Hashimoto thyroiditis	Thyroid gland
Multiple sclerosis†	Central nervous system
Myasthenia gravis	Muscle
Pernicious anemia	Gut
Sympathetic ophthalmia	Eye
Ulcerative colitis	Gut

---

*Diabetes mellitus is heterogeneous, and the extent of the role of autoimmunity in its etiology is still undetermined.
†Not proven.

---

general dysfunction of the affected organ system and sometimes to death. **Rheumatic fever**, a disease in which the heart tissue is seriously damaged, seems to occur as a result of the immune system's battle against a streptococcal infection. The streptococcal bacteria produce an antigen that is coincidentally very similar to antigens of the heart muscle cells. The immunoglobulins produced against the streptococcal antigens also react against the heart tissue, causing serious heart disease. A list of diseases that involve autoimmune responses is presented in Table 11–6.

Immunologic deficiency diseases

Several rare genetic disorders, known as **immunologic deficiency diseases**, involve the inability to respond to antigens. These defects may be autosomal or X-linked, and they may involve T cells, B cells, or both. A person who lacks B lymphocytes is highly susceptible to bacterial infections, especially in the skin and respiratory tract, though not to viral infections if the T-cell level is normal. A person who lacks T cells is especially susceptible to intracellular infection by viruses and bacteria. Injections of immunoglobulins and antibiotics help such a person live normally.

One of the most dramatic of the immunologic deficiency diseases is **Swiss-type agammaglobulinemia**, the result of an autosomal gene mutation. Patients with this disease have very few, if any, T or B cells, and they have virtually no protection against infection. The condition is commonly fatal, the child succumbing to infections within the first two to four years of life. We do not know the specific cause of the disease, though it does appear to involve enzymes that metabolize nucleic acids. Very occasionally, the condition has been successfully treated by transplants of bone marrow or thymus, and there have been several well-publicized cases of affected children who were raised in sterile tents until they were old enough to receive such transplants (Figure 11–16).

Many other diseases are associated with defects in the immune system, but in most cases we are unable to say whether they are caused by defective genes of the immune system or by some other defect that in turn causes the immune system to malfunction.

Evolution of the Immune System

It is difficult to analyze the evolutionary history of a system as complex as the immune system. But if we bear in mind some key points, we may be able to better understand how the diverse evolutionary forces favored the emergence of this increasingly complex series of genes. One point is that the immune system is essential for the survival of the individual and of the vertebrate species; it serves as a protector against infections and aberrant mutations in one's own cells. Another point worth remembering is that the great complexity of the immune system can be accounted for by mechanisms that are commonly found to occur in the natural world: gene duplication by unequal crossing over, mutation, and repair. Over time, this can increase the number of gene loci coding for immunoglobulin-type molecules and diversification of nucleotide sequences.

The HLA system of gene loci is, without exception, the most diversified gene complex known in humans. The fact that HLA antigens are associated with many

Figure 11–16
Severe immunoglobin deficiencies have kept this child, David, confined within a germ-free "bubble" for all of his eleven years. His parents and physician hope eventually to use monoclonal antibodies to cure him.

diseases emphasizes our point that the system plays a crucial role as a defense against infection by foreign agents. Indeed, systems analogous to HLA exist in all mammals, strengthening our view of the importance of the HLA gene complex.

The fact that the HLA antigens are located on the cell membrane suggests that the primary function of the HLA system is to regulate contact between the cell and the external environment, an environment that includes other cells. Many investigators think that cell-surface antigens such as the HLA antigens are critical to normal cell differentiation and development and hence to evolution. These antigens are seen as molecular signals that control cell movement and aggregation by defining cell–cell contact relationships.

Other researchers think that the primary reason for the HLA system's continued evolution and diversification was that it provided a defense against viral and bacterial attack. The cell-surface antigens may block the attachment of a virus to a human cell. Or, they may act in another way. When a virus infects a cell, reproduces, then leaves that cell to infect other cells, it commonly incorporates part of the cell membrane into its own structure. If that membrane carries cell-surface antigens, they may trigger a strong defensive reaction in the next person infected. A neat feature of this idea is the viral genome's inability to mutate to a form that is not vulnerable to attack by a host organism. The host, you see, is reacting to foreign human antigens as well as to the viral antigens. The extreme polymorphism of the HLA system ensures that, with the exception of identical twins, no two people have the same set of HLA antigens, and thus they are protected against foreign tissue.

Certain combinations of cell-surface antigens may be more effective than others in providing protection against viral and bacterial infection. Perhaps that is why certain HLA antigens are found associated together more frequently than we might expect on the basis of random associations.

The HLA proteins and the Ig molecules are much alike and probably have a common evolutionary pathway. Presumably, the Ig genes evolved as units that functioned at the cell surface and then gradually became more functionally diverse. In lower organisms, we see cell-surface antigens regulating cell–cell contact, regulating scavenger functions, and even serving as receptors in the type of cell–cell communication that operates among hormones. As Ig genes diversified, so too did their functions.

Perhaps the key event in the evolution of the Ig molecules was their capacity to first interact with an antigen and then to trigger the cell to divide. This inducement to cell division resulted in a population of cells producing the same Ig species, which could then launch a more intense attack against foreign antigens.

The functional divergence of the immune system is well integrated with the diversification of the Ig gene loci. We can postulate that the *V*-gene diversity occurred through recombinational mechanisms and the independent accumulation of mutations. The homologous flanking sequences on either side of the *V* genes would greatly facilitate this recombination.

The genes of the V and C regions exhibit many areas of homology. This similarity certainly suggests that the Ig genes evolved from a primitive gene by duplication and mutation to form a cluster of related genes. Translocational and inversional events may have led to the separation of the loci.

Summary

1 The **immune system** defends against foreign pathogens and mutations of our own cells and against the foreign tissues encountered in organ transplants.

2 The **cellular immune system** consists of white blood cells called **T lymphocytes** that circulate in the blood and lymph. T lymphocytes, which probably have antibody fragments on their surfaces, defend against viral and bacterial infections, cancer, parasitic infections, fungal growths, and tissue transplants.

3 The **humoral immune system** consists of circulating **antibodies**, or **immunoglobulins**, that defend against bacterial and viral infections and are particularly important in fighting bacteria. Antibodies are produced by white cells called **plasma cells**, which are derived from **B lymphocytes**. Plasma cells and B lymphocytes are located in the organs of the **lymph system**.

4 An **antigen** is a foreign substance, usually a large molecule or collection of molecules, that stimulates the **immune response**. The immune response consists of production of T and/or B lymphocytes directed against a specific antigen.

5 T lymphocytes, which probably have antibodies on their surfaces, stick to specific antigens and inactivate them. The circulating antibodies produced by the plasma cells bind to antigens and cause them to clump, so that they can be **phagocytized**, or engulfed, by white cells called **macrophages**. Each kind of T cell attacks a single antigen (or a group of closely related antigens); each kind of plasma cell produces one species of circulating antibody, which attacks a single antigen.

6 The key elements in the immune response are the antigen, antibody, and **complement**. The last is a collection of blood proteins that interact with the antigen–antibody complex and help to inactivate the antigen.

7 The **clonal selection theory** of antibody formation proposes that specific lymphocyte **stem cells** (ancestral cells), committed before birth to the synthesis of a specific antibody, proliferate on contact with specific antigens, forming large clones of T or B lymphocytes that produce that antibody.

8 Many antigens cause production of **memory cells**, stable lines of lymphocytes that persist for a long time after infection. When we are exposed to such an antigen for the second time, the memory cells produced in response to the first exposure proliferate very rapidly.

9 There are five immunoglobulin classes, each including specific antibodies against a variety of antigens. The largest and best understood is the class called **IgG**. Each IgG molecule has four polypeptide chains: two identical **heavy (H) chains** and two identical **light (L) chains**. Each H and L chain has a **constant (C) region** and a **variable (V) region**. The V region, which has a different amino acid sequence in each specific type of antibody, determines the affinity (i.e., attraction) of the molecule for a specific antigen. The IgG molecule has two **antigen-binding sites**.

10 Each immunoglobulin class has a distinctive type of H chain, which differs in its C region from the H chains of other classes. There are two types of L chains; both are found in all Ig classes, but only one type is found in a given molecule. The classes are distinguished by the numbers of L and H chains in each molecule, as well as by the type of H chain.

11 **Bence-Jones proteins**, found in the urine of multiple myeloma patients, consist of two L chains. Studying these proteins, each of them unique to an individual, has helped us understand the structure of immunoglobulins. More recently, development of **hybridomas**, fusion products of B lymphocytes and melanoma cells, has enabled us to produce large quantities of specific antibodies for study.

12 The **germ-line theory** of antibody variability proposes that variability is the result of a large number of distinct *V* (variable region) genes present in each lymphocytic stem cell. The theory suggests that all these *V* genes are present in the zygote and derived from the germ cells of the parents. The **somatic-mutation theory** proposes that there are few *V* genes in the zygote and that mutations occur in lymphocytic stem cells as they divide, generating diversity in the *V* genes.

13 The **gene-rearrangement theory** is based on evidence that the *V* region is coded for by two different chromosome regions, the *V* gene and the **Joiner (J) region**, each of which has many variant copies on the chromosome. The theory proposes that *V*, *J*, and *C* genes are rearranged to generate different combinations.

14 Cells have surface antigens called **histocompatibility antigens**, proteins whose combinations make each individual immunologically unique. In humans, these antigens are coded for by a gene complex called the **HLA complex**. Because the HLA complex includes many loci, each with multiple alleles, no two individuals (except identical twins or members of an inbred strain) ever have the same HLA genotype. The HLA system is crucial in organ transplantation: the more similar the HLA genotypes of donor and recipient, the greater the chances of success.

15 The blood group antigens, such as the ABO system and the Rh system, play a vital role in blood transfusion and obstetrics. The ABO, H, and secretor systems form an integrated network of blood antigen production.

16 The **A** and the **B antigens** are carbohydrates; they exist in alcohol-soluble form on the erythrocyte membrane, and in water-soluble form in the body fluids. The carbohydrate's terminal sugar determines whether it is the A or the B antigen. This sugar is formed by a specific enzyme, coded for by the *A* or the *B* gene. If both genes are present (type AB) both carbohydrates are present; if neither gene is present (type O), only the nonantigenic portion of the carbohydrate, called the **H substance**, is present.

17 The H substance, necessary for forming the A and the B antigens, is coded for by a locus separate from the ABO locus. The homozygous recessive condition (*hh*) produces the rare **Bombay phenotype**, which has no H substance and thus appears to be blood type O, regardless of the ABO genotype.

18 A third locus, called the **secretor locus**, is involved in the formation of the soluble form of the H substance. In the homozygous recessive (*se se*) state, no water-soluble H substance is made, so no A, B, or H antigens are secreted in body fluids such as saliva, regardless of the blood type.

19 The **Rh antigen** is responsible for **hemolytic disease of the newborn**. When an Rh⁻ woman gives birth to an Rh⁺ child, she develops antibodies against the Rh⁺ antigen. During a later pregnancy with an Rh⁺ fetus, the mother's antibodies may attack the child's red blood cells, causing hemolysis.

20 **Autoimmune diseases** occur when changes in a cell line prompt the body's immune system to attack the cells, or when the immune system mutates so that it identifies self-antigens as foreign.

21 In **immunologic deficiency diseases**, the immune system fails to respond to a foreign antigen. In extreme cases, a person may be born with no ability to produce either T or B lymphocytes; such a person cannot survive under normal conditions.

22 The immune system apparently evolved through gene duplication and diversification. First the genes coded for a system of cell-surface proteins that interacted with the environment. Later in evolutionary history, a primitive immune system seems to have emerged from the functions of these early cell-surface protein genes.

Key Terms and Concepts

Immune system

Lymphocyte

Antibody (immunoglobulin)

Immune response

Antigen

Cellular immune system

T lymphocyte (T cell)

Humoral immune system

Plasma Cell

B lymphocyte (B cell)

Lymph system

Stem cell

Bursa

Macrophage

Phagocytize

Clonal selection theory

Memory cell

Sensitization

Complement

Histamine

Gamma globulin

Heavy (H) chain

Light (L) chain

Disulfide bond

Constant (C) region

Variable (V) region

Antigen-binding site

Bence-Jones protein

Hybridoma

Germ-line theory

Somatic-mutation theory

Gene-rearrangement theory

Joiner (J) region

Killer T (T_K) cell

Histocompatibility antigen

Lymphocyte A complex (HLA)

Haplotype

A antigen

B antigen

H substance

Bombay phenotype

Glycolipid

Alcohol-soluble antigen

Glycoprotein

Water-soluble antigen

Secretor locus

Rh antigen

Hemolytic disease of the newborn (HDN)

Bilirubin

Self-antigen

Autoimmune disease

Sympathetic ophthalmia

Rheumatoid arthritis

Rheumatoid factor

Systemic lupus erythematosus

Rheumatic fever

Immunologic deficiency disease

Swiss-type agammaglobulinemia

Monoclonal antibody

Problems

11–1 How does the cellular immune system differ from the humoral immune system?

11–2 How do T lymphocytes differ from B lymphocytes?

11–3 What is the significance of the plasma cell?

11–4 What do macrophages do?

11–5   Describe the clonal selection theory of antibody formation.

11–6   What is the function of complement?

11–7   A person is exposed to an antigen for the second time and produces an immune response that is both more rapid and more intense than the first time. Explain this.

11–8   Describe a common view of the immune response that was prevalent before the discovery of the diversity of antibody structure.

11–9   What is the basic structure of the immunoglobulin molecule?

11–10   What is the significance of the Bence-Jones protein?

11–11   Describe and evaluate the major theories that attempt to explain antibody diversity.

11–12   What is the J region of the immunoglobulin H and L chains?

11–13   Evaluate the one gene–one polypeptide concept in light of what we understand about the immunoglobulin genes.

11–14   If $V, J,$ and C genes are not contiguous, how do they produce a single polypeptide chain?

11–15   Is there a role for unequal crossing over in the generation of antibody diversity? Explain.

11–16   Explain how recombination and repair can generate antibody diversity.

11–17   Killer-T lymphocytes appear to have two receptor sites. How do we know this and what are their specificities?

11–18   What is a haplotype? -

11–19   Except for identical twins, no two people have the same HLA alleles. True or false?

11–20   Discuss the factors that can lead to the acceptance or rejection of transplanted tissue.

11–21   What are the main types of autoimmunity a person may develop?

11–22   What are the characteristics of the immunologic deficiency diseases?

11–23   What may account for the association between HLA antigens and certain diseases?

11–24   What is the evolutionary significance of the HLA system?

11–25   What is the evolutionary significance of the Ig system?

11–26   Which of the following would provide you with the best chance of a successful organ transplant, and which would provide you with the lowest probability of success?
(a) brother, nontwin
(b) brother, fraternal twin
(c) identical twin
(d) friend, same age and sex
(e) mother
(f) uncle

11–27   Under what circumstances could Mr. Smith (page 316) be the father of the child in question?

Answers

See Appendix A.

References

Alexander, J. W., and R. A. Good. 1977. *Fundamentals of Clinical Immunology.* Saunders, Philadelphia.

Amzel, L. M., and R. J. Poljak. 1979. Three dimensional structure of immunoglobins. *Annu. Rev. Biochem.* 48:961–997.

Bodmer, W. F., and J. G. Bodmer. 1978. Evolution and function of the HLA system. *Br. Med. Bull.* 34:309–316.

Davis, M. M., S. K. Kim, and L. Hood. 1981. Immunoglobulin class switching: Developmentally regulated DNA rearrangements during differentiation. *Cell* 22:1–2.

Edelman, G. M. 1973. Antibody structure and molecular immunology. *Science* 180:830–836.

Edelman, G. M., and J. Gally. 1972. The genetic control of immunoglobin synthesis. *Annu. Rev. Genet.* 6:1–46.

Ferrara, G. B., ed. 1977. *HLA System: New Aspects.* North-Holland Publishing Co., Amsterdam.

Freedman, S. O., and P. Gold. 1976. *Clinical Immunology,* 2nd ed. Harper and Row, New York.

Fudenberg, H. H., J. R. L. Pink, A.-C. Wang, and S. D. Douglas. 1978. *Basic Immunogenetics,* 2nd ed. Oxford University Press, New York.

Golab, E. S. 1979. *The Cellular Basis of the Immune Response.* Sinauer, Sunderland, MA.

Hildemann, W. H. 1973. Genetics of immune responsiveness. *Annu. Rev. Genet.* 7:19–36.

Hood, L., J. Campbell, and S. Elgin. 1975. The organization, expression, and evolution of antibody genes and other multigene families. *Annu. Rev. Genet.* 9:305–353.

Hood, L., I. L. Weissman, and W. B. Wood. 1978. *Immunology.* Benjamin/Cummings, Menlo Park, CA.

Klein, J. 1979. The major histocompatibility complex of the mouse. *Science* 203:516–521.

McMichael, A., and H. McDevitt. 1977. The association between the HLA system and disease. *Prog. Med. Genet.* 2:39–100.

Marcu, K. B. 1982. Immunoglobulin heavy-chain constant region genes. *Cell* 29:719–721.

Marx, J. L. 1978. Antibodies (I): New information about gene structure. *Science* 202:298–299.

Marx, J. L. 1978. Antibodies (II): Another look at the diversity problem. *Science* 202:412–415.

Marx, J. L. 1981. Antibodies: Getting their genes together. *Science* 212:1015–1017.

Moscona, A. A., and M. Friedlander, eds. 1979–1980. *Immunological Approaches to Embryonic Development and Differentiation,* pts. 1 and 2. (*Curr. Top. Dev. Biol.,* vol. 13.) Academic Press, New York.

Newell, N., et al. 1980. J genes for heavy chain immunoglobins of mouse. *Science* 209:1128–1132.

Origins of Lymphocyte Diversity. 1977. (*Cold Spring Harbor Symp. Quant. Biol.,* vol. 41.) Cold Spring Harbor, New York.

Richards, F., W. Konigsberg, W. Rosenstein, R. Varga, and J. Varga. 1975. On the specificity of antibodies. *Science* 189:130–137.

Ryder, L. P., A. Sverjgaard, and J. Dausset. 1981. Genetics of HLA disease association. *Annu. Rev. Genet.* 15:169–188.

Seidman, J. G., A. Leder, M. Nau, B. Norman, and P. Leder. 1978. Antibody diversity. *Science* 202:11–17.

Sercarz, E., L. A. Herzenberg, and C. F. Fox, eds. 1977. *Regulatory Genetics of the Immune System.* Academic Press, New York.

Sercarz, E., A. Williamson, and C. F. Fox, eds. 1974. *The Immune System: Genes, Receptors, Signals.* Academic Press, New York.

Snell, G. D., J. Dausset, and S. Nathenson. 1976. *Histocompatibility.* Academic Press, New York.

Talmage, D. W. 1979. Recognition and memory in the immune system. *Am. Sci.* 67:173–177.

Watson, J. D. 1983. *The Molecular Biology of the Gene,* 4th ed. Benjamin/ Cummings, Menlo Park, CA.

Yelton, D. E., and M. D. Scharff. 1980. Monoclonal antibodies. *Am. Sci.* 68:510–516.

12 Complex Patterns of Inheritance

In our earlier discussions of Mendelism, we dealt with traits that exhibited *discontinuous variation*. Peas, for example, were either tall or short; they had seeds that were round or smooth; they had flowers that were white or purple. But inherited traits do not always appear in specific ratios, nor are they always expressed in an either-or fashion. Most of the traits we notice in people—height, build, or hair texture, for example—exhibit a pattern of *continuous variation*.

There is an important distinction between a Mendelian unit of inheritance and a trait. A trait is really an artificial construct of the human intellect. We partition a body into recognizable parts or traits such as eyes, nose, ears, teeth, skin color, height, weight, hair texture, blood pressure, electroencephalogram pattern, and susceptibility to disease. But in a very strict sense, we do not inherit these traits; rather, we inherit a genotype whose products act in various ways to produce a phenotype. The end result of the genotype's activity is the effect it has on some trait. Chromosomes obey the basic laws of inheritance; traits do not.

Sometimes a trait is **monogenic**; that is, it is the result of gene activity at a single locus. Mendel's traits fall into this category; so do dwarfism, albinism, and cystic fibrosis. But most human traits are produced by two or more gene loci interacting with each other and with the environment. A trait that is determined by several gene loci, each contributing additively to the final phenotype, is called a **polygenic trait**. The effect of each relevant gene locus is usually small and is susceptible to environmental influence. Skin color and height are both examples of polygenic traits, and both exhibit a continuous pattern of variation. For example, the range of skin color runs from very dark to very light, with all gradations in between.

We sometimes regard **multifactorial inheritance** as synonymous with polygenic inheritance. But the former term is usually reserved for traits that are determined by both genetic and environmental factors, without reference to the nature of the genetic components. The term **polygenic inheritance** indicates that several loci are involved, without reference to the environmental components. Also implicit in this term is the idea that each contributing gene locus exerts only a slight effect on the phenotype.

Many polygenic traits do not exhibit a continuous pattern of variation. For example, susceptibility to a disease may be determined by several gene loci, and the trait we call "susceptibility" may exhibit a continuous pattern of variation. But the trait "disease" is either expressed or not expressed—it exhibits a pattern of

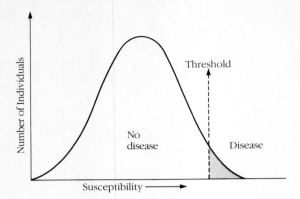

Figure 12–1
Susceptibility to a particular disease involves many factors and exhibits a normal distribution. However, only a limited segment of the population gets the disease, so disease expression is a threshold trait.

discontinuous variation. We refer to such traits as **threshold traits** (Figure 12–1). The threshold is a consequence of the complex nature of the trait.

Finally, genes often interact to produce novel or modified phenotypes. Suppose, for example, that gene *A* codes for phenotype A, and that gene *B* codes for phenotype B. When genes *A* and *B* are together, they may code for a novel phenotype C by interacting with each other. This type of gene interaction can produce modified Mendelian ratios that deviate from the classical 3:1 or 9:3:3:1 ratios. In addition to this type of interaction, an allele at one locus may modify the expression of an allele at another locus. This too can lead to a modification of Mendelian phenotypic ratios.

In this chapter, we will examine some of the more complex inheritance patterns. Even though these patterns may not exhibit classic Mendelian ratios, they do not in any way invalidate Mendelian principles; the genes and chromosomes still segregate and assort independently. That is why we must emphasize the difference between the inheritance of a gene and the inheritance of a trait.

Modifier Genes and Epistasis

A classic Mendelian inheritance pattern is often obscured by a gene locus that modifies the expression of another, nonallelic, gene. Any gene that affects the phenotypic expression of genes at other loci is called a **modifier gene**. A classic example of modifier gene activity is the trait called piebald spotting in mice. This trait is characterized by a patch of white fur on the stomach and is determined by a recessive gene (*s*). But the size of the white patch is variable; the patch ranges from a small spot to one that covers the entire coat except for a small area of pigmented fur (Figure 12–2). The basic piebald trait appears when the *s* gene is homozygous, but the extent of piebaldness is controlled by four unlinked loci whose activity somehow modifies the activity of the *s* gene locus.

Human eye color appears to be the result of several interacting gene loci. The minimal amount of pigment gives the iris of the eye a blue appearance. But when more pigment is deposited in the iris, it takes on color ranging from yellowish brown through green, hazel, and light brown to dark brown. There appears to be a continuous spectrum of eye colors, but most people simply group the vari-

ous shades into "blue" and "brown." This grouping has led to the erroneous idea that eye color is determined by a single pair of alleles, with "brown" (*B*) dominant over "blue" (*b*).

We can propose that three independently assorting gene loci determine eye color and that these loci code for melanin synthesis. A true blue-eyed person would deposit a small amount of melanin in the iris and would have the genotype *aabbcc*. One copy of a dominant allele would produce some melanin, but the eye would still look blue to most observers, even though it would be a darker blue. We can thus see how two "blue-eyed" people can produce a "brown-eyed" child. We do not have to suggest that the mother had a fling with a brown-eyed mailman. Consider the following couple:

$$\text{"Blue"} \qquad\qquad\qquad \text{"Blue"}$$

$$\frac{a}{a}\ \frac{b}{b}\ \frac{C}{c} \qquad \times \qquad \frac{A}{a}\ \frac{b}{b}\ \frac{c}{c}$$

$$\downarrow$$

$$\frac{A}{a}\ \frac{b}{b}\ \frac{C}{c} \qquad \text{"Brown" (light)}$$

The parents are classified by casual observers as blue-eyed, although a closer examination would show that their eyes are a bit darker than pure blue. The child's eyes will be light brown because of the two pigment-producing alleles. Eye color is thus an example of a polygenic trait. It might also be considered an example of a threshold trait, because people commonly classify eyes as blue or brown; and even though the genotypes form a continuous range, the phenotype is usually only one of two forms (Figure 12–3).

Sometimes one gene completely masks the phenotypic expression of a gene at another gene locus. The gene that is doing the masking is called an **epistatic gene**, and the phenomenon is called **epistasis**.

We often find epistasis operating in phenotypes that are determined by a series of genetically controlled reactions. For example, if a phenotype (x) is determined by two independently assorting pairs of alleles, with each locus coding for an enzyme in a metabolic sequence, we may observe an epistatic effect:

gene *A*: produces enzyme a
gene *a*: does not produce a functional enzyme a
gene *B*: produces enzyme b
gene *b*: does not produce a functional enzyme b

and

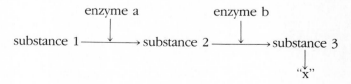

Figure 12–2
Variation in piebald spotting in *ss* (piebald) mice. The variation in spotting pattern is the result of modifier genes. A mouse carrying an *S* allele will show no piebald spotting, regardless of the modifier genes present.

According to this scheme, *Aa* or *AA* produces enzyme a, and *BB* or *Bb* produces enzyme b. Thus, any genotypes with both *A* and *B* will produce the "x" phenotype (*A-B-*). Homozygous recessive alleles at one or both loci (*A-bb*, *aaB-*, and *aabb*)

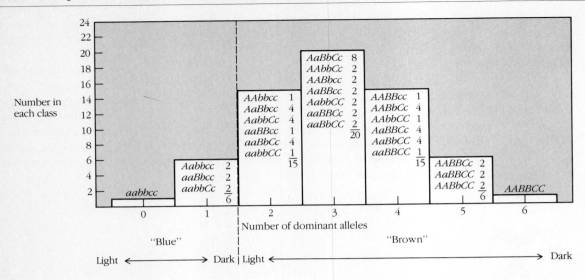

Figure 12–3
A histogram of genotypic and phenotypic classes possible from a mating between two "brown-eyed" heterozygotes such as *AaBbCc × AaBbCc*.

will result in a "non-x" phenotype. For a classic Mendelian dihybrid cross (discussed in Chapter 2), we would observe the following phenotypic ratio:

		Phenotypic ratio
9 *A-B-*	"x"	9
3 *A-bb* ⎤		
3 *aaB-* ⎬	"non-x"	7
1 *aabb* ⎦		

A modified Mendelian phenotypic ratio, in this case 9:7 instead of 9:3:3:1, is typical of epistasis. We say that *aa* is epistatic to *B-*, and *bb* is epistatic to *A-*.

Epistatic-gene effects are seen in the inheritance of albinism, a trait that occurs when the synthesis of melanin is interrupted (see Figure 10–4). In classic albinism, an enzyme is missing that is required to convert tyrosine to DOPA and dopaquinone. In this recessive condition, the homozygous allelic pair is epistatic to all other genes in the sequence, because no melanin forms even when normal gene products (enzymes) are present for all the other reactions in the sequence.

We can also see how, in this example of albinism, two albinos can produce a nonalbino child. Let's say the mother is missing tyrosinase and therefore cannot convert tyrosine to DOPA and then to dopaquinone; and the father cannot convert dopaquinone to melanin:

gene *A*: normal tyrosinase
gene *a*: no tyrosinase
gene *B*: converts dopaquinone to melanin
gene *b*: does not convert dopaquinone to melanin

The child could receive one normal gene from each parent and thus have all the enzymes necessary for the synthetic pathway:

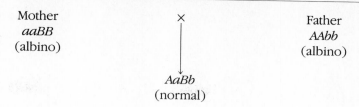

| Mother
aaBB
(albino) | ×

↓
AaBb
(normal) | Father
AAbb
(albino) |

In this case *aa* and *bb* are epistatic to all other genes in the sequence.

The Bombay phenotype is another example of epistasis. The *H* allele is responsible for producing the H substance, which is the precursor for both the A and the B antigens. A person who is *bh* does not produce the H substance and thus cannot produce A or B antigens even though the *I^A* and/or *I^B* alleles are present.

Our discussion of epistasis and modifier genes shows how gene products produced at different loci can interact to produce modified Mendelian ratios. The environment can also modify these ratios.

Interaction of Genotype and Environment

Genotypes are frequently sensitive to the modifying influences of the environment. For example, a child who is born with phenylketonuria can live a reasonably normal, healthy life if his or her environment is modified to exclude the amino acid phenylalanine, or at least greatly reduce its presence. In much of the rest of this chapter, we will explore some of the ways that the environment interacts with the genotype. Most of the phenotypes we observe are the result of gene–environment interaction.

The environmental factors that can modify a genotype's expression are numerous and complex. We are usually unable to characterize or quantify these influences, but they include at least the following internal and external factors: (1) temperature, (2) light conditions, (3) nutrition, (4) sex, (5) age, and (6) availability of substrates.

Consideration of the environment can affect our interpretation of the inheritance pattern of a pair of alleles. Suppose that a pair of alleles has segregated in Mendelian fashion to give a 1:2:1 genotypic ratio. If each genotypic class were phenotypically distinct and were not subject to extensive environmental modification, we would observe a 1:2:1 phenotypic ratio. If, on the other hand, each genotypic class were subject to environmental modification, the phenotypic classes could so overlap that there would be a continuous distribution of phenotypes, and in no way could we be certain that two alleles were segregating (Figure 12–4).

Some people have suggested that the behavioral disorder schizophrenia is the result of a dominant allele with incomplete penetrance and variable expressivity. Schizophrenia does indeed exhibit a wide range of characteristics, so if this genetic model is applicable, then the AA and Aa phenotypes will overlap extensively and produce a continuous pattern of variation ranging from extremely disordered behavior to normal.

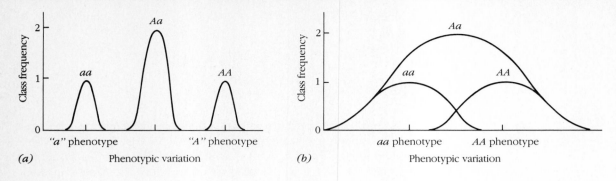

Figure 12–4
Phenotypic variation when genotypic expression is (*a*) minimally and (*b*) extensively modified by the environment.

Phenotypes may vary to the degree that the gene activity controlling their expression can be modified. That is, if a genotype could be placed in many different environments, a certain range of phenotypic expressions would be observed, and that range of expression would vary from one genotype to another. The **norm of reaction** is the entire range of phenotypes expressed by a specific genotype in all environments. This norm can be narrow or quite broad. The ABO blood groups, for example, show essentially no variability, irrespective of the environment; the norm is narrow. Intelligence, on the other hand, varies widely with the environment; the norm is broad.

Obvious environmental influences

The role of the environment in influencing a phenotype can be obvious or complex. In the obvious category are the extreme forms of expression modification. In one form, the environment modifies a normal genotype so that it is expressed as a mutant phenotype. This simulated phenotype is, of course, not inherited, and it is called a **phenocopy** because it resembles a known genetically caused phenotype. There are many examples of phenocopy induction.

One of the most tragic instances of induced phenocopy occurred in the late 1950s and early 1960s, when deformed children resulted from the use of the tranquilizer thalidomide by women in their early pregnancy. Some 6000 children, most of them born in West Germany and Great Britain, expressed a previously known phenotype called **phocomelia** (Greek: *phoke*, "seal," + *melos*, "limb"), in which one or more limbs are reduced to flipperlike appendages (Figure 12–5). Thalidomide, the common factor in most of these malformations (Table 12–1), had been deemed safe by the manufacturer and was commonly dispensed without a prescription. During the sixth week of development, the fetus is especially sensitive to this drug, which interferes with normal limb formation and results in a phenocopy of a rare dominant mutation.

In another form of obvious environmental influence, mutant genotypes are expressed as normal phenotypes. Diabetes (technically, diabetes mellitus, the most common form of diabetes) is an example of this kind of influence. Diabetes is an inherited disorder, controlled either by several genes or by a single pair of alleles with low penetrance. It is characterized by a deficiency or defectiveness of the pancreatic hormone insulin. Diabetics cannot utilize glucose properly, so they must depend on an external source of insulin. Insulin injections (or, in some

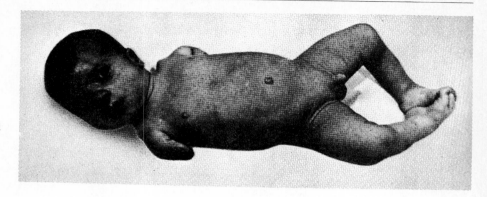

Figure 12–5
Phocomelia as an induced phenocopy in a newborn child whose mother took thalidomide during early pregnancy.

cases, oral drugs or even dietetic controls alone) adjust the internal environment so that the mutant genotype is not expressed and the normal phenotype is produced.

In the genetic disease hemophilia A, blood is unable to clot because a specific protein required for normal clotting is either defective or scant. The blood of hemophiliacs clots normally if they receive injections of this factor; thus, a mutant genotype produces a nonmutant phenotype. Unlike insulin, which can be obtained readily from nonhuman animals, and now through recombinant DNA technology, the antihemophilic factor can be gotten only from humans. It is therefore difficult to obtain and extremely expensive.

Complex environmental influences

The relationship between the genotype and the environment is not always so obvious. Traits such as intelligence, certain mental illnesses, and some forms of cancer show very complex gene–environment interactions. For such traits, the

Table 12–1   Thalidomide as a common factor in an induced phenocopy

	1960	1961	1962 Jan–July	1962 Aug–Oct	Total
Total births	19,052	19,917	13,326	5,542	57,837
Malformations resembling phocomelia	28	60	40	2	130
History					
Thalidomide taken	13	46	33	2	94
No evidence of intake	0	5	1	0	6
No history available	15	9	6	0	30

relative influence of genes and environment is difficult to determine. Certain forms of cancer demonstrate the complex interaction of the genotype and environment.

Cancer is characterized by abnormal cell division and tissue growth. Cancer in general is probably the result of altered gene functions, and certain types of cancer even appear to follow a Mendelian pattern of inheritance. These are usually the tissue-specific types, such as colon cancer. Extensive analysis of families in which cancer of a specific type appears indicates that certain combinations of

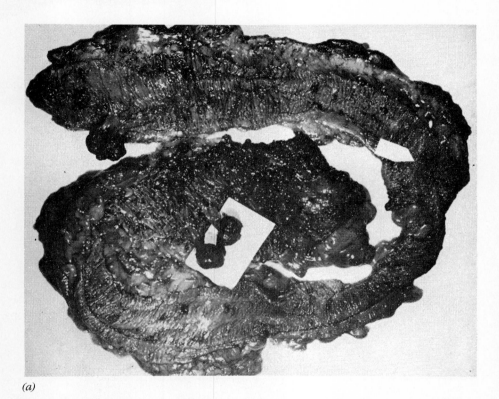

Figure 12–6
(*a*) Intestinal polyps.
(*b*) Pedigree of a family expressing Gardner syndrome.

(*a*)

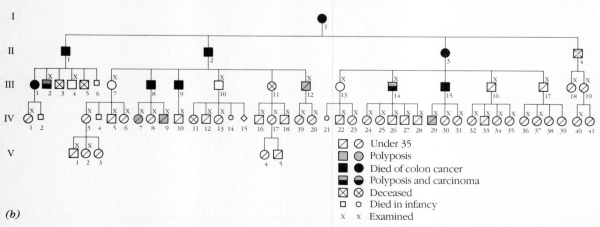

(*b*)

genes create a potential for expressing a specific type of cancer. But this potential may be realized only if the environment is conducive to the expression of the genotype. Here we have an example of a *threshold trait*, for the person may be phenotypically healthy even though he or she is carrying the genotype for susceptibility to the disease.

An interesting example of a genetically based cancer that has a clear but complex environmental component is Gardner syndrome, a form of colon cancer accompanied by numerous polyps (lumps of tissue) that protrude into the intestine (Figure 12–6a). This cancer seems to be the result of a single dominant gene located on chromosome 2, and the role of the environment in the syndrome is small. The pedigree (Figure 12–6b) shows that eight members of one family died of colon cancer (I1, II1, II2, II3, III1, III8, III9, III15). Since one of the persons examined (III12) had polyps but no cancer, there may be two steps in the development of the disease: the formation of polyps and their subsequent conversion to malignant growths. Of 69 family members, 4 had nonmalignant polyps (III12, IV7, IV9, IV29). Later examination of the 4 who had the nonmalignant polyps revealed that 2 had developed malignant growths. It seems that the genotype causes (or permits) the polyps to develop, and something in the environment then converts the polyps into malignant growths in about 90% of the cases. We have no idea what causes this conversion.

The environment of a gene is the sum total of all factors external to it, both genetic and nongenetic. Since genes commonly code for enzymes whose activity is controlled by the environment, it seems reasonable that a change in the environment might alter the activity of enzymes and thus modify a phenotype. Environmental forces act in similar ways on structural proteins.

Polygenic Traits

When Mendel's work was unearthed at the turn of this century, a very obvious question that soon arose was how widely his laws could be applied. Mendel's experiments showed that for the traits he examined, the variation was discontinuous and his laws were valid. But did these laws apply equally to traits that are genetically controlled but exhibit a continuous pattern of variation, such as human height?

In the infancy of genetics, there were two major schools of thought regarding continuous variation. One school argued that continuous traits followed a different set of laws. According to this school, Mendelism was the exception rather than the rule, since most traits are not so clearly discontinuous as those Mendel studied. Supporters of this position proposed a **blending model of inheritance**, in which inheritance units from each parent mix and blend in their offspring. Producing children, they thought, is analogous to mixing red and white paint to get pink; a child's traits represent an averaging of parental traits.

The blending model of inheritance is rooted in a premise that is clearly contradictory to Mendelism. It says that an individual receives half of his inheritance from each parent, a fourth from each of his four grandparents, an eighth from each of his eight great-grandparents, and so on. This form of inheritance is called the **law of ancestral inheritance** (Figure 12–7), and it produces the blending patterns we mentioned. What makes this model incompatible with

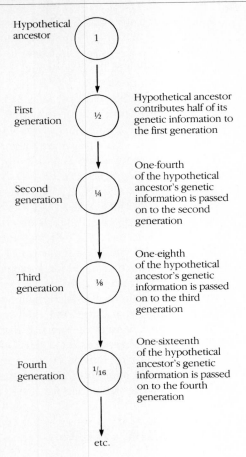

Hypothetical ancestor

1

First generation

½

Hypothetical ancestor contributes half of its genetic information to the first generation

Second generation

¼

One-fourth of the hypothetical ancestor's genetic information is passed on to the second generation

Third generation

⅛

One-eighth of the hypothetical ancestor's genetic information is passed on to the third generation

Fourth generation

$^1/_{16}$

One-sixteenth of the hypothetical ancestor's genetic information is passed on to the fourth generation

etc.

Figure 12–7
Law of ancestral inheritance. Each generation, no matter how far removed from a hypothetical ancestor, passes on a specific proportion of that ancestor's genetic information.

Mendelism is its assertion that one carries a specific fraction of genetic material from every ancestor, no matter how remote. Mendelism, on the other hand, asserts that there is a certain *probability* that one will carry a specific ancestor's chromosome, but it is by no means a sure thing. For example, the propositus (the person who serves as the basis for a study) indicated in Figure 12–8 should, according to the law of ancestral inheritance, carry one-fourth of the genetic material of each grandparent, yet we see that the person carries no chromosomes at all from grandparents X and R.

The alternative to the blending model of inheritance and the law of ancestral inheritance was *polygenic inheritance*. According to this school of thought, all inherited traits can be interpreted within the framework of Mendel's principles. Continuous traits, in this view, are determined by several pairs of alleles, by incomplete dominance, and by the interaction among genes and between genes and the environment.

After extensive experimentation, the Mendelian view has prevailed. The first major step toward its acceptance was the careful partitioning of the phenotype into genetic and environmental components. This was begun by Wilhelm Johannsen in 1909. This very perceptive experimenter demonstrated that in a homozygous strain of broad beans, or fava beans (*Phaseolus vulgaris*), variations in bean

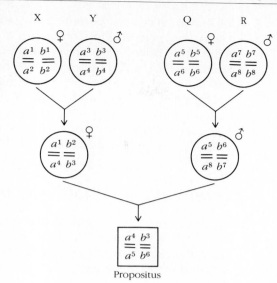

Figure 12–8
Mendelism argues against the law of ancestral inheritance. The propositus does not carry any genetic information from grandparents X and R.

size could be attributed almost exclusively to environmental influences (Figure 12–9). Under field conditions in which the plants were freely interbreeding, the beans varied continuously in weight from 150 to 750 mg. By selecting only the largest or smallest beans and inbreeding these plants for several generations,

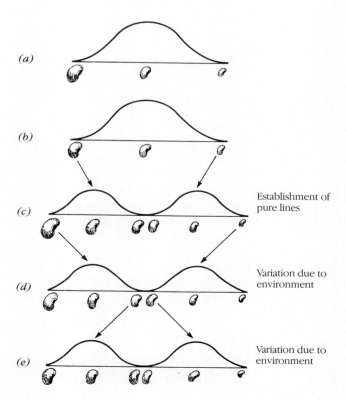

Figure 12–9
Johannsen's analysis of genotypic and environmental components of the phenotype bean size. An unselected, medium-sized bean (*a*) repeats the normal distribution, as demonstrated in (*b*). Pure lines are established (*c*) when either large or small beans are selected and inbred over several generations. Whether the largest bean from the small line or the smallest bean from the large line is selected (*d*), the distribution is still normal (*e*).

Johannsen established pure lines that were homozygous at most gene loci. Yet within these pure lines, he continued to find variation, and further selection from within them failed to alter the phenotypic curve, whether he selected the largest bean from the small line or the smallest bean from the large line. Since the plants within any one line were all the same genetically, Johannsen concluded that the variation he observed must be due to the environment. His work thus led to the realization that a phenotype has two components, genotypic and environmental. As a matter of fact, Johannsen coined the terms *gene, genotype,* and *phenotype* as a result of his studies (though Mendel clearly had the basic concepts in his original paper).

Confirmation of polygenic inheritance

In 1909, Herman Nilsson-Ehle demonstrated clearly that more than one gene pair can influence a given trait. Crossing a strain of white-kerneled wheat with a strain that had dark red kernels, he found that the F_1 was intermediate (Figure 12–10). Self-fertilization of the F_1 produced an F_2 phenotypic ratio of

> 1 dark red
> 6 moderate red
> 15 red
> 20 intermediate red
> 15 light red
> 6 light light red
> 1 white

Nilsson-Ehle proposed that this ratio is the result of three pairs of independently assorting alleles, each pair showing incomplete dominance. He did additional crosses to verify his hypothesis. This one experiment indicated the existence of polygenic inheritance. His experiments and subsequent experiments confirmed this important hypothesis.

In situations where the phenotypic classes are not clear-cut, we can sometimes estimate the number of allele pairs contributing to a trait by determining the proportion of F_2 individuals that fall into the parental classes. For example, if we have a continuously varying trait in which the parental phenotype makes up $\frac{1}{16}$ of the total F_2 offspring, we may conclude that two pairs of alleles control this trait. A parental extreme constituting $\frac{1}{64}$ of the total F_2 suggests three allelic pairs; $\frac{1}{256}$ suggests four allelic pairs, $\frac{1}{1024}$ suggests five allelic pairs, and so on. These fractions represent the completely homozygous dominant or recessive condition, just as they do in the case of discontinuous variation. But they represent only an approximation, because there may be other gene loci influencing the trait located on the same chromosome and thus not assorting independently of genes on that chromosome. For example, because three closely linked loci will generally be transmitted as a unit, they may be labeled as a single Mendelian unit.

In humans, we do not have the convenience of pure lines when we study a trait showing continuous variation, so we usually cannot be sure of the exact number of allelic pairs involved. However, we are able to make good estimates for certain kinds of traits, and the trait of skin color comes the closest to serving as

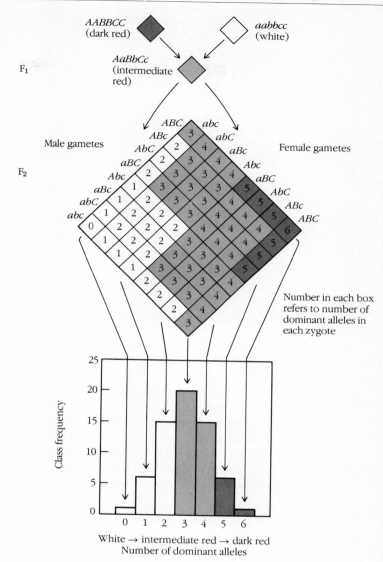

Number in each box refers to number of dominant alleles in each zygote

Figure 12–10
Nilsson-Ehle's analysis of the inheritance of grain color in wheat, showing that the trait is controlled by three independently assorting gene loci.

White → intermediate red → dark red
Number of dominant alleles

a model for proper analysis. Skin color is inherited as a polygenic trait, the genes having an additive effect. Authors of some early studies suggested that skin color is the result of two independently assorting gene loci and incomplete dominance, but these studies could be faulted for the arbitrary way in which the researchers determined skin-color classes. More modern studies have attempted to classify skin color in a more precise quantitative fashion by using the ability of the skin to reflect light of a specific wavelength (Figure 12–11). When we use skin reflectance as our criterion, we obtain a continuous distribution, not a series of discrete classes.

Analysis of the offspring of hundreds of different mating combinations indicates that 3–5 allelic pairs control skin color.

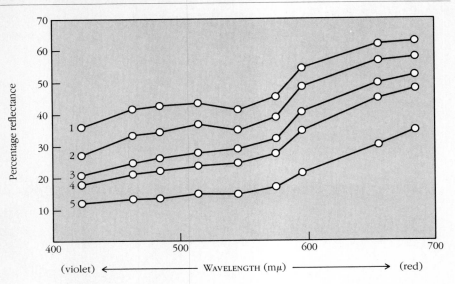

Figure 12–11
Mean reflectance from the skins of Europeans, Africans, and various hybrid groups, as measured with a reflectance spectrophotometer. 1 = European; 2 = offspring of marriage between Europeans and F₁ hybrids of European × West African; 3 = offspring of marriage between Europeans and West Africans; 4 = offspring of marriage between West Africans and F₁ hybrids of European × West African; 5 = West African.

Patterns of variation

Polygenic inheritance sometimes produces F_2 phenotypes that exceed the parental extremes. This **transgressive variation** occurs when the parents do not represent the extreme classes. For example, crosses between very dark-skinned people and very light-skinned people generally produce an F_1 that is intermediate in skin color. But matings between unrelated F_1 intermediates occasionally produce children who are either lighter than the light grandparents or darker than the dark grandparents. In such cases, we can infer that several allelic pairs are assorting independently and producing a cumulative effect. When the grandchildren (F_2) are lighter or darker than either of the grandparents (P), in all probability the grandparents were not of the most extreme genetic classes:

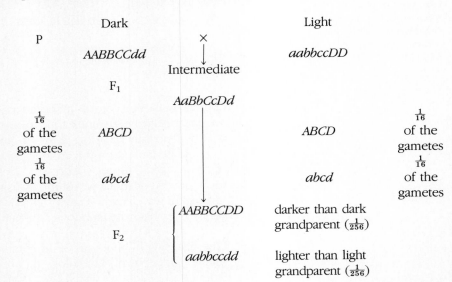

Box 12–1   *Calculating the Variance for the Human Height Sample*

Individual Values	Mean	Difference Squared	Difference Squared Divided by $N - 1$	
157	− 173 =	$(-16)^2$ =	256/12	= 21.3
161	− 173 =	$(-12)^2$ =	144/12	= 12.0
162	− 173 =	$(-11)^2$ =	121/12	= 10.1
163	− 173 =	$(-10)^2$ =	100/12	= 8.3
169	− 173 =	$(-4)^2$ =	16/12	= 1.3
174	− 173 =	$(1)^2$ =	1/12	= 0.1
176	− 173 =	$(3)^2$ =	9/12	= 0.8
177	− 173 =	$(4)^2$ =	16/12	= 1.3
179	− 173 =	$(6)^2$ =	36/12	= 3.0
180	− 173 =	$(7)^2$ =	49/12	= 4.1
181	− 173 =	$(8)^2$ =	64/12	= 5.3
183	− 173 =	$(10)^2$ =	100/12	= 8.3
186	− 173 =	$(13)^2$ =	169/12	= 14.0
			89.9 = variance	

Transgressive variation is less common than the contrasting phenomenon of **regressive variation** (or **regression toward the mean**), which is the tendency of offspring to approach the mean of the population rather than to exceed the parental extremes. For example, the children of very tall parents tend (on the average) to be shorter than their parents: they are closer than their parents to the mean for the population. Conversely, the children of very short parents tend to be taller than their parents—to approach the populational mean. Regression is primarily the result of increased heterozygosity:

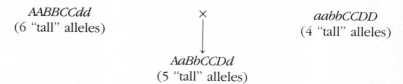

AABBCCdd	×	*aabbCCDD*
(6 "tall" alleles)		(4 "tall" alleles)

AaBbCCDd
(5 "tall" alleles)

The pattern of variation we observe in a population is to some extent controlled by the mating pattern. For some traits, people exhibit **random mating**. For example, we do not select our partners on the basis of their blood type or the form of acid phosphatase they have. On the other hand, we do select our partners on the basis of intelligence, body shape, skin color, hair color, and many other traits. The tendency for males of a particular type to mate with females of a particular type is called **assortative mating**. If the two mates are more alike than is to be expected by chance, they illustrate **positive assortative mating**. Examples are intelligent people tending to marry intelligent people and light-skinned people tending to marry light-skinned people. If the two mates are less alike than is to be expected by chance, they illustrate **negative assortative mating**. Both transgressive and regressive variations are the result of the way mating pairs form in a population, a point we shall return to in Chapter 14.

Analysis of polygenic inheritance patterns

A continuous distribution pattern does not invariably indicate polygenic inheritance. Mendelian traits—that is, those determined by a single pair of alleles—are often more complex than was once believed and may sometimes exhibit continuous distribution.

For example, acid phosphatase, an enzyme found in the red blood cells, is produced by a multiple allelic series at one locus: P^A, P^B, P^C. Each allele produces an enzyme form with a different activity; P^C is the most active, and P^A is the least active. The six possible genotypes (AA, AB, AC, BB, BC, CC) are distinguishable from one another on the basis of their electrophoretic differences, but this determination is possible only under carefully controlled conditions. Usually there is so much overlap in functioning among the six genotypes that they are indistinguishable from one another on the basis of their function (Figure 12–12). The overall activity distribution presents a continuous pattern of variation. Here we have a Mendelian trait obscured by a continuous-variation pattern.

More often, however, a continuous-distribution pattern is the result of polygenic inheritance and is impossible to resolve into discrete phenotypic or genotypic classes. We will consider here some of the methods used in analyzing quantitative traits.

Geneticists formerly thought that the ability to taste phenylthiocarbamide (PTC) is determined by a single dominant allele (T), and that the homozygous recessive condition results in the inability to taste PTC:

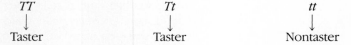

TT	Tt	tt
↓	↓	↓
Taster	Taster	Nontaster

But on examining the tasters more carefully, we find a considerable amount of variation in response to different PTC concentrations. One study of PTC tasting

Figure 12–12
Distribution of acid phosphatase activities in red blood cells. The dashed curve shows a continuous distribution of the enzyme in the general population. The other curves show activity values for five separate genotypes. Because of their overlapping distributions, these genotypes cannot be distinguished on the basis of their enzyme activity alone.

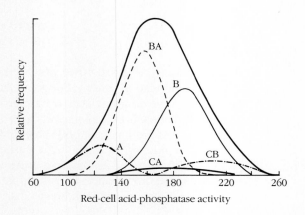

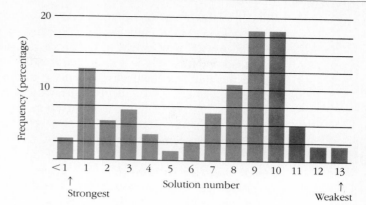

Figure 12–13
Distributions of taste thresholds for PTC in a population.

ability (Figure 12–13) tested different concentrations of PTC; solution 1 was the strongest and 13 the weakest. People sampled each solution until they reached a concentration they could taste, called the **threshold solution**. In this particular study, solution 5 was arbitrarily assigned as the breakpoint between "tasters" and "nontasters." That is, people who could not taste any solution weaker than solution 4 were classed as "nontasters"; those who could taste solution 6 or weaker ones were classed as "tasters." Note, however, that there is a gradation in tasting capabilities, so that "nontasters" at one concentration are "tasters" at another. We also find that different populations have different average thresholds and produce different distribution patterns. PTC tasting, though it is determined mainly by a single allelic pair, is evidently influenced by other genes.

Many traits are determined by an unknown number of gene loci. Recently, for example, it has been estimated that there may be as many as 7000 gene loci controlling or influencing the process of aging in humans. Such complex traits are impossible to analyze and organize into distinct genotypic and phenotypic classes. Human height and IQ (Figure 12–14) are clear examples of traits that are

Figure 12–14
Distribution of IQ and height among two different European populations.

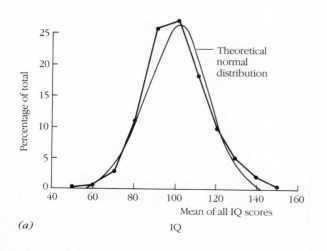

(a) IQ

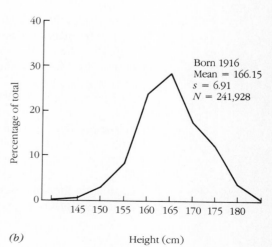

(b) Height (cm)

difficult to organize into separate classes, except in an artificial way. These traits are polygenic and are also very susceptible to environmental influences.

It is usually best to begin the analysis of a quantitative trait by describing it in quantitative terms. We ordinarily take random samples from a population, describe the trait quantitatively from these samples, and then extrapolate to the whole population. Values calculated from samples are **statistics**; values calculated from the whole population are **parameters**. It is most common to analyze quantitative traits on the basis of statistics, and we shall now look at how some of those statistical measurements are made.

Mean, median, and mode   In the analysis of a sample of people expressing a quantitative trait, it is important for us to know where a "typical" value is on the value scale of the trait. In other words, we need values that give us locations in our sample. Suppose, for example, we were to pick 13 males from your class, measure their height (in centimeters), and arrange the values in sequence:

$$157 \quad 161 \quad 162 \quad 163 \quad 169 \quad 174 \quad 176$$
$$177 \quad 179 \quad 180 \quad 182 \quad 183 \quad 186$$

Location points that help us characterize this sample are the mean and the median. The **mean ($\bar{x}$)** (or average) is the sum of an array of quantities divided by the number of items in the sample. In the foregoing sample, the mean is

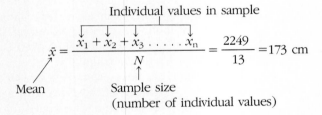

$$\bar{x} = \frac{x_1 + x_2 + x_3 \ldots \ldots x_n}{N} = \frac{2249}{13} = 173 \text{ cm}$$

The **median** is the middle value in a group of values arranged in order of size. The median here is 176 cm: the seventh, and thus the middle, value of the 13 observations.

Though this is a small sample, we often organize the data into classes and form a **frequency distribution** (Table 12–2). The **modal class**, or **mode**, is that class containing more individuals than any other class in the frequency distribution. In this particular example, the modal class is the 176–180 class. If the sample is large enough, a frequency distribution can provide useful information, especially when converted to graph form (Figure 12–15).

Variance and standard deviation   The location points we just described tell us nothing about the variation, that is, how variable or divergent the measurements are among individuals in the sample. In other words, we do not know how the values are dispersed. One way of characterizing dispersion in a sample is by calculating the **variance**. When all values in a sample are expressed as plus and minus deviations from the mean, the variance is the mean of the squared deviations:

Table 12–2   **Calculation of the frequency distribution of the height sample**

Class	Values for Individuals in Each Class	Class Mean	Frequency
156–160	157	—	1
161–165	161, 162, 163	162	3
166–170	169	—	1
171–175	174	—	1
176–180	176, 177, 179, 180	178	4
181–185	181, 183	182	2
186–190	186	—	1

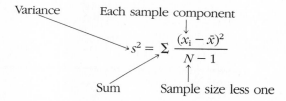

Variance   Each sample component

$$s^2 = \Sigma\ \frac{(x_i - \bar{x})^2}{N - 1}$$

Sum   Sample size less one

Thus, as we see from Box 12–1, the variance is 92.2 cm².

The variance can be used by itself as the variability measure, but since it is expressed in square units, we usually take its square root, which returns us to the original linear scale of measurement. The square root of the variance is called the **standard deviation** (s). Thus, from our height sample we obtain

$$s^2 = 92.2 \text{ cm}^2$$
$$s = \sqrt{92.2} = 9.6 \text{ cm}$$

The larger the standard deviation, the larger the variability (i.e., the more the observations are dispersed from the mean) in our sample. The importance of

Figure 12–15
Histogram of the class frequencies for height, as discussed in text. The modal class is 176–180.

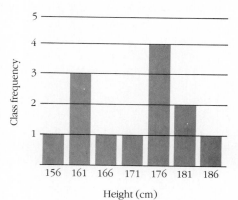

Table 12–3	Two sets of scores that have the same mean but differ in variability	
	Set A	Set B
	32	46
	32	42
	31	38
	31	34
	30	30
	30	30
	29	26
	29	22
	28	18
	28	14
	$\Sigma X = 300$	$\Sigma X = 300$
	$N = 10$	$N = 10$
	$\bar{x} = \dfrac{300}{10} = 30$	$\bar{x} = \dfrac{300}{10} = 30$
	Range = $32 - 28 = 4$	Range = $46 - 14 = 32$

knowing about the dispersion of values in a sample is apparent in Table 12–3, where the two samples shown have identical means but differ greatly in their dispersion properties.

Standard deviations enter into one other important aspect of patterns of continuous variation. If we were to graph most of the continuously varying traits in a population, we would see that they are distributed in a bell-shaped pattern called the **normal distribution**, or **Gaussian distribution**, in which the distribution of values is symmetrical about the mean (Figure 12–16). In such a normal distribution, about two-thirds of all individuals fall between one standard deviation above and one standard deviation below the mean. Ninety-five percent of all individuals fall between two standard deviations on either side of the mean. In Figure 12–16, the mean is zero, and the standard deviation is 1.

Figure 12–16
Standard deviation in distributions having a normal shape. In both graphs the scale of the horizontal axis makes for a mean of zero and a standard deviation (s) of 1. About one-third of all individuals in the distribution fall between the mean and one standard deviation above the mean. As the distribution is symmetrical, some two-thirds of the individuals fall between one standard deviation below the mean and one above it. Most (95%) of all individuals in the distribution fall between the standard deviations on either side of the mean.

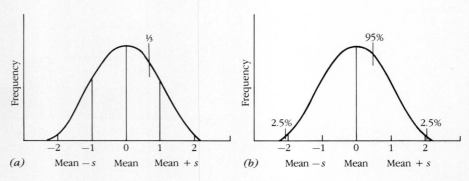

Figure 12–17
Fingerprint patterns.
(*a*) Arch with no triradius. With no line count, the TRC is zero. (*b*) Loop with one triradius. The TRC along the line from the triradius to the loop center is 13. (*c*) Whorl with two triradii. The TRC along the two lines is 17 and 8. (We use the higher count.)

(*a*)   (*b*)   (*c*)

Fingerprint ridges: Analysis of a polygenic trait

All of us are familiar enough with fingerprints to know that they are so specific in their pattern that no two people have the same pattern except identical twins, and even here there are some minor differences. Fingerprint patterns are genetically controlled and are formed early in embryonic development.

Fingerprints are classified by three main characteristics: (1) their pattern of **loops, whorls, and arches**; (2) their 10-digit **total ridge count (TRC)**; and (3) the number of their **triradii**, or points from which three ridge systems course in three directions at 120° angles. These characteristics are summarized in Figure 12–17. Notice that the arch pattern has no triradii; the loop has one, and the whorl has two. The TRC is a count of the ridges that lie along a line drawn from the core of the pattern to the triradial point. In the loop pattern shown, there are 13 ridges, and in the whorl pattern one ridge count has 17 ridges and the other has 8. To determine the TRC in whorl patterns, we use the higher ridge count. The average TRC (total for all fingers) in British males is 145, with a standard deviation of 51; for females it is 127, with a standard deviation of 52. This means that two-thirds of all the males are in the 94–196 range, and two-thirds of all females are in the 75–179 range.

To analyze the role that genes play in the determination of TRC, we compare the TRC in pairs of individuals with different degrees of genetic relationship between them. The **correlation coefficient** measures phenotypic similarity in continuously varying traits such as TRC. For a group of paired individuals, the correlation coefficient is equal to 1 if the traits are always equal or if the trait in one person varies in exact proportional to the same trait in the other person. It is zero if the traits vary independently in all pairs. Correlation coefficients are a measure of relatedness—genetic relatedness in our discussion. A parent and child have an expected correlation coefficient of 0.50 for genetically determined traits; siblings also have an expected genetic correlation coefficient of 0.50 because the members of each pair have 50% of their genes in common. Identical twins have an expected genetic correlation coefficient of 1.00.

S. B. Holt studied the correlation of TRC with genetic relatedness, and found that there was almost a perfect correlation (Table 12–4). A nearly perfect agreement between theoretical and observed TRC patterns indicates that the trait we call TRC is determined almost entirely by the genotype. Further, according to other studies, the genes apparently have an additive effect on the trait (dominant

Table 12–4   Correlations between relatives for total dermal ridge count

Relationship	Number of pairs	Observed Correlation Coefficient	Theoretical Correlation Coefficient Between Relatives Shown
Parent–child	810	0.48	0.50
Mother–child	405	0.48 ± 0.04	0.50
Father–child	405	0.49 ± 0.04	0.50
Father–mother	200	0.05 ± 0.07	0.00
Midparent–child	405	0.66 ± 0.03	0.71
Sibling–sibling	642	0.50 ± 0.04	0.50
Monozygotic twins	80	0.95 ± 0.01	1.00
Dizygotic twins	92	0.49 ± 0.08	0.50

Source: S. B. Holt, Quantitative genetics of fingerprint patterns, *Br. Med Bull.* 17:247–250.

genes do not seem to be involved). This analysis does not, however, allow us to estimate the number of genes that are involved in the TRC trait.

Fingerprint patterns are often diagnostic in cases of chromosomal anomalies. For example, in Turner syndrome (XO), the TRC tends to be much higher than normal. This may be because the XO embryo is edematous (swollen with fluids), even in the fingertip area. The expanded fingertips mean that more ridges form and the patterns are larger. Table 12–5 summarizes some of this information.

Use of twins in the analysis of multifactorial traits

The relationship among phenotype, genotype, and environment is not always obvious. Traits such as intelligence, personality, certain mental disorders, and some forms of cancer show various degrees of gene–environment interaction. For such traits, the precise effects of genes and environment on the phenotype are difficult to determine. Studies of twins have been used extensively to try to measure the relative contribution of each set of influences.

Francis Galton, in the latter part of the nineteenth century, was among the first to recognize the difference between identical and fraternal twins and to use twins in the analysis of multifactorial traits. He concluded that identical twins are **monozygotic (MZ)** (come from a single fertilized egg) and are therefore genetically identical. Fraternal twins are **dizygotic (DZ)** (come from two independently fertilized eggs) and are no more alike than nontwin siblings with respect to their genotypes (Figure 12–18).

Twin studies are one of our most valuable tools in analyzing the "nature versus nurture" problem, as Galton called it. Using MZ twins raised apart, we can study the effect of different environments on identical genotypes; using DZ twins, we can assess the consequences of identical environments on different genotypes. When we compare the expression of specific traits in MZ and DZ twins, we get an indication of the genetic contribution to a phenotype.

Table 12–5	Fingerprint patterns associated with various chromosomal anomalies	
Chromosome Abnormality	Fingers
5p⁻	Many arches; low TRC
Trisomy 13	Many arches; low TRC
Trisomy 18	6–10 arches (also on toes); very low TRC
Down syndrome	Many ulnar loops (usually 10); radial loop on fourth and/or fifth digits
45,X	Large loops or whorls; high TRC
XXY	Many arches; low TRC
XYY	Normal
Other syndromes with extra X and Y chromosomes	Many arches; low TRC; the more sex chromosomes, the lower the TRC

Complications of studies of twins   Galton said that any differences found to exist between MZ twins had to be the result of the environment because these twins were genetically identical. But we must exercise caution in analyzing data from MZ-twin studies. For one thing, to make sure that the twins are indeed MZ, we must do extensive testing of blood antigens and other genetic markers. In addition, it is common for twins to experience different environments in utero, so

Figure 12–18
Monozygotic twin pair.

Box 12–2   *Of Politics, Polygenes, and the Social Scene*

It seems well-nigh inevitable. Whenever social movements press for political and social equality for an underprivileged group, a spate of controversy quickly ensues over whether that group is really "biologically equal" to the perceived demands of the social positions to which they aspire. Data are produced in profusion on both sides of the issue; accusations and rebuttals proliferate. Data that seem to support the status quo are immediately viewed by some as having been contrived, or at least publicized, to serve the interests of those already "in power," and data that "demonstrate" apparent biological equality are challenged as biased in favor of the underprivileged group. The person attempting an objective analysis is often inundated by statistics and overwhelmed by the intricacies of experimental design, the nuances of logical inference, and highly emotional charges and countercharges.

So why is this matter appropriate for a genetics text? Because nearly always, the arguments over "biological equality" concern themselves with human characteristics and abilities which, if indeed they are genetically based, are polygenic (multigenic) in nature. From a purely social point of view, it may be that a solid understanding of polygenic systems has more relevance than does mastery of the simpler Mendelian systems previously discussed in this text. "Polygenic" describes those systems in which traits are controlled by many pairs of genes operating together, each with small effects. The more widely known systems—albinism, hemophilia, PKU, and so on, where a given trait is largely controlled by one or at most a very few pairs of genes—are sometimes designated as "oligogenic" ("few genes") or "Mendelian" for distinction.

But polygenic systems are unquestionably the more difficult to analyze. Not only are they very susceptible to phenotypic alterations by influences of the environment, but there is yet no rigorous definition of what "a" polygene *is;* the term really designates a phenomenon, a type of genetic control and behavior, but the concept itself fails to carry with it any real explanation of the general causal mechanisms involved.

Let us take, for example, the question of "equality" between men and women. The current popular literature contains arguments (variously backed by evidence) for a host of differences between males and females. Women are said to have a stronger immune system, greater sensitivity to taste, quicker responses to flashes of light, more acute hearing in the higher ranges, better night vision, and a better sense of touch. They are said to be more "helping," more nurturant, and more moved by emotion. They are supposedly more sensitive to "body language" and hence better judges of character. They are acclaimed to have better memories for faces, to be less aggressive and violent, less competitive, and to be superior in verbal skills to men. But we are also informed that women are inferior to men in terms of analytical skills, the ability to handle mathematics, and cognitive and spatial abilities. The latter deficiencies, we sometimes hear, explain the relatively low percentage of women in professions requiring mathematical and scientific expertise.

We do not yet know, really, whether the perceived differences between men and women are real. But any such differences are so susceptible to misinterpretation and political exploitation that we cannot avoid thinking through the implications they carry for the legitimate roles of the two sexes in our society.

Suppose, for example, that the mental differences often claimed turn out to be not real after all. Then we would have no justification at all for denying women equal opportunities and remunerations. But suppose the mental differences *are* real? Are they necessarily genetic in nature? We often hear that the mental differences are due to subtle environmental differences in the way we raise and educate children. If so, we may wish to seek ways to maximize or minimize the environmental effects as deemed desirable (we shall leave unexamined the question of who should decide what is desirable, and how such decisions should be implemented). Alternatively, we are often told that "hormones" may cause the differences. Hormones, of course, are produced and regulated by genes, even though a hormonal system may be altered by a variety of nongenetic factors. Then what happens to the ideal of equality if genetic differences in mental abilities separate men and women? When we examine the issue more thoroughly, using our understanding of polygenics, we see that the issue of "biological equality," like the issue of equality itself, is broader in scope than we first perceived it. *Are* all women different from *all* men in these characteristics?

The answer is no—not even in the data of the most ardent proponents. For example, it does appear that males as a group score slightly higher on standardized exams than do females. But it would be silly in the extreme to suggest that all males score better than do all females. Indeed, the available evidence suggests that males have a broader range of performance than do females: they not only score higher, they also score lower. Males generally score better than females on math tests only in average, or mean, scores. And there is far more difference between different males (or different females) than there is between the means of the two groups. In other words, the *inter*-group difference is far less than the *intra*-group differences.

In short, one cannot make any valid judgment at all about the mathematical ability of any given person just by knowing his or her gender. Any social policy or judgment that discriminated in any way between males and females and claimed differences in mathematical ability as its justification would seem to be on shaky grounds indeed.

The same kind of situation occurs, with some variations, for virtually all the other traits possibly based on polygenic differences—and especially for those dealing with mental or intellectual abilities. It appears, then, that regardless of whether the currently claimed differences in mental ability are real, or regardless of their causes *if* real, they cannot be used as the basis for discriminating public policy. Opportunities and rewards in society should be available on the basis of individual personal merit, and not on the mean performance of any particular group to which one may belong.

that development may proceed differently for each twin, although the resulting phenotypic differences are usually very subtle.

Mutations may also complicate the analysis of MZ twins. When the zygote first divides into two separate units, we assume that mitosis proceeds normally and that there are no mutations; the units should then be genetically identical and programmed to give rise to two identical individuals. But this assumption is not necessarily true. Some mitotic divisions may be inaccurate and result in aneuploid cells that in turn may produce a clone of cells that have lost or gained a chromosome. Somatic mutations may also occur independently in the twins, so that there are patches of tissue that are not genetically identical in the two. Thus, identical twins may be harboring subtle genetic differences that are not visible to the observer but are influencing the trait being studied.

Problems are also inherent in studies of DZ twins. Though these twins are presumed to arise from the independent fertilization of two eggs, produced by meiosis in two different oocytes, it is possible that an ovum might divide in two (mitotically) after meiosis to yield two genetically identical oocytes. Each egg would then be fertilized by a different sperm, but the maternal components of the zygotes would be the same. The resulting DZ twins would be genetically more different than MZ twins, but more alike than normal DZ twins. Though this kind of twinning is probably very rare and has never been proven, we must consider it a possibility when we label twins as DZ. These and other problems suggest that there are certain limitations to what we can know about the genetic makeup of twins.

Studies of twins may also make certain unwarranted assumptions about environmental constancy, or the lack of it. MZ twins, because of their inherent similarities, tend to seek out very similar environments. So we are not always justified in presuming that very different environmental influences are at work on twins raised apart. And even MZ twins reared together will share an environment more uniform than that of DZ twins reared together.

T. Bouchard and his colleagues at the University of Minnesota are currently engaged in a large-scale analysis of MZ twins who have been raised apart. The study, not yet complete, is showing some very surprising results. The twins are much more similar than expected. In one case, male twins were separated at four weeks of age and reunited when they were 39. Both turned out to be involved in police work, and both vacationed in Florida, drove Chevrolets, had dogs named Toy, married and divorced women named Linda, remarried women named Betty, and named their sons James Alan and James Allen. In school they were both good at math and poor in spelling. They both enjoy carpentry and mechanical drawing, have similar smoking and drinking habits, and bite their fingernails down to the quick; both put on 10 pounds at the same time in their lives. And this is an incomplete list of their similarities. The total picture of this and other twin pairs studied so far shows remarkable resemblances, far more than can be ascribed to coincidence (see handwriting samples from a twin pair in Figure 12–19). The most reasonable interpretation of all this is that identical genotypes tend to seek out similar environments—a situation that adds even more twists to studies of MZ twins raised apart.

DZ twins are genetically more different than MZ twins and are often even of the opposite sex, so they tend to seek out different environments. As DZ twins age, they grow more dissimilar, but as MZ twins age, they remain remarkably similar. All these contrasts tend to distort our interpretation of the data—we do not know to what extent older DZ twins differ because of inherent differences and to what extent their differences are due to social influences. To cut down some-

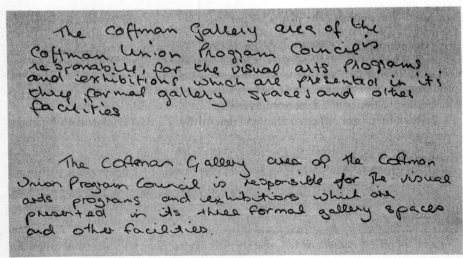

Figure 12–19
Handwriting for a twin pair.

what on these problems, we should compare MZ twins only with DZ twins of the same sex and age.

Despite all these problems, studies of twins are very valuable as research tools as long as we exercise caution in our reliance on the data we derive from them. When controlled breeding experiments are not possible, and they certainly are not in humans, studies of twins offer the best means of estimating how much of the variability observed between different persons is due to heredity and how much is due to environmental differences encountered during each person's development. These studies tell us something about the genetic predisposition to express a particular trait, though they do not tell us how many gene loci are involved, how the genes function, or what their transmission patterns are.

Concordance in monozygotic and dizygotic twins   An elegantly simple and effective way to assess differences between twins is to examine traits that are either present or absent in the members of the twin sets. If both members of a pair of twins express or fail to express a given trait or characteristic, they are **concordant** for that phenotype. If they differ in the phenotype, they are **discordant** for it. To assess the roles of heredity and environment, we figure out the degree of concordance and discordance in both MZ and DZ twins. For a monogenic dominant trait with complete penetrance, we expect the concordance to be 50% in DZ twins because each twin has a 50% chance of receiving the gene for it. We expect the concordance for the same trait to be 100% in MZ twins because each twin has the same gene. If the trait is recessive and both parents are carriers, the concordance for DZ twins is expected to be 25%; for MZ twins, the concordance is expected to be 100%.

With incomplete penetrance and polygenic inheritance, the analysis becomes more difficult. In looking at eye color, we find that 99% of all MZ twins are concordant—that is, both members of the pair have the same eye color. By contrast, 28% of DZ twins are concordant, and thus 72% are discordant—that is, the eye color of one twin is different from the eye color of the other. These data clearly indicate that the eye-color trait is determined largely by the genotype. Measles, on the other hand, has a concordance rate of 95% in MZ twins and 87% in DZ twins. This is not surprising, since measles is caused by an environmental agent, a virus. The slight difference in concordance and discordance for measles may indicate an inherited predisposition for measles, such as cell membrane attachment site proteins, but such a predisposition, if it exists, is slight. Concordance and discordance values for some selected traits are summarized in Figure 12–20.

Effects of Genotype and Environment on Quantitative Traits

For well over 100 years the debate over nature versus nurture for various human traits has been raging, and it still shows no signs of letting up. The debate is really about the relative influence of genes and environment on particular traits. Emotions can often run high when it comes to discussing phenotypes such as intelligence, a trait we shall examine in the next chapter in more detail.

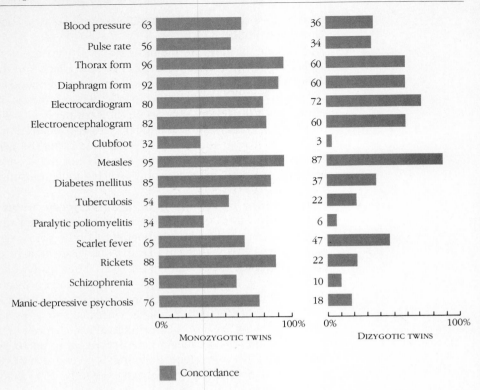

Figure 12–20
Concordance and discord-
ance in twins for a variety
of traits.

A population of schoolchildren raised in an academically enriched environment will score higher on IQ tests than a population of children who are educationally deprived. We obviously cannot conclude from this that the well-educated group is genetically more intelligent than the poorly educated group. The environments here have had a decided impact. Yet intelligence, like many other traits, is influenced by genes as well as the environment. Realizing this, can we determine the extent to which a given trait is determined by genes and the extent to which it is determined by the environment?

This question is, in fact, poorly phrased. I purposely phrased it this way because it reflects a common inaccuracy. No trait develops in a vacuum. Genes *and* environment interact continuously to produce phenotypes. Genes do not act independently of the environment. A better and more accurate way to phrase the question is as follows:

> To what extent is the variation we observe in a population due to the genetic differences among individuals, and to what extent is it due to environmental differences?

When we set out to study the "nature versus nurture" question, we must phrase it in this fashion.

Perhaps the single most important concept employed to study the nature–nurture issue is **heritability**. This concept gives us a quantitative estimate of the fraction of the phenotypic variation that is the result of genetic differences among individuals. To obtain an estimate of heritability, we use the following symbolism:

H = heritability
V_T = the total phenotypic variance observed in a trait
V_G = the fraction of the phenotypic variance due to genetic differences among individuals
V_E = the fraction of the phenotypic variance that is due to differences in environmental conditions to which the individuals were exposed

The relationship between these components is

$$V_T = V_G + V_E$$

and

$$H = \frac{V_G}{V_G + V_E} = \frac{V_G}{V_T}$$

Finding V_T is relatively easy. We use the same procedure for obtaining variance that we discussed earlier. But finding values for V_G and V_E is difficult, especially under conditions where controlled breeding experiments are not possible.

One technique used to estimate heritability for human traits is called the twin method. This method takes advantage of total phenotypic variance in groups with known degrees of genetic relatedness. We can show how it works in the following way:

$V_{T(MZ)}$ = total phenotypic variance for a trait among identical twins
$V_{T(DZ)}$ = total phenotypic variance for a trait among nonidentical twins of the same sex

Since MZ twins are genetically identical, we can assume that any differences between them are the result of environmental differences. Thus,

$$V_{T(MZ)} = V_E$$

DZ twins have half their genes in common. The phenotypic variance we observe in this case is the result of both genetic and environmental differences; but since they have half their genes in common, the total genetic variance (V_G) will be half that of unrelated individuals:

$$V_{T(DZ)} = \tfrac{1}{2} V_G + V_E$$

By finding the difference between $V_{T(MZ)}$ and $V_{T(DZ)}$, we are able to estimate, for a given trait, half of the total V_G:

$$V_{T(DZ)} - V_{T(MZ)} = \tfrac{1}{2}V_G + V_E - V_E = \tfrac{1}{2}V_G$$

Table 12–6	Heritabilities for certain traits in humans (estimated by studies of twins)	
Trait		**Heritability**
Stature		0.81
Sitting height		0.76
Weight		0.78
Cephalic index		0.75
Binet mental age		0.65
Binet IQ		0.68
Otis IQ		0.80
Verbal aptitude		0.68
Arithmetic aptitude		0.12
Aptitude for science		0.34
Aptitude for history and literature		0.45
Spelling ability		0.53
Foot-tapping speed		0.50

Two times this value gives us the V_G component. Once we have calculated V_T, we can determine heritability:

$$H = \frac{V_G}{V_T}$$

In calculations of this type, we need to make some simplifying assumptions that may or may not be warranted. For example, we assume that V_E is the same for both MZ and DZ twins, but it may not be true. The environment for MZ twins is probably more similar than that for DZ twins. If so, the V_E for DZ twins is greater than the V_E for MZ twins. This difference will make the V_G value in the preceding formula larger than it should be and lead to an overestimate of H.

The twin method for calculating heritability requires knowing the total phenotypic variability of the trait in question. But say we know only the variance between twins. We can still estimate heritability by using the following formula:

$$H = \frac{V_{DZ} - V_{MZ}}{V_{DZ}}$$

For example, as shown in Table 12–7, we determine the IQ differences for each twin member among 10 pairs of DZ and 10 pairs of MZ twins, and we find the variance to be 6.10 for the former and 3.84 for the latter. From these data, we are able to estimate the heritability:

$$H = \frac{6.10 - 3.84}{6.10} = 0.37$$

Table 12–6 presents a summary of heritability estimates for selected human traits. There are, of course, some problems that make accurate estimations of

Pair	MZ Twins (Same Sex; Raised Together)	DZ Twins (Same Sex; Raised Together)
Table 12–7	**Number of IQ differences for each twin member among pairs of MZ and DZ twins**	
1	3	13
2	6	5
3	6	9
4	3	8
5	5	9
6	1	12
7	7	13
8	6	11
9	3	10
10	6	9
Mean	4.6	9.9
Variance	3.84	6.10

heritability very difficult, but these extend beyond the scope of this book. In Chapter 13, we will examine some of the pitfalls in the use of heritability for the IQ phenotype.

Summary

1 Phenotypes result from the interaction of gene products with one another and with the environment. Genotypes, not traits, are inherited.

2 **Polygenic inheritance** refers to traits that are determined by several gene loci, each contributing additively to the final phenotype. Such traits are usually influenced by environmental factors. **Multifactorial inheritance** is nearly the same as polygenic inheritance, but it emphasizes the role of environmental components and does not assume gene additivity.

3 A **threshold trait** is a polygenic trait that is either expressed or not expressed. A **quantitative trait** is one that displays continuous variation.

4 **Modifier genes** modify the expression of other, nonallelic, genes. **Epistatic genes** completely mask the expression of other, nonallelic, genes.

5 The **norm of reaction** is the range of phenotypes that can be expressed by a specific genotype in a variety of environments.

6 The environment can be so manipulated that a nonmutant genotype expresses a mutant phenotype, or a mutant genotype expresses a nonmutant phenotype. An environmentally induced abnormal phenotype that resembles a phenotype due to gene mutation is called a **phenocopy**.

7 In certain forms of cancer, we can give a quantitative estimate of the role of the genotype and the role of the environment in producing the phenotype.

8 The **law of ancestral inheritance** represents an unsuccessful early attempt to explain continuous patterns of variation. This theory gave way to the multifacto-

rial and polygenic models of inheritance in both of which several Mendelian units contribute to the expression of one trait.

9 Selection experiments with the fava bean showed that continuous traits have both a genetic and an environmental component.

10 **Transgressive variation** means that offspring are phenotypically more extreme than their ancestors. **Regressive variation**, which is much more common, means that offspring approach the populational mean for a trait more closely than did their parents. Both phenomena are explained by the independent assortment of polygenes.

11 Quantitative traits are statistically analyzed and described in terms of the **mean, median, mode, variance**, and **standard deviation** for a population. Mean, median, and mode describe average or typical values; variance and standard deviation describe the distribution of values within the population.

12 Most continuous traits have a **normal distribution**, or **Gaussian distribution**, in which the distribution of values is symmetrical around the mean.

13 Fingerprint-ridge analysis allows us to apply the tools of quantitative analysis to a continuously varying trait.

14 Studies of twins have provided us with valuable information on quantitative traits. By comparing **monozygotic twins** raised separately with monozygotic twins raised together and with **dizygotic twins**, we obtain concordance values that indicate how much the genotype contributes to the phenotype for a given trait.

15 **Heritability** is a measure of the degree to which phenotypic variation is genetically determined and, thus, the degree to which it can be changed by selection. The value of *H* (heritability) for a given trait depends on the specific population and the specific environment in which the variation occurs.

Key Terms and Concepts

Monogenic trait

Polygenic trait

Multifactorial inheritance

Polygenic inheritance

Threshold trait

Modifier gene

Epistatic gene

Epistatis

Normal of reaction

Phenocopy

Phocomelia

Blending model of inheritance

Law of ancestral inheritance

Transgressive variation

Regressive variation (regression toward the mean)

Random mating

Assortative mating

Positive assortative mating

Negative assortative mating

Threshold solution

Statistics

Parameter

Mean

Median

Frequency distribution

Modal class (mode)

Variance

Standard deviation

Normal distribution (Gaussian distribution)

Loops, whorls, and arches

Total ridge count

Triradii

Correlation coefficient

Monozygotic twins

Dizygotic twins

Concordant

Discordant

Heritability

Problems

12–1 Distinguish between the terms *epistasis* and *dominance*.

12–2 How would you distinguish between a phenotype controlled by multifactorial inheritance and one controlled by a single gene pair with variable expressivity?

12–3 In a classroom, the following distribution of grades was obtained in two sections:

Section 1		*Section 2*	
Range	Frequency	Range	Frequency
80–89	4	90–99	2
70–79	10	80–89	6
60–69	18	70–79	12
59–59	9	60–69	20
40–49	5	50–59	10
30–39	2	40–49	2

(a) Estimate the means of sections 1, 2, and the entire class. (Use midpoints in each range as values.)

(b) Estimate the variance and standard deviation of the entire group.

12–4 Nilsson-Ehle studied the inheritance of kernel color in wheat. In one cross, he crossed a true-breeding red variety to a true-breeding white variety. The F_1 was red, as was the F_2. When he self-fertilized the 78 F_2 plants, he obtained the following results:

Number of F_2 Plants	Progeny from Self-Fertilized Plants
50	All red
15	15 red:1 white
8	3 red:1 white
5	63 red:1 white
78	

(a) How would you interpret the data with respect to the number of gene loci involved in this trait?

(b) What are the parental and F_1 genotypes?

12–5 The following table gives data on the variances of two phenotypic traits in sparrows (wing span and beak length):

Wing Span		Beak Length	
V_T	271.4	V_T	627.8
V_E	71.2	V_E	107.3

Calculate the heritability for each trait.

12–6 A certain trait is determined by five unlinked pairs of alleles, with each dominant allele having an additive effect on the phenotype. What phenotypic classes would you predict, and in what frequencies, from a cross between two people who are heterozygous for all five loci?

12–7 In the following relationships between a person and others in his family, estimate the proportion of genes they have in common:

person
monozygotic twin
parent
child
first cousin
great-grandparent
uncle

12–8 On the assumption that a trait is determined by a polygenic series with additive effects, how would you explain regressive variation?

12–9 A trait is determined by alleles at three independently assorting gene loci. Assuming that two alleles exist for each locus (*Aa*, *Bb*, *Cc*), and that uppercase alleles have an additive effect, how many possible genotypes can have four uppercase letters and what are they?

12–10 In one family line, the trait deaf-mutism appears to be inherited as an autosomal recessive trait. Normal parents produce, on occasion, a deaf-mute child. Marriages between two deaf-mutes produce all deaf-mute children. However, in one marriage between two unrelated deaf-mutes, several children were produced, all of whom were normal. Assuming that all the children were legitimate, how would you explain this?

12–11 Tuberculosis is a debilitating respiratory disease caused by a bacterium. In studies of concordance for this disease, we find that concordance is much higher for MZ twins than it is for DZ twins. How would you explain this difference in concordance values for a trait that is caused by a bacterial infection, not genes?

12–12 What are the subtle distinctions between multifactorial inheritance and polygenic inheritance?

12–13 Explain how it is possible for two brown-eyed parents to produce a blue-eyed child.

12–14 What genetic model might account for the continuous pattern of variation in the behavioral disorder schizophrenia?

12–15 Explain what a phenocopy is and give three examples of its occurrence.

Answers

See Appendix A.

References

Ayala, F. J., and J. A. Kiger. 1980. *Modern Genetics.* Benjamin/Cummings, Menlo Park, CA.

Bulmer, M. G. 1970. *The Biology of Twinning in Man.* Oxford University Press, Oxford.

Cavalli-Sforza, L., and W. F. Bodmer. 1971. *The Genetics of Human Populations.* Freeman, San Francisco.

Falconer, D. S. 1960. *Introduction to Quantitative Genetics.* Oliver and Boyd, Edinburgh.

Fraser, F. C. 1976. The multifactorial/threshold concept: Uses and misuses. *Teratology* 14:267–280.

Holt, S. B. 1961. Quantitative genetics of fingerprint patterns. *Br. Med. Bull.* 17:247–250.

Hrubec, Z. 1973. The effect of diagnostic ascertainment in twins on the assessment of the genetic factor in disease etiology. *Am. J. Hum. Genet.* 25:15–28.

Jenkins, J. B. 1979. *Genetics,* 2nd ed. Houghton Mifflin, Boston.

MacGillivray, I., P. P. S. Nylander, and G. Corney. 1975. *Human Multiple Reproduction.* Saunders, Philadelphia.

Mather, L., and J. L. Jinks. 1975. *Introduction to Biometrical Genetics.* Cornell University Press, Ithaca, New York.

Newman, H. H., F. N. Freeman, and J. K. Holzinger. 1937. *Twins: A Study of Heredity and Environment.* University of Chicago Press, Chicago.

Penrose, L. S. 1969. Dermatoglyphics. *Sci. Am.* 221:72–85 (July).

Stern, C. 1970. Model estimates of the number of gene pairs involved in pigmentation variability of the Negro American. *Hum. Hered.* 20:165–168.

13 The Genetics of Human Behavior

To many people, scientists and nonscientists alike, the genetics of human behavior is the most important and interesting aspect of human biology. It is hardly a coincidence that no other aspect of biology prompts such emotional reactions: any theory of human behavior has social and political implications, and these are the reasons for our passionate interest and partisanship. More than a century after Galton first raised the question of nature versus nurture, controversy still rages over the respective contributions of the genes and the environment.

Some nonscientists, believing that a person can become whatever his or her environment permits, refuse to accept the current idea that many behavioral traits are genetically influenced—sometimes very strongly so. Others believe that genes are almost solely responsible for such mental qualities as intelligence, personality, mental stability, and certain patterns of social behavior. Few, if any, modern scientists accept either of these extreme positions; nearly all would agree with the statement that human behavior is influenced by the environment as well as the genotype. Nevertheless, scientists continue to argue furiously about the extent of the genetic contribution and the nature of the interactions between genes and environment. The nature–nurture debate will not be resolved for a very long time, if, indeed, it ever can be.

The genetics of human behavior suffers from experimental hardships. We can grow human cells in culture and study their chromosomal structure and gene products by using microscopic and biochemical techniques, but such methods are of no help in analyzing behavioral phenomena unless a behavior is the result of a specific, detectable chromosomal or protein defect. Known instances of such behaviors are rare, yet we must be able ultimately to associate specific gene functions with specific behaviors. We must also be able to resolve a behavior pattern into its genetic and environmental components, and to do so without any bias. This last requirement—looking at ourselves in the same detached way that we look at bacteria or laboratory mice—is the most difficult of all to fulfill. Many who have reflected on this problem believe that no one can have an unbiased approach to human behavior.

Genes Influence Behavior

Behavior is a specific phenotype that refers to the actions or reactions of organisms (including humans) under specified circumstances. In this chapter we will discuss various forms of human and nonhuman behavior from a genetic perspective, including pathologies such as schizophrenia and manic-depressive psychosis as well as normal behaviors such as intelligence.

Many proteins are involved in the structure and functioning of the nervous system. When any one of them is altered, there is every reason to expect some alteration in behavior. Thus, genes, through their products, have the capacity to influence a behavioral trait. Proteins, of course, do not cause a complex specific behavior pattern. That is, there is no gene product that causes a person to hoist glasses of whiskey repeatedly and by so doing become an alcoholic. There are, however, genes that function in the metabolism of ethanol, and some of these genes may produce proteins that make a person more sensitive to ethanol. This sensitivity, coupled with certain personality types and certain environmental stresses, may produce alcoholism.

We have already presented many examples of single gene mutations that influence behavior. Recall, for instance, that Huntington's disease, an autosomal dominant trait, causes a slow disintegration of the nervous system's functioning and, consequently, a variety of bizarre behaviors. Lesch–Nyhan syndrome, a sex-linked recessive disorder, is characterized by hyperexcitability of the nervous system, causing nearly continuous movements and a compulsive tendency to self-mutilation. This behavioral anomaly is associated with an enzyme defect, as discussed in Chapter 4. Phenylketonuria is an autosomal recessive trait that also results from an enzyme defect and that is characterized by severe mental dysfunction. There are literally hundreds of single gene loci that in one way or another influence human behavior.

Clearly, genes influence behavior and in some cases control it. But genes do not act in isolation: the environment can modify the activity of the genotype (Figure 13–1). What makes behavior so difficult to study, in fact, is the prolonged interaction between the genotype and the environment. In simple genetic anomalies such as polydactyly, there is a specific developmental stage that is most sensitive to the forces of the environment; in behavior, *all* developmental stages are sensitive, both prenatal and postnatal. But as complex as the genotype–environmental interactions are, they can sometimes be untangled, and we will look at some examples of how this is done.

Genetics of Behavior in Some Nonhuman Animals

Because behavior is so difficult to study in humans, geneticists interested in the genetics of human behavior employ a strategy used commonly in genetics research: they analyze inherited behavior patterns in simpler, nonhuman systems. We will take somewhat the same route by considering first some examples of such studies.

Hygienic behavior in bees

Like humans, honeybees are susceptible to bacterial infections. A bacillus bacterium infects bee larvae as they develop inside their cell and then kills them. This

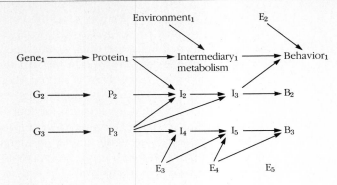

Figure 13–1
The genotype influences behavior indirectly by influencing the physiological system.

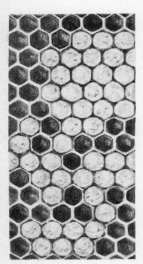

Figure 13–2
Honeybee comb. The empty cells have been uncapped and infected larvae have been cleaned out by *uurr* (hygienic) bees.

disease is known as American foulbrood. Bees of certain strains are able to combat this infection by **hygienic**, or **nest-cleaning behavior**, which is characterized by two procedures: uncapping the cells of infected larvae and then cleaning them out (Figure 13–2). Bees of other strains are unable to rid their hive of the infected larvae.

We find that this hygienic behavior is controlled by two pairs of independently assorting genes. When bees from a hygienic strain are crossed with those of a nonhygienic strain, the F_1 bees are unable either to uncap the cells containing infected larvae or to remove infected larvae from their cells if they are artificially uncapped. When the F_1 strain is crossed back to the hygienic strain, we are able to see the components of the nest-cleaning behavior sorting out:

$\frac{1}{4}$ of the bees do not express any hygienic behavior
$\frac{1}{4}$ uncap the cells of infected larvae but do not remove them
$\frac{1}{4}$ remove the infected larvae, but only after the cells have been first artificially uncapped
$\frac{1}{4}$ express the complete hygienic behavior pattern

We can interpret this remarkable partitioning of behavior in the following manner:

U = no uncapping
u = uncapping
R = no removal of larvae
r = removal of larvae

Nonhygienic *UURR* × *uurr* Hygienic

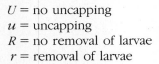

UuRr × *uurr* (backcross)

$\frac{1}{4}$ *UuRr* nonhygienic
$\frac{1}{4}$ *Uurr* no uncapping; removal of larvae
$\frac{1}{4}$ *uuRr* uncapping; no removal of larvae
$\frac{1}{4}$ *uurr* complete hygienic

It is not known how the gene products, so far unidentified, control the two behavioral components. Perhaps there is an odor in infected cells that only hygienic bees can recognize, or perhaps the color of infected cells is different and is perceived only by hygienic bees.

Genetic dissection of Drosophila behavior

The fruit fly, *Drosophila*, which has been so instrumental in the more classic areas of biology and genetics, is proving to be a marvelous organism to use in studies of behavior. Among its genetically controlled behaviors are its courtship ritual, its movement toward light (positive phototropism), and its movement away from the center of gravity (negative geotropism). A wide variety of behavioral mutants have been isolated in *Drosophila*. Among these are the "drop-dead" type, which lives normally for a couple of days, then just drops dead; the hyperkinetic type, which has nervous, rapid, and jerky movements of the legs when anesthetized; and the fruity male, which courts other males and does not copulate with females.

Special techniques have been developed for studying behavior in *Drosophila*, some of which may be applicable to humans. One of the most interesting is the construction of a **genetic mosaic**. The mosaic, a type of organism we discussed in Chapter 6, is used here for X-linked behavioral genes (Figure 13–3). The mosaic flies have XO cells along with XX cells in the same body. We create this fly by using females who carry an unstable X chromosome that has been broken and formed into a ring, and that tends to get lost as cells divide. When the chromosome is lost, the cell and all its descendants are XO.

Figure 13–3
The genetic mosaic.
(*a*) The loss of the ring X to form a mosaic organism.
(*b*) The distribution pattern of normal and mutant cells depends on the plane of cleavage at the first division.
(*c*) Different cleavage planes and their consequences.

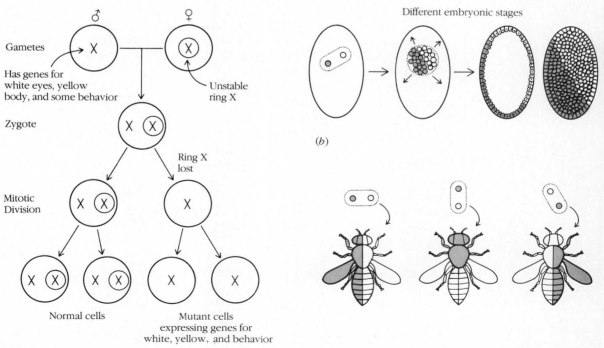

Figure 13–4
The types of genetic mosaics used in the analysis of the hyperkinetic mutant (*Hk*). S = strong shaking effect; black = XX tissue, heterozygous for *Hk*; white = XO tissue, hemizygous for *Hk*.

A specially constructed genetic mosaic is used to reveal which parts of a fly's body control mating behavior. For example, if a fly's head is all XO (which in flies determines maleness) and the rest of the body is XX (normal female), it behaves as a normal male by chasing females around the bottle. We can tell which body parts are XO male and which are XX female because the normal chromosome carries recessive genes for different body and eye colors (Figure 13–3). When both X chromosomes are present, the unstable X carries dominant alleles for body and eye color and the cell line is phenotypically normal; if the unstable X is absent, the cell line expresses the recessive alleles for body and eye color.

The dominant mutation called hyperkinetic (*Hk*) has been located to the X chromosome. When females are homozygous and males are hemizygous for this mutant allele, they shake their legs arrhythmically when anesthetized with ether. Heterozygous females do the same thing but do not shake their legs as vigorously. The *Hk* mutants show another strange behavior: when an object moves near them—say, a hand waves over the vial housing them—they jump and fall over. Using different kinds of mosaics (Figure 13–4), we find that each leg behaves autonomously and that the focal point for the mutant defect lies in clusters of nerves in the insect's chest region. Thus, if the tissue containing a nerve cluster is mutant ($X^{Hk}O$), the leg controlled by that nerve cluster will shake. There are six legs and thus six focal points.

In addition to *Hk*, we have discovered other movement aberrancies in *Drosophila*. Shaker (*Sk*) mutants are similar to *Hk*, but their shaking pattern is quite different. A paralyzed (*para*) mutant moves normally at 22°C, but when the temperature is raised to 29°C, it is paralyzed. The paralysis disappears when the temperature is lowered to 22°C. There is some hope that studies of these behavioral mutants will one day shed light on some human pathological conditions.

The construction of genetic mosaics, especially mosaics that involve the brain, has enabled us to locate with relative precision the somatic sources of behavior in the fly's body. The mosaic is really a mix of normal and abnormal parts that can help us better understand behavior genetics.

Genetically influenced behavior in mice

The mouse has been and continues to be a favorite animal for studying the genetics of behavior. Mice can be systematically and rapidly bred with adequate control matings; and, because they are mammals with nervous systems similar to ours, it is reasonable to cautiously extrapolate the results of mouse experiments to humans.

Neurological disturbances Many mouse behavior patterns are controlled by a single gene locus. Classic examples are the behavioral anomalies known as "waltzing" and "twirling." The so-called waltzing mice, or waltzers, far from being graceful on their feet, shake their heads and run in tight circles. They are also very irritable. The inheritance pattern for the waltzing trait suggests that it is an autosomal recessive trait:

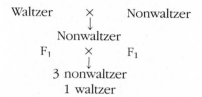

$$\text{Waltzer} \quad \times \quad \text{Nonwaltzer}$$
$$\downarrow$$
$$\text{Nonwaltzer}$$
$$F_1 \quad \times \quad F_1$$
$$\downarrow$$
$$\text{3 nonwaltzer}$$
$$\text{1 waltzer}$$

The twirling trait is similar to the waltzing trait in that the mouse is irritable, shakes its head, and runs around in circles. However, twirling seems to be the result of a dominant allele (T) that is lethal when homozygous:

$$\text{Twirler } (Tt) \times \text{Twirler } (Tt)$$
$$\downarrow$$
1 dead within 24 hours of birth (TT)
2 twirlers (Tt)
1 normal (tt)

In twirlers and waltzers we can point to abnormalities in brain structure, though we are unable to pinpoint a defective gene product.

Obesity Obesity is a problem for some mice as well as for some people. Our knowledge of one cause for obesity in mice unfortunately offers little comfort to humans. In mice we have discovered an "obese" gene (ob), that functions as an autosomal recessive. Mice that are ob/ob have a ravenous appetite, get quite fat, and become relatively inactive. The cause of the overeating may be a genetically controlled defect in the hypothalamic region of the brain that controls appetite satisfaction. A similar situation may well exist in humans, especially in cases where obesity runs in families. That is, the genotype may control obesity in humans as it occasionally does in mice.

Alcoholism Many behavioral patterns in mice appear to be controlled by polygenic inheritance. These include seizures at the sound of loud noises, running speed, learning ability, alcohol preference, and emotionality. We will look at alcohol preference as an example of a complex behavior that is influenced by several gene loci. Alcohol preference in mice is genetically controlled, though we have not yet identified specific genes. Our research into human alcoholism also indicates a genetic influence.

Different strains of mice differ in their preference for alcohol (Figure 13–5). When mice are given their free choice of water or 10% alcohol (about the same

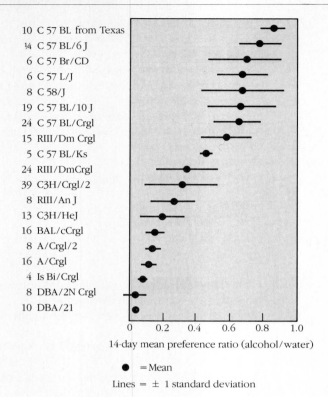

10	C 57 BL from Texas
¼	C 57 BL/6 J
6	C 57 Br/CD
6	C 57 L/J
8	C 58/J
19	C 57 BL/10 J
24	C 57 BL/Crgl
15	RIII/Dm Crgl
5	C 57 BL/Ks
24	RIII/DmCrgl
39	C3H/Crgl/2
8	RIII/An J
13	C3H/HeJ
16	BAL/cCrgl
8	A/Crgl/2
16	A/Crgl
4	Is Bi/Crgl
8	DBA/2N Crgl
10	DBA/21

14-day mean preference ratio (alcohol/water)

● = Mean

Lines = ± 1 standard deviation

Figure 13–5
Differences in alcohol preference in some inbred mouse strains.

alcoholic content as wine), at one extreme are strains that consume over 80% of their daily fluid from the alcohol solution, and at the other extreme are strains that avoid the alcohol nearly all the time.

Not only are there strain-specific preferences for alcohol, but there are also strain-specific differences in the effects of alcohol. In general, strains with the highest alcohol preference are the least susceptible to alcohol's effects. For example, alcohol-induced sleep is shortest in strains with the highest alcohol preference. Apparently the strain differences are due to both the brain's susceptibility to alcohol and the liver's ability to metabolize the alcohol via the enzyme called ethanol dehydrogenase. These findings have clear implications for human alcohol preferences. In humans, alcoholism is probably the result of genetic peculiarities in the metabolic functions of the brain and liver, coupled with the individual's response to environmental stresses.

Discontinuous Traits in Humans

Human-behavior genetics is certainly the hardest area of human genetics to study properly. It is made difficult by the complexity of the behavioral phenotype, which is often hard to define; by the complexity of the genotype–environment interaction; by our inability to perform controlled crosses while controlling the environment; and by the absence of highly inbred lines. Nevertheless, there are adventurous souls who relish the challenges of a difficult problem and tackle it enthusiastically. Thanks to the untiring efforts of such people, we are beginning to have a much deeper understanding of the relationships between the genotype and the behavioral phenotype. We will first examine discontinuous, or threshold, behavioral traits—traits that are either present or absent, but may have a continuously varying pattern of susceptibility.

Schizophrenia

No behavior disorder is more important or more difficult to analyze than **schizophrenia**, in which the affected person has trouble distinguishing between reality and his or her imaginings. Part of the problem for geneticists is that schizophrenia (literally, split personality) is really an array of behavioral aberrancies that may vary extensively. The schizophrenic disorder is in general characterized by

1 Delusions
2 Bizarre, illogical responses to stimuli
3 Hallucinations
4 Loss of interest in life's daily activities
5 Loss of desire for normal activity
6 Loss of capacity to experience even the simplest of normal pleasures

Some or all of these characteristics may appear in a patient at any one time or over a period of time. We now estimate that nearly 2% of all hospitalized patients are schizophrenics. About 1% of the U.S. population either has had or currently has schizophrenia.

There is clearly a genetic basis for schizophrenia. Over five decades of studying concordance rates in monozygotic and dizygotic twins has given us convincing evidence that the genotype has a critical role to play in the development of this disorder. Thirteen studies carried out between 1928 and 1969 indicated the average concordance values to be 52% (in a 15–76% range) for 571 pairs of MZ twins, and 10% (in a 0–17% range) for 1281 pairs of DZ twins. These studies also indicated that while genetic factors are crucial, even essential, to the development of schizophrenia, the environment is also important. In other words, a person who has a genotype that predisposes him or her to develop schizophrenia may not express the disorder unless certain environmental factors are present.

The mode of inheritance of schizophrenia is a matter of speculation. One model suggests the disorder is caused by a single autosomal dominant gene with nearly complete penetrance. This idea is supported by data showing a correlation between the degree of genetic relatedness and the proportion of relatives affected with schizophrenia or schizoid behavior (Figure 13–6). Schizoid behavior, or preschizophrenia, is characterized by withdrawn and frightened behavior, undeveloped defenses, a sense of worthlessness, and babylike actions. As seen from the figure, the proportion of persons with schizophrenia or schizoid behavior is about what we would expect if the disorder were caused by a single dominant gene interacting with the environment.

A second model views schizophrenia as the result of polygenic inheritance. According to this model, a person inherits genes that define a predisposition to the disorder, but the disorder will develop only if precipitating environmental stresses are present. It is proposed that several genes affect various aspects of

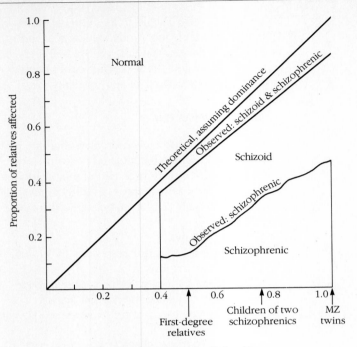

Figure 13–6
Observed and expected frequencies of schizophrenia based on the theory that the trait is caused by a single dominant gene. Examples of first-degree relatives are parents, their children, and siblings ($Aa \times aa \rightarrow \frac{1}{2}Aa, \frac{1}{2}aa$). They share 50% of their genes. Two schizophrenics (Aa) would be expected to produce offspring of which 75% would carry *at least* one dominant allele for schizophrenia.

intermediary metabolism. There can be wide variation in the metabolic effects for two reasons: different genes affect different parts of the metabolic network, and multiple alleles may affect a single pathway differentially. The potential for interaction among different gene loci complicates matters even more.

A third model for schizophrenia takes into account the possible relationships among gene loci as well as those between the genotype and environment (Figure 13–7). The genotype generates metabolites that, if their levels are high enough, can induce schizophrenic behavior. According to the model, environmental stresses cause the level of steroid hormones to rise, and steroid hormones are known to be inducers of gene activity. The schizophrenic genotype may be more susceptible to the gene-activating effects of steroids, or it may cause a person to be more susceptible to stress and thus to produce more steroids, or both. The increased activity of certain genes may raise the levels of enzymes responsible for the formation of chemicals such as dimethyltryptamine (DMT) and mescaline. The body normally produces both these substances, but at very low levels. At higher levels, they are powerful hallucinogenic agents, capable of inducing schizophrenic symptoms. Increased gene activity may also raise the level of the neurotransmitter known as dopamine. This elevated level may result in a hyperaroused, confused nervous state, which may be the cause of some schizophrenic symptoms. Monoamine oxidase (MAO), which normally inactivates neurotransmitters such as dopamine, may be mutant, leading to increased levels of neurotransmitters and therefore to schizophrenic symptoms.

This third model helps explain the great variability we find in the symptoms of schizophrenia, for it suggests that there is variability both in the genotypes

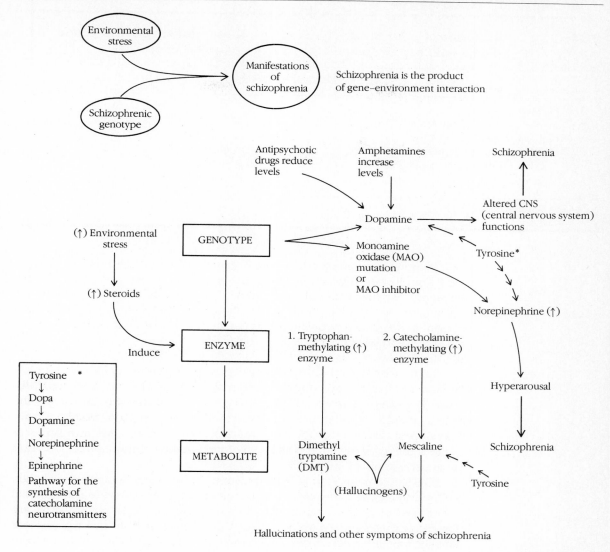

Figure 13–7
Model for the causes of schizophrenia.

leading to the schizophrenia phenotype and in the environmental stresses that may precipitate the disorder.

Manic-depressive psychosis

The puzzling mental disorder known as **manic-depressive psychosis** involves extreme behavior anomalies that come and go, usually in regular cycles. The extremes range from mania (exaggerated elation) to deep depression. Manic-depressives are not usually hospitalized because they generally recover from

Table 13–1	**Risk estimates for manic-depressive psychosis and schizophrenia for the general population and for relatives of manic-depressives**	
	Manic-Depressive (%)	Schizophrenic (%)
General population	0.7	0.9
Parents of manic-depressives	7.6	0.6
Siblings of manic-depressives	8.8	0.8
Children of manic-depressives	11.2	2.5
Children of two manic-depressive parents	33.0	—

Source: After Rosenthal, 1970.

their extreme sways in mood. However, suicide is common in those who experience a prolonged depression phase. In the general population, the estimated frequency of manic-depressive psychosis is in the broad range of 0.07–7%, depending on how narrowly one defines the illness. The median risk factor for a person in the general population is about 1%. The genotype is clearly important in manic-depressive psychosis. The concordance rate is 76% for MZ twins and 18% for DZ twins.

Some have suggested that manic-depressive psychosis grades into schizophrenia and that the two disorders are really variants of the same general disorder. This does not appear to be true. The frequency of the two disorders in the general population and in relatives of manic-depressives suggests that the two are independent (Table 13–1). Schizophrenia occurs no more often in relatives of manic-depressives (with the possible exception of their children) than it does in the general population.

There is some indication from pedigree studies that a dominant, sex-linked gene, situated close to the genes for red–green color blindness (deutan) and red color blindness (protan) may be a causative factor in manic-depressive psychosis. However, this evidence must be weighed in the light of other observations that show the trait transmitted from father to son. A recent study by L. R. Weitkamp and his colleagues identified a gene that caused the state of depression to develop. This gene is reportedly linked to HLA genes (discussed in Chapter 11) on chromosome 6, but this conclusion is debatable. The gene's function is not known, and its exact location on the chromosome is being investigated. It appears that the genetic basis for this disorder is more complex than a sex-linked dominant gene.

Manic-depressive psychosis is one of two main types of cyclic depressive psychoses, and it is commonly referred to as *bipolar* because of its two extremes. The other type, called **unipolar depression**, involves cyclical swings between normality and depression, with no swing into mania. More and more data suggest that the two forms are independently inherited. Unipolar depressives tend to have unipolar relatives, and bipolar depressives tend to have bipolar relatives. But these relationships do not always hold up. Although unipolar persons rarely have bipolar relatives, bipolar persons sometimes have unipolar relatives. So perhaps some segregation of genetic factors accounts for the two forms of the disorder.

The unipolar and bipolar forms are symptomatically and pharmacologically distinct from each other, but we are unable to describe the genetic basis for either type, except to say that they are to a large extent genetically controlled.

Continuous Traits in Humans

Manic-depressive psychosis and schizophrenia are both discontinuous, or threshold, traits. The tendency to develop these disorders is perhaps controlled by polygenic inheritance, but the disorders themselves are either present or absent. The traits we discuss next are different in the sense that they are always present, but in continuously varying forms. As you read the discussion of intelligence, keep in mind that there is no such thing as "no IQ" or "IQ = 0."

Intelligence

Intelligence is a complex trait determined by the genotype and the environment. It is really a constellation of related abilities that includes:

1 Defining and understanding words
2 Thinking of words rapidly
3 Analyzing mathematical relationships
4 Analyzing spatial relationships
5 Memorizing and recalling information
6 Perceiving similarities and differences among objects
7 Formulating rules, principles, or concepts for solving problems or understanding situations

Some have taken the rather extreme position that intelligence is determined almost entirely by the genotype, but this view has not been widely accepted. It is reasonable to suggest that there is an innate general reasoning ability, but certain mental skills, such as musical ability, verbal ability, or quantitative ability, may be unrelated to general reasoning. Intelligence tests (e.g., those devised by Alfred Binet and revised by L. M. Terman, M. A. Merrill, and others) employ a single number, the **intelligence quotient**, or **IQ**, and are intended to measure a person's innate intelligence. The IQ is derived by taking a person's mental age (as determined by the test), dividing it by his or her actual age, and then multiplying by 100. Thus, a 10-year-old who scored as high on the test as an average 12-year-old would have an IQ of $\frac{12}{10} \times 100$ or 120. The mean score for the U.S. population is 100. Whether the IQ score is an accurate measure of intelligence is an issue for considerable debate, particularly when it is used for comparisons between different social classes and/or ethnic groups.

IQ and degree of relatedness   Attempts to quantify the relative contributions of genotype and environment to intelligence are complicated by disagreement in defining intelligence and by skepticism of the tests designed to measure it and the sampling procedures used. Nevertheless, it is generally agreed that intelligence is a polygenic trait subject to environmental modification. Studies of twins reveal that MZ twins raised together are more concordant for IQ than either MZ twins

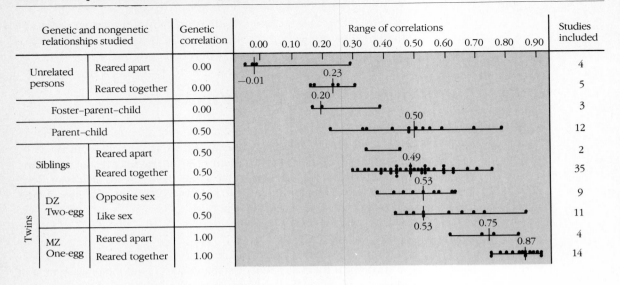

Genetic and nongenetic relationships studied		Genetic correlation	Range of correlations	Studies included
Unrelated persons	Reared apart	0.00	−0.01 ... 0.23	4
	Reared together	0.00	0.20	5
Foster–parent–child		0.00		3
Parent–child		0.50	0.50	12
Siblings	Reared apart	0.50	0.49	2
	Reared together	0.50	0.53	35
Twins — DZ Two-egg	Opposite sex	0.50		9
	Like sex	0.50	0.53	11
Twins — MZ One-egg	Reared apart	1.00	0.75 ... 0.87	4
	Reared together	1.00		14

Range of correlations axis: 0.00 0.10 0.20 0.30 0.40 0.50 0.60 0.70 0.80 0.90

Figure 13–8
Correlations of scores on intelligence tests taken by pairs of people with varying degrees of hereditary and environmental similarity. Line lengths = ranges of the correlations in the studies. Dots = correlations in each study. Vertical slashes = average correlation in each study.

raised separately or DZ twins raised together. The lower concordance for MZ twins raised apart indicates that environmental factors can modify the expression of intelligence. Studies of MZ twins separated at birth have shown that the greater the environmental differences between them, the greater the IQ differences—up to a point. The greatest difference found was 24 points in one pair; the differences between the members of other pairs were generally smaller. MZ twins separated at birth and raised in similar environments showed IQ differences of only about six points. In subsequent studies, the IQs of adopted children were compared with the IQs of their biological mothers and of their adoptive mothers. Invariably, a child's IQ was found to be more closely correlated with the IQ of the biological mother than with that of the adoptive mother. These data suggest that the environment is limited in its influence on IQ.

Current studies of intelligence show that the genotype has a greater influence on IQ than do environmental factors. This statement is based largely on studies of twins. Figure 13–8 compares the IQs of persons with different degrees of genetic relatedness, ranging from totally unrelated persons to identical twins. To assess how much of the variability observed between individuals is due to genetic differences, we use the **correlation coefficients** discussed in Chapter 12. Recall that a value of 1 tells us that there is perfect agreement between members of a pair and that a value of zero indicates no agreement. A value of −1 is a perfect negative correlation, meaning that an increase in one score is accompanied by an equal decrease in the other.

The study on which the figure is based is a summary of several separate studies, and it points out clearly the strength of the genetic component of IQ. As the genetic relatedness diminishes between pairs of individuals, the IQ correlation also goes down. For example, compare unrelated persons raised together (such as two boys or two girls adopted into the same family) with MZ twins raised together. The former correlation is 0.23, and the latter is 0.87. With such information available, it is hard to deny the importance of the genotype to the IQ trait.

The hereditarian argument has been challenged by geneticists and psychologists who think that the environment is more important than the genotype. These people criticize the studies we have mentioned as being too small and not properly sampled. They also criticize the IQ testing procedure. The debate goes on about the relative importance of the genotype and environment to human intelligence, and it is destined to continue as long as we remain in the dark about the genes that influence the various mental processes we call intelligence.

IQ and race   Our discussion of intelligence and its genetic and environmental roots is pertinent to a controversial study published by Arthur Jensen in 1969. Jensen's thesis is that there are inherent differences in the average intelligence of different racial groups. At the core of his argument is the syllogism (or deductive scheme of reasoning) that if races differ genetically, and if intelligence is to a large extent a genetically determined trait, then perhaps we can expect to observe genetically based differences in intelligence among racial groups.

Jensen carefully examined studies that show blacks as a group scoring about 15 points lower than whites on IQ tests, and he contends that the difference in scores is largely due to genetic factors. In his original study, he argued further that about 80% of the variation in IQ observed among members of a population is determined by the genotype and that there are inherent black–white group differences.

For obvious reasons, Jensen's ideas have been hotly contested. Critics point to the difficulties in defining intelligence, to the argument that IQ tests are culturally biased, and to the difficulty of defining a human race. Most investigators think that interracial IQ differences can be explained by environmental differences, such as home life and educational opportunities, and by the way sampling and testing procedures were done. Another issue that clouds the picture of IQ differences is the reliability of the often-cited twin-study data of Cyril Burt, which showed a very strong genetic component to the IQ score. We now know that these data, formerly used in many studies of intelligence, were manipulated, if not entirely fraudulent. Their exclusion from current reviews has resulted in a general lowering of estimates of the heritability of IQ. Not only has Jensen revised his heritability estimates downward in the light of Burt's trickery, but his own studies in rural Georgia schools support this downward revision.

The idea that there may be racial differences in intellectual capabilities is explosive. If accepted, it could drastically alter our social and educational programs and lead to racial segregation at an official level. School systems could justify segregated schools or classrooms; indeed, some legislators have tried to invoke this sneaky form of racism by misinterpreting the conclusions of the Jensen study. The potential dangers of trying to show that there are genetically rooted intellectual differences between races are enormous. They can be lessened only by an understanding of the methodology used, the assumptions made, and the data collected. Certainly a major problem with studies of this type is the concept of race. As a biological term, it is limited in value, and for humans it is nearly useless. There is no such thing as a pure race. Different human groups have different frequencies of genes, but there is so much overlap that we cannot make clear distinctions between those groups on genetic grounds. Another major problem with this type of study is the use of the concept of heritability, which we will discuss shortly.

Box 13–1   A Genetically-Influenced Stress Behavior?

Twice before in this text we have referred briefly to Lesch-Nyhan syndrome, an X-linked trait characterized by self-mutilation in at least some of the patients. But its bizarre manifestations (Fig. 4-14) merit still further discussion.

Just what perverse trick of nature is it that drives a person to chew off his own lips and fingertips? It is certainly not involved with a lack of sensation, for the unfortunate victims are often seen shrieking in pain at the very moment they are tearing at their own flesh.

Biochemically, we know many things about the trait. For example, we know it is associated with nucleic acid metabolism. To keep our cells actively dividing and functioning, the body needs a constant supply of the nitrogeneous base making up the genetic code. The two major purine bases, it will be remembered, are adenine and guanine. The body has the ability to manufacture both these bases from simpler compounds, but it can also "recycle" them from nucleic acids that are being dismantled as part of the normal processes of cell activity. This pathway of recycling is called the "salvage pathway," and is normally mediated by HGPRT, the enzyme lacking in Lesch-Nyhan patients. This means that the bases which would ordinarily be reused by the body are wasted in Lesch-Nyhan patients and are converted to uric acid, which then builds up to abnormally high levels in the blood and urine. But we do not yet know how these abnormal levels of chemicals cause the unusual behavior.

At birth, the Lesch-Nyhan babies appear perfectly normal. They will usually manifest continuing bouts of colic, but colic is so common among babies that it has no diagnostic value at all for the syndrome. Most of the babies will excrete uric acid in their urine, where it will appear in the diapers as a red-brown or orange "sand," but this trait is not alarming enough to cause most mothers any particular concern. Lesch-Nyhan babies often vomit, too—but what baby doesn't, on occasion?

The first real clues that something is wrong seem to appear at about 6–8 months of age,

when the babies are clearly not making "normal" physical progress. Indeed, there may be actual regression of motor skills, which will become more pronounced as the child grows older. The muscles grow abnormally tight, and the baby exhibits uncontrolled writhing motions and spasms of the trunk and limbs. The legs may ultimately "scissor," making walking virtually impossible without assistance. The poor muscle control may also inhibit speech, toilet training, or the development of other motor skills. Many patients exhibit periodic seizures.

At about two or three years of age, behavior for which Lesch-Nyhan syndrome is most noted begins: the compulsive biting and tearing of the victim's own lips, mouth, fingers, toes, etc. Only a minority of the patients exhibit this phenomenon, and a few have not begun it until their teen years. Such variation has led some researchers to suspect genetic heterogeneity—that is, more than one basic genetic lesion with common phenotypes. For our purposes, however, we shall confine ourselves to the "pure" expression of the trait.

These children do not always confine their "aggression" to themselves; they will often bite, hit, or scratch others who are not sufficiently wary. They are often delighted at their ability to snatch and fling away the eyeglasses of their nurses or doctors with lightning speed. Curiously, however, the children are often the favorites of their attendants in public institutions, despite their occasional aggressive acts, because of their usual alertness, genuine humor, and comparatively bright demeanor. Though their IQ scores range characteristically between 30 and 65, classifying them as severely retarded, many consider their scores to be artificially depressed because of the difficulty many have in speaking. And often, after an attack of biting or hitting an attendant, the patient will apologize profusely, all the while sporting an innocent, good-natured grin.

Self-mutilation is sporadic and often bafflingly inconsistent. It may be only the right hand, for instance, which will inflict damage elsewhere on the body and which must be restrained by a tie or mittens; the left hand may be perfectly harmless. Some patients may go for months without need of restraints or protection, then ask for such because they sense that they are "going out of control." When asked why they mutilate themselves, some will reply "Because I can't help it" or "I do it when I get mad." It is clear, however, that most of them are not doing it just to get attention: they are most at risk when they are alone and unattended.

Most patients exhibit chronic thirst and drink copious amounts of water. This is probably fortunate, in terms of survival at least. Most patients die during their teen years, though a few have survived into their twenties or even thirties. Death is often associated with kidney failure and/or kidney stones from the urate crystals associated with their disease. It is quite likely, in the minds of some researchers, that such kidney problems would appear even earlier if only normal amounts of water were consumed.

Current research pursues a number of different approaches. Work into the direct action of the abnormal chemical levels on nerve cells continues. Other drugs (such as caffeine in high doses) are known also to cause the mutilating behavior; we are searching for the mechanisms involved. There are now a variety of tests to identify heterozygous female carriers before they produce affected sons (affected, homozygous females are unknown); their histories and behavior are of great research interest. Considering that heterozygous females will be subject to random X-chromosome inactivation (Lyonization), one would predict that heterozygous females affected with various levels of uric acid build-up will exist and can be found. How will the levels of uric acid (or related chemical imbalances) relate to their behavior, at what ages, etc? Will these provide a clue for understanding and treating the affected males?

A particularly interesting line of study involves the production of mice whose bodies are mosaics of tissues, some of which are of cells lacking the HGPRT enzyme, others of which are normal. We have means to detect where the deficient cells are located in the body; it is hoped by this means to identify which tissues are critical to the production of the syndrome (which tissues house the necessary gene function), and thereby to dissect further the steps involved in the production of the syndrome. Knowing those, we may eventually be able to devise means of therapy and treatment.

A final thought: Several researchers point out that the major mutilation done by Lesch-Nyhan patients is the biting of fingertips and lips. The patients engage in such activity much more during emotional stress. But many normal people under stress can be observed to chew their lips or bite their fingernails. Is Lesch-Nyhan syndrome only an extreme manifestation of a relatively common form of stress-induced behavior? And if so, just what are the implications of that?

Before proceeding with a discussion of heritability and IQ, let us clarify one other issue. Jensen has often and unfairly been portrayed as a bigot—a sort of intellectual Archie Bunker. His objective from his 1969 paper onward has *not* been to foster notions of racial superiority, but rather to suggest educational remedies that could provide maximum educational benefits for *all*. If there are interracial differences in IQ scores, and there appear to be, we would be slipshod in ignoring them and failing to restructure our educational system to overcome them.

Heritability and IQ Let us consider a series of hypothetical situations that illustrate the problems associated with the concept of heritability (H), especially in IQ studies. These situations will show that for the IQ trait (and other traits as well), we can arrive at a number of different H values.

In the first hypothetical situation, we place several genetically identical persons into extremely heterogeneous environments. The environments vary extensively in diet, intellectual stimulation, cultural enrichment, and other important ways. We determine the IQ of each person. Since the genotypes are identical ($V_G = 0$), all the variation in IQ (whether great or small) is the result of environmental variation. The heritability of IQ in this case would be 0.

If we now take several genetically unrelated persons and place them all in the same environment, we will come up with a different heritability estimate of IQ. Suppose the environment is an optimal one for intellectual growth and development. In this instance, the *H* value would approach 1, because the environmental variability (V_E) is 0, and all IQ variation is the result of genetic variation. If we were to take these same genetically unrelated persons and place them in a uniformly unenriched environment, we would find that they all performed relatively more poorly on the IQ test. But the heritability estimate would still be 1, because all the variability would still be of genetic origin. In other words, the variation within the group might be small, given a uniform environment, but the heritability would still be high. Finally, if we were to take these same genetically unrelated persons and place them into heterogenous environments, we would find the *H* value of IQ to be somewhere between 0 and 1.

Three very important points emerge from these situations. First, they emphasize that *H* does not express the extent to which a trait is determined by genes. It does express the proportion of phenotypic *variation* that is the result of genetic variation.

Second, *H* measures only *a specific population in a specific environment*. In our examples, we obtained *H* values ranging from 0 to 1 for the same trait. The value varied because we focused on specific populations in specific environments, and it is clear that an *H* estimate from one population cannot be directly extrapolated to other populations.

Third, these situations show that just because *H* is high, say 1, it does not follow that observed variation must be the result of genetic differences. We derived high *H* values from the two situations in which genetically unrelated persons were placed into either a deprived or an enriched environment, yet it is obvious that the IQ difference *between* the two groups was the result of environmental, not genetic, differences. The *H* value does not tell us whether this environmentally caused difference is greater or smaller than the genetically caused differences within each group.

These examples illustrate a major problem with some discussions of IQ. Some traits with a high *H* value, such as IQ, show interracial differences, but we cannot necessarily conclude that those differences are the result of genetic differences. We must exercise extreme caution in using *H* estimates.

Primary mental abilities   We said earlier that one of the weaknesses of the IQ test was the single numerical characterization it presented of a person's intelligence. No single number can adequately embrace the complexity and diversity of intellectual capabilities. The Chicago Primary Mental Abilities Test (PMAT) is designed to measure separate mental abilities, such as verbal skills, space perception, quantitative skills, general reasoning, word fluency, and memory scores (Figure 13–9).

VERBAL REASONING

Each of the five sentences in this test has the first word and the last word left out. You are to pick out words that will fill the blanks so that the sentence will be true and sensible.

Example X.   . . . . . . is to water as eat is to . . . . . .

     A continue — drive
     B foot — enemy
     C drink — food
     D girl — industry
     E drink — enemy

NUMERICAL ABILITY

This test consists of five numerical problems. Next to each problem there are five answers. You are to pick out the correct answer and mark its letter on the Practice Answer Sheet.

Example X.   Add

                A 14
     13   **B** 16
     <u>12</u>   **C** 25
                D 59
                N none of these

SPACE RELATIONS

This test consists of five patterns which can be folded into figures. To the right of each pattern there are four figures. You are to decide which **one** of these figures can be made from the pattern shown. The pattern always shows the **outside** of the figure. Here is an example:

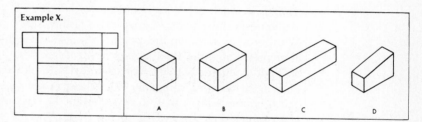

ABSTRACT REASONING

Each row consists of four figures called Problem Figures and five called Answers Figures. The four Problem Figures make a series. You are to find out which one of the Practice Answer Figures would be the next (or the fifth one) in the series of Problem Figures.

Example X.

PROBLEM FIGURES   ANSWER FIGURES

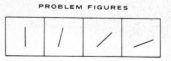 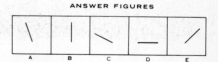

MECHANICAL REASONING

This test consists of a number of pictures and questions about those pictures.

Figure 13–9

Sample items from an intelligence test. The answer to the Verbal Reasoning question is C; Numerical Ability, C; Space Relations, C; Abstract Reasoning, D; and Mechanical Reasoning, B.

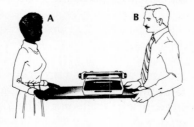

Example X.

Which person has the heavier load?
(If equal, mark C.)

Table 13–2   Heritability estimates based on MZ and DZ studies of the Chicago Primary Mental Abilities Test

Test Item	Study				Average of Studies b, c, and d
	a	*b*	*c*	*d*	
	Blewett (1954)	Thurstone et al. (1955)	Vandenberg (1962)	Vandenberg (1964)	
Verbal	0.68	0.64	0.62	0.43	0.56
Space	0.51	0.76	0.44	0.72	0.64
Number	0.07	0.34	0.61	0.56	0.50
Reasoning	0.64	0.26	0.29	0.09	0.21
Word fluency	0.64	0.60	0.61	0.55	0.59
Memory	—	0.38	0.21	—	0.30

Several studies have examined the PMAT and attempted to determine heritabilities for each of the six abilities noted (Table 13–2). We see that the genotype may be especially significant in verbal ability, space relations, and word fluency. All the studies agree on this point. They also agree that heredity is less important in memorizing skills. But the American studies (T. G. Thurstone and S. G. Vandenberg) do not agree with the British study (D. B. Blewett) on the issue of numerical skills and reasoning capability. The British study suggests heredity is not important in numerical skills but is important in reasoning capability, whereas the American studies suggest just the opposite. The differences are probably the result of substantial differences in educational backgrounds, philosophies, and socioeconomic experiences. These studies do point out that complex mental functions can be separated and that some of them seem to be strongly influenced by the genotype. We can go no further in our assessment of the genetic control of these mental abilities.

Criminality

Are people predisposed to live as criminals in this world? Are there such things as "criminal genes"? Few topics pack as much emotional wallop as the genotype's influence on criminal or antisocial behavior. As with IQ and other behavioral traits, there is no specific gene that causes the specific behavior we call criminality. But there is evidence that criminality is related to such things as body build, electroencephalogram (EEG) patterns, IQ, personality type, and psychopathology, all of which are genetically influenced.

Several studies have centered on the concordance for criminal behavior among MZ and DZ twins. They all suggest a stronger concordance for criminality in MZ twins than in DZ twins, though the concordance rates do vary (Figure 13–10). This variation is the result of different definitions of criminality. The 1929 German study considered only the more brutal crimes; the 1968 Danish study

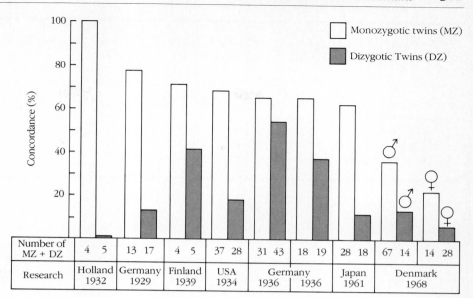

Figure 13–10
Concordance rates of MZ and DZ twins for criminality. Criminals are defined by court convictions.

Number of MZ + DZ	4 5	13 17	4 5	37 28	31 43	18 19	28 18	67 14	14 28
Research	Holland 1932	Germany 1929	Finland 1939	USA 1934	Germany 1936	Germany 1936	Japan 1961	Denmark 1968	

embraced a much wider range of criminal acts. Although these studies suggest a genetic influence on behavior patterns that lead to the commission of antisocial or criminal acts, they are all ambiguous about the role of the environment. For example, from looking at these studies, we are unable to assess the influence of the social interaction between members of twin pairs.

Adoption studies have strengthened the view that antisocial behavior is influenced by the genotype. Children whose biological mothers were convicted of criminal acts and who had been adopted by unrelated people were examined for antisocial behavior patterns. They were matched against adopted children whose biological mothers were not criminals (or at least if they were, they had not been caught!). The adoptive home environments were comparable in both instances. The results show that the children whose mothers exhibited antisocial behavior were much more likely to exhibit that behavior than those whose mothers did not. The adoption studies are somewhat flawed because the children were not always adopted at ages young enough to be free of the biological mother's influence and because the samples are not perfectly age-matched.

Though we cannot single out specific genetic elements as the causative agents in criminality, we are able to point to some genetically influenced characteristics that in turn are important in the development of criminality:

1 Criminals tend to have lower IQs than noncriminals.
2 Criminals tend to have abnormal brain-wave patterns, as determined by EEGs.
3 Criminals tend to be mesomorphic or more athletically oriented than noncriminals (Table 13–3).

But none of these tendencies is sufficiently reliable to offer a predictive value for criminality.

Table 13–3	Correlation coefficients between physique and temperament		
Body Build	Relaxed, Easy To Get Along With	Aggressive, Active, Athletic, Dominating	Secretive, Self-conscious, Introverted
Endomorphy (heavy, rounded)	0.79	−0.29	−0.32
Mesomorphy (husky, muscular)	−0.23	0.82	−0.58
Ectomorphy (linear, delicate)	−0.40	−0.53	0.83

We mentioned in Chapter 6 that an extra Y chromosome (XYY) may predispose a person to overly aggressive or antisocial behavior. Whether it does or not has not yet been conclusively demonstrated. It is certainly tantalizing to speculate on the possible link between the Y chromosome and a specific behavior. It may be worthwhile for you to go back to Chapter 6 and review this issue before proceeding.

Homosexuality

Though genetic mechanisms that may lead to homosexual or heterosexual behavior are not known, studies of twins suggest a strong genetic influence on homosexual behavior. There are two arbitrary levels of homosexual behavior: high-grade and low-grade, which are based on Alfred Kinsey's 1943 study of the sexual behavior of human males. When both high- and low-grade forms were considered together, the concordance rate was 100% for MZ twins and 26% for DZ twins (Table 13–4). For the high-grade form only, the concordance rate was 70% for MZ twins and 4% for DZ twins. According to Kinsey, homosexual behavior of the low-grade type is common in the American population. But even if we discount the low-grade form, we are still faced with evidence for a strong genetic influence.

The environment must not be an underrated factor in these studies. It may be, for example, that one twin induces the other to engage in homosexual behavior. However, studies of separated twins show that homosexual behavior patterns developed independently in the two members of each twin pair. Further, the

Table 13–4	Homosexuality in MZ and DZ twins			
Number	Concordance (High-Grade)	Concordance (Low-Grade)	Discordance	Concordance
MZ: 44	31 (70%)	13 (30%)	—	100%
DZ: 51	2 (4%)	11 (22%)	38 (74%)	26%

specific patterns of homosexual activity were similar in the twin pairs. Though the environment may be important to the development of homosexual behavior, it is difficult to avoid the conclusion that the genotype exerts a strong influence.

Personality and life-style

This chapter has dealt with behavioral traits (or groups of traits) that are fairly unusual, like schizophrenia, or clearly distinguished from other aspects of behavior, like intelligence. But personality consists of a host of more or less common, interacting characteristics that are hard to define, and the problem of sorting out the genetic and environmental influences that have shaped a personality is formidable. Studies of twins have provided evidence, however, that many ordinary behaviors and preferences are strongly influenced by genes—recall some of the remarkable "coincidences" turned up by the University of Minnesota study we discussed in Chapter 12.

A recent newspaper story provides a striking example of parallel development in genetically identical people. MZ male triplets were separated at birth and none was ever told of his brothers. Two of the brothers recently enrolled at the same college in New York. When they first encountered each other, they were shocked at how much alike they were. They soon discovered that they were in fact twins. The third triplet saw their picture in the newspaper, noted the similarities, and called them up to suggest that perhaps he too was a brother. The adoption agency later confirmed that they were identical triplets. The three brothers had never seen or heard about one another until college, yet they all drove the same type of car, had the same preferences for food, dressed alike, and showed many other commonalities. Clearly, genetic factors are involved in the development of personality and life-style, but we cannot say what these factors are.

Summary

1 Genes produce proteins, and since proteins are important to the structure and function of the nervous system, it follows that genes can influence behavior. Examples of single genes controlling behavior are Huntington's disease and Lesch–Nyhan syndrome. Some genetic disorders, such as PKU, affect intelligence.

2 **Hygienic behavior** in bees is controlled by two gene loci. One locus controls the uncapping of infected cells, and the other controls the removal of the infected larvae. These loci are inherited independently and a defective gene at either locus prevents the hygienic behavior. This is a simple example of polygenic behavioral regulation.

3 X-linked genes in *Drosophila* have been shown to control certain behavior patterns. The mode of these genes' activity is studied through the construction of genetic mosaics that can reveal where in the body the control center for a behavior is located.

4 Behavior studies using mice can be more reliably extrapolated to human behavior problems than can those using bees. In mouse studies we find that traits such as obesity and alcoholism have a genetic basis.

5 **Schizophrenia** is a discontinuous, or threshold, trait controlled to a large extent by the genotype. We are unable to say for sure whether the condition is the result of an autosomal dominant allele with incomplete penetrance or of polygenic inheritance, though most evidence favors the latter idea.

6 Bipolar manic-depressive psychosis and **unipolar depression** are both genetically influenced psychopathologies.

7 Several continuously varying behavior traits have genetic components. These include intelligence, antisocial behavior, and homosexuality.

Key Terms and Concepts

Behavior

Hygienic (nest-cleaning) behavior

Schizophrenia

Manic-depressive psychosis

Unipolar depression

Problems

13–1 Pairs of MZ and DZ twins were studied for their ability to learn certain procedures. Learning was determined for each member of a pair, and the differences in learning speed (in minutes) between MZ and DZ twin pairs were calculated. The results are shown below. (Each point represents one twin pair.) What do you conclude from these data?

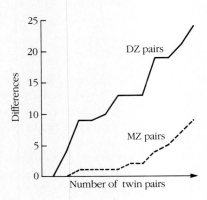

13–2 Discuss enzyme defects that lead to mental aberrancies.

13–3 A morbidity risk is an estimate of the chance of being affected with a trait. The table below shows the median morbidity risks for the general population and for relatives of schizophrenic index cases. On the basis of your knowledge of schizophrenia and degrees of genetic relatedness, match the groups with the risk estimates given at the end of the table.

General population

First-degree relatives
  Parents of schizophrenics
  Siblings of schizophrenics
  Children of schizophrenics
  Second-degree relatives (i.e., grandchildren)
  Third-degree relatives (i.e., great-grandchildren)
    9.7; 4.2; 2.1; 0.9; 7.5; 1.7

13–4 Two zigzag mice were crossed. They produced 84 zigzag offspring and 58 normal offspring. What can you conclude from this cross?

13–5 Can two nonhygienic bees produce hygienic offspring? Explain.

13–6 What is a genetic mosaic and how has it been used to locate the somatic sources of behavior?

13–7 Discuss two *Drosophila* mutants that have genetically caused movement aberrancies.

13–8 You find a mouse that is irritable, shakes its head, and runs around in circles. How would you determine whether this mouse was a waltzer or a twirler?

13–9 Discuss the model for schizophrenia that considers the interactions among gene loci as well as those between the genotype and the environment.

13–10 How do the two main types of depressive illnesses differ from each other?

13–11 Evaluate the problems associated with heritability, especially as it applies to IQ studies.

Answers

See Appendix A.

References

Childs, B. 1972. Genetic analysis of human behavior. *Annu. Rev. Med.* 23:373–406.

Childs, B., J. M. Finucci, M. S. Preston, and A. E. Pulver. 1976. Human behavior genetics. *Adv. Hum. Genet.* 7:57–97.

Crowe, R. R. 1974. An adoption study of antisocial personality. *Arch. Gen. Psychiatry* 31:785–791.

Ehrman, L., and P. A. Parsons. 1976. *The Genetics of Behavior*. Sinauer, Sunderland, MA.

Fuller, J. L., and W. R. Thompson. 1960. *Behavior Genetics*. Wiley, New York.

Fuller, J. L., and W. R. Thompson. 1978. *Foundations of Behavior Genetics*. Mosby, St. Louis.

Gottesman, I. I., and J. Shields. 1972. *Schizophrenia and Genetics: A Twin Study Vantage Point.* Academic Press, New York.

Hutchings, B., and S. Mednick. 1975. Registered criminality in the adoptive and biological parents of registered male criminal adoptees. In *Genetic Research in Psychiatry,* Fieve, R. R., D. Rosenthal, and H. Brill, eds. Johns Hopkins University Press, Baltimore.

Jensen, A. R. 1969. How much can we boost IQ and scholastic achievement? *Harvard Educ. Rev.* 39:1–123.

Kagan, J. S., J. M. Hunt, J. F. Crow, C. Bereiter, D. Elkind, L. J. Cronback, and W. F. Brazziel. 1969. How much can we boost IQ and scholastic achievement? A discussion. *Harvard Educ. Rev.* 39:273–356.

Kallmann, F. J. 1953. *Heredity in Health and Mental Disorder.* Norton, New York.

Mendlewicz, J., and J. D. Ranier. 1974. Morbidity risk and genetic transmission in manic depressive illness. *Am. J. Hum. Gen.* 26:692–701.

Mendlewicz, J. 1977. Adoption study supporting genetic transmission in manic depressive illness. *Nature* 268:327–329.

Plomin, R., J. C. DeFries, and G. E. McClearn. 1980. *Behavioral Genetics, A Primer.* Freeman, San Francisco.

Shah, S. A. 1970. *Report on the XYY chromosome abnormality.* U.S.P.H.S. Publication no. 2103. U.S. Government Printing Office, Washington, DC.

Schulsinger, F. 1972. Psychopathology, heredity and environment, *Int. J. Ment. Health* 1:190–206.

Vandenberg, S. G., ed. 1968. *Progress in Human Behavior Genetics.* Johns Hopkins University Press, Baltimore.

Weitkamp, L. R., H. C. Stancer, E. Persod, C. Flood, and S. Guttormseu. 1981. Depressive disorders and HLA: A gene on chromosome 6 that can affect behavior. *New Engl. J. Med.* 305:1301–1307.

Wender, P. H. 1972. Adopted children and their families in the evaluation of nature–nurture interactions in the schizophrenic disorders. *Annu. Rev. Med.* 23:355–372.

14 Populations and Evolution

We humans are a genetically diverse collection of individuals, organized into populations, spanning the globe, and occupying all habitats. Our diversity, indeed our very existence, is the result of millions of years of evolutionary history. We have an understandable fascination with our origins and the complex forces that acted on prehuman and early human populations to eventually give rise to modern human beings in all their diversity.

About 6 million years ago, there were four or five, or possibly more, groups of hominids (humanlike organisms) wandering the East African plains. All these groups except one became extinct. That one group went on to evolve into modern humans. We have, unfortunately, far more questions about this evolution than we have answers. Among the questions we will consider in this chapter are: How does the genetic composition of a population change over time, and how do the genetic changes that accumulate in a population lead ultimately to the emergence of a new species?

Populations and Genetic Variability

The human species is a community of actually or potentially interbreeding individuals, all of whom share a common **gene pool**—the sum total of all genes in a population. Because each of us has the potential for mating with any other person of the opposite sex, we are all considered to be members of the same species. But belonging to the same biological unit, the species, does not mean that we are all genetically the same. Even the most casual observer knows that peoples from around the world vary in hair and skin color, height, weight, hair texture, head shape, blood type, and hundreds of other ways (Figure 14–1). Many of these differences are the result of genetic variability.

To understand the evolution of humans, we first need to understand the genetic composition of human populations. Until recently, it was widely thought that there was relatively little genetic variation in the human population. According to this **classical model**, for each genetic locus there is an allele that is normal, or "wild type." This allele predominates at that locus, which means that most individuals are homozygous at most loci. Alleles other than the wild type exist, but because they are not usually beneficial, they are not very common. An alternative

Figure 14–1
Variation in humans. In this photo, we see variation in facial features, skin color, eye shape, hair color, sex, and other features.

hypothesis now seems more likely to be true. This is the **balance model**, which argues that there is really no single wild-type allele but rather a variety of alleles that exist at each locus (Figure 14–2). The balance model view predicts that there is heterozygosity at many loci since there is no single favored allele for a given locus. Key questions here are: What is the extent of genetic variation in a population, and what are its causes?

Evidence supporting the view that there is a tremendous amount of heterozygosity is overwhelming, not only for humans but for other species as well. The most direct evidence for genetic variation in populations emerges from studies of enzyme variation. Variation in enzyme structure is studied by means of electrophoresis (Figure 14–3), a technique described in Chapter 10. Electrophoresis separates proteins on the basis of their total electrical charge, a property that varies according to their amino acid composition. For example, in comparing the amino acid sequence of a specific enzyme in two persons, we may find that there is a single amino acid difference. Person A has lysine at a particular position in the

Figure 14–2
Two models of the genetic structure of populations. The hypothetical genotypes of three typical persons are shown according to the two models. According to the classical model, there is a wild-type allele at most loci (designated by a plus sign), with an occasional mutant allele (such as C^2, B^4, and O^3). According to the balance model, there is no single wild-type allele for each locus; rather, individuals are heterozygous at most loci.

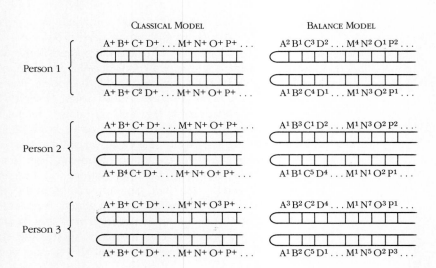

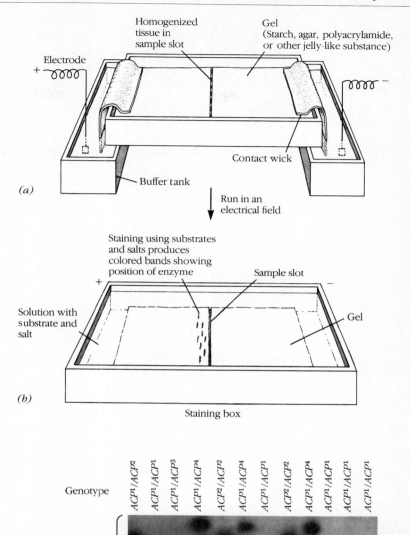

Figure 14–3
Techniques of gel electrophoresis (*a*) and enzyme assay (*b*) used to measure genetic variation in natural populations. The enzymes migrate according to their electrical charge. The allelic variation is determined by the band positions. (*c*) The result of electrophoresis for a sample of 12 persons tested for acid-phosphatase variation. This enzyme is constructed of two polypeptide chains, so the heterozygote shows three different bands: two of the bands represent the enzyme with identical polypeptides, and one band represents the enzyme with two different polypeptides. There are four alleles in evidence from this gel: ACP^1, ACP^2, ACP^3, and ACP^4.

enzyme, and Person B has glycine at the same position. Lysine has a net positive charge, whereas glycine is electrically neutral. The net result of this single amino acid difference is a charge difference, detectable by electrophoresis, between the two variant enzyme molecules. Since the amino acid sequence in a protein is genetically determined, the difference we observe here is the result of allelic differences.

Using electrophoresis to analyze enzyme variation, we have come to the conclusion that in humans, at least 30% of all gene loci vary within the whole species. But this percentage may well be an underestimate, for we know that electrophoresis detects differences only in structural genes in cases where amino acid charge differences exist, and that, for example, substituting one neutral amino acid for another in an enzyme molecule does not result in a charge difference.

How is all this genetic variation maintained, how does it change over time, and what are its adaptive advantages? The rest of this chapter will be devoted to discussing these questions.

Genotypic and phenotypic frequencies

A study of the human population shows us that for any given gene locus, particular proportions of homozygotes and heterozygotes appear. We need to know if these observed proportions are significant in any way. Are certain genotypic proportions more advantageous than others? Are certain gene combinations more advantageous than others? To answer these questions, we must first determine how genotypic and phenotypic frequencies are calculated.

In a sample of 200 newborns, we find that 100 are blood type MM, 82 are MN, and 18 are NN. Assuming that this is a random sample of the population, and that mating in the population is random with respect to M or N blood type, the genotypic and phenotypic frequencies in the population are:

Blood Type	Frequency
MM	$\frac{100}{200} = 0.50$
MN	$\frac{82}{200} = 0.41$
NN	$\frac{18}{200} = 0.09$

In dealing with the genetics of populations, it is important that the population be freely interbreeding with respect to the trait in question. Granting this, we can determine frequencies on the basis of samples and evaluate the forces that alter those frequencies over time. In the example just given, for our sample to be representative of the entire population, parental pairs had to form without regard for the M and N alleles present. MM people could not preferentially mate with MM people or preferentially avoid NN people. If they did, the sample would not accurately reflect the genotypic or phenotypic composition of the population.

Gene frequencies and random mating

We can determine if the population just mentioned is randomly mating by determining its gene frequencies and the predicted outcome of random fertilization. Gene frequencies are determined by counting up alleles and then figuring out their proportion. Returning again to our blood type sample, 100 newborns were MM and thus carried 200 M alleles; 82 children were MN and thus carried 82 M and 82 N alleles; and 18 children were NN and carried 36 N alleles. The total number of alleles in this sample was $200 \times 2 = 400$, because each person carries two alleles. Of the 400 alleles, 282 are M and 118 are N. The gene frequencies are therefore

$$M \text{ frequency} = \frac{282}{400} = 0.7$$
$$N \text{ frequency} = \frac{118}{400} = 0.3$$

If *M* and *N* alleles combined at random, we would expect the following genotypes:

$$M \times M = 0.7 \times 0.7 = 0.49 \; MM \text{ newborns}$$
$$M \times N = 0.7 \times 0.3 = 0.21 \; MN \text{ newborns}\Big\}$$
$$N \times M = 0.3 \times 0.7 = 0.21 \; MN \text{ newborns}\Big\} \quad 0.42 \; MN$$
$$N \times N = 0.3 \times 0.3 = 0.09 \; NN \text{ newborns}$$

The observed frequency of 0.50 *MM* , 0.41 *MN*, and 0.09 *NN* matches this prediction exactly.

If the population was not randomly mating, the *observed* genotypic proportions would not match the proportions expected on the basis of gene frequency determinations. For example, suppose our sample of 200 newborns was composed of the following genotypes:

$$MM = 120 = 0.6$$
$$MN = 40 = 0.2$$
$$NN = 40 = 0.2$$

Using the procedure outlined before, we again find that the frequency of gene *M* is $0.7(240 + \frac{40}{400} = 0.7)$, and the frequency of *N* is $0.3(40 + \frac{80}{400} = 0.3)$. But now when we assume random mating, we find that observed and expected genotypes *do not match*:

$$0.60 \; MM \text{ versus } 0.49$$
$$0.20 \; MN \text{ versus } 0.42$$
$$0.20 \; NN \text{ versus } 0.09$$

These genotypes are not the result of random mating. Either our sampling procedures were wrong or—a most unlikely proposition—the parents of these children paired up using the *M* and *N* allele as a criterion.

Though the notion of humans mating randomly may seem absurd at first glance, you can see from all this that it is not. Mating *is* random with respect to certain traits: we do not choose mates on the basis of blood type or enzyme variants. We do, however, choose mates on the basis of personality, intellect, hair color, skin color, size, and so on. For these traits, mating is not random.

Hardy–Weinberg equilibrium

Calculating the gene and genotypic frequencies brings us to the single most important concept in evolutionary genetics, the **Hardy–Weinberg principle of genetic equilibrium**. Working independently, the English mathematician G. H. Hardy and the German physician Wilhelm Weinberg responded to an erroneous statement made by U. Yule in 1908. Yule had said that a dominant gene, in the course of time and in the absence of counteracting forces, would produce a 3 to 1 phenotypic ratio in a freely interbreeding population. Because he did not observe this ratio for such dominant traits as brachydactyly, he said that Mendelism was open to criticism.

In Hardy's response, published in 1908, he sounds like a kindly old professor gently reprimanding an irrepressible young student who has gone off half cocked.

What Yule had omitted from his analysis, Hardy pointed out, was the frequency of the particular alleles in the population. Hardy (and also Weinberg) considered a population in which there was a pair of alleles, *A* and *a*, with frequencies of *p* and *q*, respectively. They further assumed that mating was random with respect to the genotypes being considered and that there were no external forces acting on the genotypes to change their frequencies. By "no external forces" they meant that

1 All genotypes are equally viable and equally fertile.
2 The population is large enough so that statistical fluctuations do not occur.
3 Mutations (*A* to *a*, and *a* to *A*) and migrations are rare enough to be ignored.

Given these conditions, genotypic frequencies are expressed by the formula

$$(p + q) \times (p + q) = (p + q)^2$$
$$= p^2 + 2pq + q^2$$

That is, the heterozygote (*Aa*) frequency will be *2pq*, and the homozygote frequencies will be p^2 and q^2 for *AA* and *aa*, respectively. Further, the gene and genotypic frequencies will remain stable over time, a phenomenon called genetic equilibrium. In the *M* and *N* examples discussed earlier, the *M* and *N* frequencies (*p* and *q*) will remain stable at 0.7 and 0.3, respectively, over an indefinite period. The *MM*, *MN*, and *NN* frequencies ($p^2 + 2pq + q^2$) will also remain stable indefinitely. The Hardy–Weinberg principle, or law, says that these genetic ratios will remain constant in the population over an infinite period as long as the forces that could disrupt this equilibrium are absent.

In dismantling Yule's theory, Hardy and Weinberg patiently showed that a dominant gene would be expressed in 75% of the population only if its frequency were 0.5:

$$\left.\begin{array}{l} p^2 = (0.5)^2 = 0.25 \; AA \\ 2pq = (2)(0.5)^2 = 0.50 \; Aa \end{array}\right\} \begin{array}{l} \text{75\% with the} \\ \text{dominant trait} \end{array}$$
$$q^2 = (0.5)^2 = 0.25 \; aa$$

If a dominant gene is introduced into a population, it will reach an equilibrium that is a function of its initial frequency. Suppose, for example, that 10 *AA* persons migrate into a population of 90 *aa* persons.

$$p(A) = 0.1$$
$$q(a) = 0.9$$

If members of this population now randomly mate, 19% of their offspring will have the dominant trait:

$$\left.\begin{array}{l} p^2 = 0.01 \; AA \\ 2pq = 0.18 \; Aa \end{array}\right\} \; 0.19$$
$$q^2 = 0.81 \; aa$$

These genotypic frequencies will remain stable unless they are disturbed by such outside forces as mutation, migration, or natural selection (meaning that one genotype survives better than the others and leaves more offspring behind). The Hardy–Weinberg principle of genetic equilibrium is really very simple, but its impact on biology has been enormous.

The Hardy–Weinberg principle and recessive-gene frequencies

The Hardy–Weinberg principle often allows us to calculate recessive-allele frequencies with great ease. For example, we find that the frequency of phenylketonuria (PKU, a metabolic disease discussed earlier) in the U.S. population is about 1 in 10,000. This is a recessive trait, so we can say that

$$\text{PKU} = aa = q^2 = \frac{1}{10,000} = 0.0001$$

Assuming random mating in the population (*AA* and *Aa* have no specific preferences for each other), we can estimate the gene and genotypic frequencies:

$$q^2 = 0.0001$$
$$2pq = ?$$
$$p^2 = ?$$

$$q^2 = 0.0001$$
$$q = \sqrt{0.0001} = 0.01$$

since $p + q = 1$,

$$p = 1 - q$$
$$= 1 - 0.01 = 0.99$$

Therefore,

$$AA = p^2 = (0.99)^2 = 0.9801$$
$$Aa = 2pq = (2)(0.99)(0.01) = 0.0198$$
$$aa = q^2 = (0.01)^2 = 0.0001$$

This calculation presents us with an interesting fact. In a population of 10,000 people with a low frequency of a deleterious recessive allele, the vast majority of these alleles are "hidden" in carrier heterozygotes (198 carriers in our example). In other words, it would be futile to try to eliminate a deleterious allele from a population by preventing the affected homozygote from mating.

Extending the Hardy–Weinberg principle

Multiple alleles We can employ the same basic strategy to calculate gene and genotypic frequencies for a multiple allelic series. For example, in a sample of newborns, we find the following blood type frequencies:

$$A = 0.45$$
$$B = 0.13$$
$$O = 0.36$$
$$AB = 0.06$$

Given this information, we can calculate the frequencies of the three alleles (I^A, I^B, i) and of the six genotypes ($I^A I^A$, $I^A i$, $I^B I^B$, $I^B i$, ii, $I^A I^B$). To do this, we let

$$p = \text{frequency of } I^A$$
$$q = \text{frequency of } I^B$$
$$r = \text{frequency of } i$$

The genotypic frequencies are

A	$I^A I^A = p^2$
	$I^A i = 2pq$
B	$I^B I^B = q^2$
	$I^B i = 2qr$
O	$ii = r^2$
AB	$I^A I^B = 2pq$

Now, since blood type O is always genotype ii, we know that the frequency of genotype ii is the same as that of type O phenotype, or 0.36. Therefore,

$$r^2 = 0.36$$
$$r = \sqrt{0.36} = 0.6 = \text{frequency of } i$$

Using this r value, we can solve for one of the other alleles:

$$\text{type A} + \text{type O} = I^A I^A + I^A i + ii = 0.45 + 0.36 = 0.81$$
$$\quad\quad\quad\quad\quad\quad\quad (p^2) \;\; (2pr) \;\; (r^2)$$

Therefore,

$$p^2 + 2pr + r^2 = 0.81$$
$$(p + r)^2 = 0.81$$
$$p + r = \sqrt{0.81} = 0.9$$
$$p = 0.9 - r$$
$$p = 0.9 - 0.6 = 0.3 = \text{frequency of } I^A$$

Since $p + q + r = 1$,

$$q = 1 - (p + r)$$
$$q = 1 - (0.3 + 0.6) = 0.1 = \text{frequency of } I^B$$

Using these gene frequencies, we can generate the genotypic frequencies:

$$\left.\begin{array}{l} I^A I^A = p^2 = 0.09 \\ I^A i = 2pr = 0.36 \end{array}\right\} \quad A = 45\%$$

$$
\left.
\begin{aligned}
I^B I^B &= q^2 = 0.01 \\
I^B i &= 2qr = 0.12
\end{aligned}
\right\} \qquad \text{B} = 13\%
$$

$$
\begin{aligned}
ii &= r^2 = 0.36 && \text{O} = 36\% \\
I^A I^B &= 2pq = 0.06 && \text{AB} = 6\%
\end{aligned}
$$

Sex-linked alleles   Determining the frequencies of sex-linked alleles is more complicated, because females have twice as many X-linked alleles as males. The Hardy–Weinberg frequencies for X-linked alleles A and a are

$$
\begin{aligned}
\text{Males} \qquad A &= p \\
a &= q \\
\text{Females} \qquad AA &= p^2 \\
aa &= q^2 \\
Aa &= 2pq
\end{aligned}
$$

In females the genotypic frequencies are the same as for those involving autosomal genes ($p^2 + 2pq + q^2$), but in males the genotypic frequencies are the same as the frequencies of the alleles ($p + q$). The second ratio is also, of course, the ratio of the phenotypes in males.

Red–green color blindness occurs in about 8% of the male population but only in about 0.6% of the female population. This makes sense, given the information we have just discussed. Assuming that the frequency of the color-blind allele is the same in both sexes, the male hemizygous recessive frequency is the same as the gene frequency: 0.08. The *aa* genotypic frequency in females is expected to be $(0.08)^2$, or 0.0064, and this is just about the frequency of color-blind females we find.

The rarity of the sex-linked recessive allele for hemophilia makes it easy to understand why females almost never have this affliction. The frequency of the trait in males is about 1 per 10,000 (10^{-4}). This is the q value. So the frequency of the trait in females would be q^2, or 1 per 100,000,000!

Mating frequencies

In an earlier discussion, we said that Yule was incorrect in his assertion that a dominant allele would eventually be expressed in 75% of the population. He was incorrect because he confused a specific ratio produced by a specific pair of mating individuals with ratios produced by a freely interbreeding population. Yule was correct in regard to mating between heterozygotes ($Aa \times Aa$), but he failed to consider the other mating pairs that could form. In our sample of M and N blood types, the following nine mating pairs are possible:

Male		Female
MM	×	*MM*
MM	×	*MN*
MM	×	*NN*
MN	×	*MM*
MN	×	*MN*
MN	×	*NN*
NN	×	*MM*
NN	×	*MN*
NN	×	*NN*

In a randomly mating population, these mating frequencies are predictable and are based on genotypic frequencies:

$$
\begin{array}{ll}
MM \times MM & = 0.50 \times 0.50 = 0.2500 \\
MN \times MN & = 0.41 \times 0.41 = 0.1681 \\
NN \times NN & = 0.09 \times 0.09 = 0.0081 \\
\left.\begin{array}{c} MM \times MN \\ \text{or} \\ MN \times MM \end{array}\right\} & = 0.50 \times 0.41 \times 2 = 0.4100 \\
\left.\begin{array}{c} NN \times MM \\ \text{or} \\ MM \times NN \end{array}\right\} & = 0.50 \times 0.09 \times 2 = 0.0900 \\
\left.\begin{array}{c} MN \times NN \\ \text{or} \\ NN \times MN \end{array}\right\} & = 0.41 \times 0.09 \times 2 = 0.0738 \\
& \overline{1.0000}
\end{array}
$$

The last three combinations must be multiplied by 2 because there are two ways to form these mating pairs.

From the mating frequencies, we can predict the phenotypes of the offspring. If this is a randomly mating population (and we showed earlier that it is), the offspring will have the same genotypic frequencies as the parents. Table 14–1 shows this to be the case.

Inbreeding

For the Hardy–Weinberg equilibrium to maintain, we must have random mating in a population, and we do for many traits. But mating patterns are not always random. As we discussed in Chapter 12, pairs frequently mate on the basis of phenotypes, a behavior we call *assortative mating*. Positive assortative mating, you'll recall, occurs when like persons tend to mate and negative assortative mating occurs when unlike persons tend to mate; in both cases the frequencies are greater than would be expected on the basis of randomness. **Inbreeding**, or **consanguineous mating**, is a form of positive assortative mating between persons whose genetic relation is closer than average. Inbreeding increases the chances that the mates will have like genotypes. **Outbreeding** is a form of negative assortative mating between persons who are less closely related than average.

Positive assortative mating is very much in evidence in humans when we consider such factors as income, social status, religion, interests, education, and intelligence. It is also clear that humans tend to form mating pairs from within the same subgroups, such as racial or ethnic groupings. Positive assortative mating accentuates differences between groups, a factor that has undoubtedly been important in human evolution.

Inbreeding can have serious consequences, depending on how closely related the mates are. This is because consanguineous matings increase the likelihood that deleterious recessive alleles will be brought into a homozygous condition. For this reason, or for other social reasons, virtually all societies have prohibited incestuous matings by law or custom, although they have differed in the degree of consanguinity considered incestuous.

Table 14–1	**Random mating frequencies and offspring frequency expected if *MM* = 0.49, *MN* = 0.42, and *NN* = 0.09**			

| | | Offspring Frequency | | |
Mating	Frequency	*MM*	*MN*	*NN*
MM × *MM*	0.2401	0.2401	0	0
MN × *MN*	0.1764	(1) 0.0441	(2) 0.0882	(1) 0.0441
NN × *NN*	0.0081	0	0	0.0081
MM × *MN*	0.4116	(1) 0.2058	(1) 0.2058	0
MM × *NN*	0.0882	0	0.0882	0
MN × *NN*	0.0756	0	(1) 0.0378	(1) 0.0378
Total	1.0000	0.4900	0.4200	0.0900

Incest is a phenomenon that can cut to the very core of our psyche, as Freud went to great lengths to point out. In Chapter 4 we discussed a case in point: that of the newlyweds who discovered after their marriage that they were half-siblings, since they had the same father. Caught in this dilemma are not only the highly distraught couple, who are frightened about the possibility of producing abnormal children, but also the bride's mother, who failed to reveal the secret of their parentage until they had wed; the father, who has been unwilling to cooperate with the genetic counselors; and the genetic counselors themselves, who are stymied by the lack of detailed information about the father's genetic history. The obvious solution to the dilemma is unacceptable to the couple: they refuse to part. Confused and bitter, they are receiving psychiatric counseling to help them come to grips with this situation.

In a sense, all humans are relatives, but in the case of inbreeding we normally consider lines of descent over four generations: propositus → parents → grandparents → great-grandparents. Figure 14–4 shows some of the more important types of consanguineous matings.

To predict the consequences of inbreeding, it is necessary to evaluate the precise relationship between two persons. We do this by calculating the **inbreeding coefficient (F)**, which is the probability that at a specific locus a person receives two genes that are identical by descent (i.e., they are replicas of a single gene carried by a common ancestor).

We can calculate the inbreeding coefficient for a mating between a brother and sister. In other words, we can estimate the probability that a child produced by a mating of siblings (or sib mating, as geneticists say) will have identical alleles at a given locus. Consider a locus with four alleles: *a*, *b*, *c*, and *d*. If the parents of the sibs in question carry all four alleles between them, the mating of the parents can be represented as follows:

$$\frac{\underline{a}}{b} \times \frac{\underline{c}}{d}$$

$$\tfrac{1}{4}\frac{\underline{a}}{c} \quad \tfrac{1}{4}\frac{\underline{a}}{d} \quad \tfrac{1}{4}\frac{\underline{b}}{c} \quad \tfrac{1}{4}\frac{\underline{b}}{d}$$

Parents (ab) (cd)

Sibs $\tfrac{1}{4}\,ac$ $\tfrac{1}{4}\,ad$ $\tfrac{1}{4}\,bc$ $\tfrac{1}{4}\,bd$

We now can calculate the probability that a child produced by a mating between two of the offspring will be *aa*, *bb*, *cc*, or *dd*. In other words, the child will carry a pair of alleles that are identical by descent from a common grandparent.

After looking at all possible mating pairs that can be formed from these four sibling genotypes, we determine the genotypes possible among their offspring. This information is presented in Table 14–2. You can see that one-fourth of the children of all possible sib matings will be homozygous for one or the other of the four alleles of the locus in question. We therefore say that the inbreeding coefficient for sibling mates is 0.25.

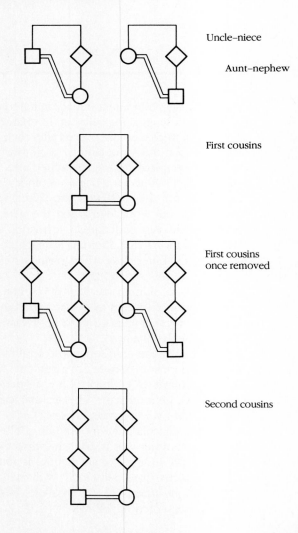

Uncle–niece

Aunt–nephew

First cousins

First cousins once removed

Second cousins

Figure 14–4
Pedigrees of consanguine-ous matings. The double bar connects the mating pair.

Table 14–2   **Calculation of the inbreeding coefficient (F) for the progeny of a mating between sibs, illustrated with the use of one locus with four alleles (a, b, c, and d)**

Father (ab)$=$Mother (cd)

Sibs $\frac{1}{4}$ ac $\frac{1}{4}$ ad $\frac{1}{4}$ bc $\frac{1}{4}$ bd

Random Mating Between Sibs

Second Sib

First Sib	$\frac{1}{4}$ ac	$\frac{1}{4}$ ad	$\frac{1}{4}$ bc	$\frac{1}{4}$ bd	Homozygotes*
$\frac{1}{4}$ ac	$\frac{1}{16} \times \begin{cases} \frac{1}{4}\,aa^* \\ \frac{1}{2}\,ac \\ \frac{1}{4}\,cc^* \end{cases}$	$\frac{1}{16} \times \begin{cases} \frac{1}{4}\,aa^* \\ \frac{1}{4}\,ac \\ \frac{1}{4}\,ad \\ \frac{1}{4}\,cd \end{cases}$	$\frac{1}{16} \times \begin{cases} \frac{1}{4}\,ab \\ \frac{1}{4}\,ac \\ \frac{1}{4}\,bc \\ \frac{1}{4}\,cc^* \end{cases}$	$\frac{1}{16} \times \begin{cases} \frac{1}{4}\,ab \\ \frac{1}{4}\,ad \\ \frac{1}{4}\,bc \\ \frac{1}{4}\,cd \end{cases}$	$\frac{2}{64}\,aa$ $\frac{2}{64}\,cc$
$\frac{1}{4}$ ad	$\frac{1}{16} \times \begin{cases} \frac{1}{4}\,aa^* \\ \frac{1}{4}\,ac \\ \frac{1}{4}\,ad \\ \frac{1}{4}\,cd \end{cases}$	$\frac{1}{16} \times \begin{cases} \frac{1}{4}\,aa^* \\ \frac{1}{2}\,ad \\ \frac{1}{4}\,dd^* \end{cases}$	$\frac{1}{16} \times \begin{cases} \frac{1}{4}\,ab \\ \frac{1}{4}\,ac \\ \frac{1}{4}\,bd \\ \frac{1}{4}\,cd \end{cases}$	$\frac{1}{16} \times \begin{cases} \frac{1}{4}\,ab \\ \frac{1}{4}\,ad \\ \frac{1}{4}\,bd \\ \frac{1}{4}\,dd^* \end{cases}$	$\frac{2}{64}\,aa$ $\frac{2}{64}\,dd$
$\frac{1}{4}$ bc	$\frac{1}{16} \times \begin{cases} \frac{1}{4}\,ab \\ \frac{1}{4}\,ac \\ \frac{1}{4}\,bc \\ \frac{1}{4}\,cc^* \end{cases}$	$\frac{1}{16} \times \begin{cases} \frac{1}{4}\,ab \\ \frac{1}{4}\,ac \\ \frac{1}{4}\,bd \\ \frac{1}{4}\,cd \end{cases}$	$\frac{1}{16} \times \begin{cases} \frac{1}{4}\,bb^* \\ \frac{1}{2}\,bc \\ \frac{1}{4}\,cc^* \end{cases}$	$\frac{1}{16} \times \begin{cases} \frac{1}{4}\,bb^* \\ \frac{1}{4}\,bc \\ \frac{1}{4}\,bd \\ \frac{1}{4}\,cd \end{cases}$	$\frac{2}{64}\,bb$ $\frac{2}{64}\,cc$
$\frac{1}{4}$ bd	$\frac{1}{16} \times \begin{cases} \frac{1}{4}\,ab \\ \frac{1}{4}\,ad \\ \frac{1}{4}\,bc \\ \frac{1}{4}\,cd \end{cases}$	$\frac{1}{16} \times \begin{cases} \frac{1}{4}\,ab \\ \frac{1}{4}\,ad \\ \frac{1}{4}\,bd \\ \frac{1}{4}\,dd^* \end{cases}$	$\frac{1}{16} \times \begin{cases} \frac{1}{4}\,bb^* \\ \frac{1}{4}\,bc \\ \frac{1}{4}\,bd \\ \frac{1}{4}\,cd \end{cases}$	$\frac{1}{16} \times \begin{cases} \frac{1}{4}\,bb^* \\ \frac{1}{2}\,bd \\ \frac{1}{4}\,dd^* \end{cases}$	$\frac{2}{64}\,bb$ $\frac{2}{64}\,dd$

*Total homozygotes $= \frac{4}{64}\,aa + \frac{4}{64}\,bb + \frac{4}{64}\,cc + \frac{4}{64}\,dd = \frac{16}{64} = \frac{1}{4} = F.$

Using the same principle we just discussed, we are able to calculate the F value for other consanguineous matings (Figure 14–5). Child G in the figure is the result of a mating between first cousins. Her parents carry genes that are identical by descent because they have common ancestors (A and B). How do we use this pedigree to calculate the inbreeding coefficient for G (F_G)? In other words, we are asking here for the probability that, for any particular autosomal locus, the alleles are identical by descent. To be identical by descent, the alleles have to be transmitted down both sides of the pedigree and then joined in Child G.

We'll assume that Persons A and B are heterozygous at a particular locus and that the four alleles are all different (a_1, a_2, a_3, and a_4). If child G's alleles are identical by descent, four genotypes are possible: a_1a_1, a_2a_2, a_3a_3, and a_4a_4. Each

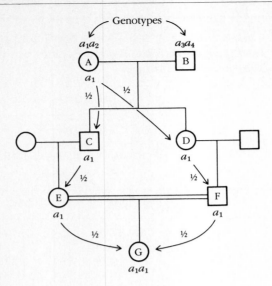

Figure 14–5
A pedigree showing a mating between first cousins. Superimposed on this pedigree is the pathway taken by the a_1 allele in order for it to be present in a homozygous state in child G. The probability of G being a_1a_1 is $(\frac{1}{2})^6$, or $\frac{1}{64}$.

of these four possibilities is equally likely. This means that we need only to calculate the probability of one and then simply multiply it by 4.

What is the probability that the child is a_1a_1? To have this genotype, A would have to pass the a_1 allele to both C and D. The probability of this occurring is $\frac{1}{2} \times \frac{1}{2}$. C would have to pass this allele to E, and D would have to pass it to F. The probability of this occurring is $\frac{1}{2} \times \frac{1}{2}$. For G to be a_1a_1, she would have to get the a_1 allele from each of her parents (E and F). The probability of this occurring is $\frac{1}{2} \times \frac{1}{2}$. The overall probability that G is a_1a_1 is obtained by multiplying all of the $\frac{1}{2}$'s together:

$$\tfrac{1}{2} \times \tfrac{1}{2} \times \tfrac{1}{2} \times \tfrac{1}{2} \times \tfrac{1}{2} \times \tfrac{1}{2} \times = \tfrac{1}{64}$$

Remember that this same probability applies to the other three genotypes as well, which means that we must multiply by 4:

$$\tfrac{1}{64} \times 4 = \tfrac{4}{64} = \tfrac{1}{16}$$

The inbreeding coefficient of a child produced by a mating between first cousins is $\frac{1}{16}$.

The inbreeding coefficient has an important meaning. In this instance it means that in Child G, 1/16th of the time the alleles at a particular locus will be identical by descent. It can also mean that in this child 1/16th of *all* loci have alleles that are identical by descent. This degree of homozygosity can have serious consequences for the health and well-being of the child, especially when you consider that homozygosity can also exist at other loci between alleles that are not identical by descent.

A more direct way to calculate the *F* value from a pedigree is to trace the paths and multiply probabilities (Figure 14–6). A summary of inbreeding coefficients is given in Table 14–3.

Uncle–niece;
aunt–nephew

First cousins
once removed (1½)

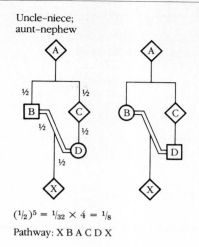

$(½)^5 = ¹/₃₂ × 4 = ⅛$

Pathway: X B A C D X

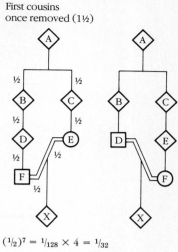

$(½)^7 = ¹/₁₂₈ × 4 = ¹/₃₂$

Pathway: X F D B A C E X

Second cousins

Third cousins

Figure 14–6
Inbreeding coefficients as determined by pathway analysis. For each pedigree, we follow the path of a specific allele through the common ancestor, then back to the inbred child. At each step we multiply by ½. Finally, we multiply by 4, because there are 4 possible alleles that can be identical by descent in ◇.

$(½)^8 = ¹/₂₅₆ × 4 = ¹/₆₄$

Pathway: X F D B A C E G X

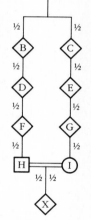

$(½)^{10} = ¹/₁₀₂₄ × 4 = ¹/₂₅₆$

Pathway: X H F D B A C E G I X

In human populations, the frequency of inbreeding is usually very low because of the customs or laws that prohibit it. When inbreeding does occur, it usually involves first or second cousins. For instance, if we sample 1000 couples in a population, we may find that 42 couples are first cousins ($F = \frac{1}{16}$) and 64 couples are second cousins ($F = \frac{1}{64}$). The remaining 894 couples are unrelated. For this population, the population inbreeding coefficient is

$$\frac{42}{1000} × \frac{1}{16} = 0.00262$$
$$\frac{64}{1000} × \frac{1}{64} = 0.00100$$

Total inbreeding coefficient = 0.00362

Table 14–3	Coefficients of inbreeding for various types of consanguineous matings	
Type of Mating		**F**
Selfing		$\frac{1}{2}$
Full-sibs		$\frac{1}{4}$
Uncle × niece, aunt × nephew, or double first cousins		$\frac{1}{8}$
First cousins		$\frac{1}{16}$
First cousins once removed		$\frac{1}{32}$
Second cousins		$\frac{1}{64}$
Second cousins once removed		$\frac{1}{128}$
Third cousins		$\frac{1}{256}$

In human populations, the inbreeding coefficient is on the average 0.001. It is higher in certain small isolated groups such as the Old Order Amish religious sect in Lancaster County, Pennsylvania. The *F* value in this group is estimated to be 0.02, which results in the expression of an enormous number of recessive traits because of the relatively high degree of homozygosity. The data in Figure 14–7 summarize inbreeding coefficients for various world populations.

The consequences of inbreeding can be quite severe. It increases the incidence of homozygous recessives in the population, and this commonly leads to

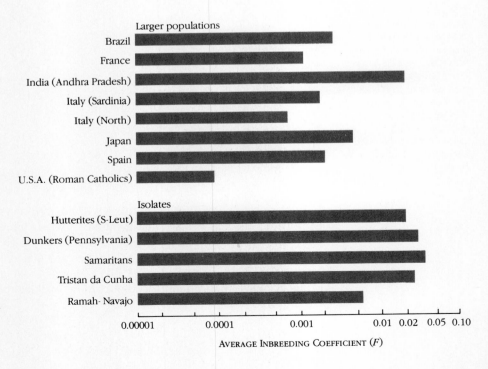

Figure 14–7
Average inbreeding coefficient in some isolated human populations (log scale).

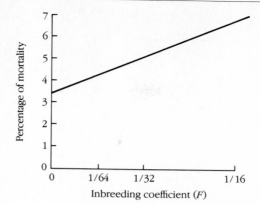

Figure 14–8
Percentage of mortality among children as a function of inbreeding. These data are based on two different Japanese populations.

an increase in mortality. In general, there is a direct relationship between the inbreeding coefficient of a population and the percentage of child mortality (Figure 14–8).

Evolution

In 1831, Charles Darwin, a young Englishman of 22, boarded the HMS *Beagle*, prepared for a long voyage, and eagerly anticipated the natural wonders of distant regions of the earth. He held the then common view that God had created each species in its present form. When Darwin returned to England five years later, he was a changed man. The data he had collected and the observations he had made led him to conclude that there was a natural and understandable explanation for the diversity of the living world and that new species were constantly being formed.

Once back in England, Darwin spent several months thinking about his data and trying to organize them into a coherent theory. He had grasped the idea of evolution but not the mechanism. About two years after his return, the pieces fell into place, and Darwin had his theory. The theory is based on three deductions from four primary observations (Figure 14–9).

Darwin's fundamental theory, perhaps the single most important theory in biology, suffered from one major weakness. It did not explain how the variation in a population originated. Darwin was unaware of Mendel's work in genetics, though they were contemporaries. Even if he had known of Mendel's work, it is unlikely that he would have incorporated Mendel's principles into his theory because the contrasting forms of the traits Mendel studied were too great: tall versus dwarf; wrinkled versus smooth; terminal flowers versus axial flowers. Darwin was looking for a mechanism to explain the very gradual changes, the variation patterns we would consider to be quantitative. It is doubtful that Darwin would have been able to extend Mendel's ideas to traits that showed a continuous range of variation.

Darwin ultimately settled for an explanation that he was not very satisfied with. He proposed a Lamarckian-type idea of how variation originates known as the **theory of acquired characteristics**, or the **theory of use and disuse**.

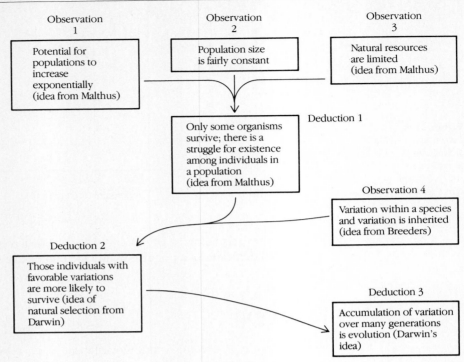

Figure 14–9
The observations and deductions that form Darwin's evolutionary theory.

Jean Baptiste de Lamarck suggested that the environment directly influenced the characteristics an organism inherits. According to Lamarckism, the giraffe's neck is long because of the animal's continuous need to stretch to reach the leaves it feeds on. In other words, the more a structure is used, the more prominent it becomes; the less it is used, the less prominent it becomes. The environment is said to convey information to the genotype and alter it in directed ways. This directed change is then transmitted to the offspring. The theory of acquired characteristics has been shown to be wrong, but at the time Darwin was writing, it seemed reasonable.

We now understand mutational processes. In addition, we are able to integrate Darwin's theory into the existing structure of modern genetics so that we have a strong basis for appreciating the evolutionary forces that act to change populations over time and lead to the emergence of new species.

Evolutionary forces

Modern evolutionary theory embraces the principles of genetics we have been discussing. Evolution is the transformation over time of the genotypic and phenotypic characteristics of populations of organisms that are related by descent, the change occurring over the course of successive generations. Let us look at how this change occurs.

Mutation A population in Hardy–Weinberg equilibrium is one in which the genes either do not mutate or mutate at such a slow rate that we can ignore this factor. But genes do mutate and by doing so provide variation, the raw material of evolution. Earlier we discussed the mechanisms of mutation; now we will consider the role of mutations in evolution and how mutations disrupt the Hardy–Weinberg equilibrium.

Although mutations are rare events in humans, we can nevertheless estimate the rate at which a normal gene mutates to a dominant mutant allele. For dominant traits such as achondroplastic dwarfism, we identify a new mutation by screening for nondwarf parents who have an achondroplastic dwarf child. The affected child is usually the result of a new mutation. We can estimate the mutation rate of a to A by looking at the number of newborns with normal parents in some total number of births:

T = total number of births
M = number of affected children with unaffected parents
m = mutation rate of a to A per generation
$m = \frac{1}{2}(M/T)$

We need to multiply by $\frac{1}{2}$ because each affected child receives two alleles, one normal and one mutant. Nonaffected children have, of course, two normal alleles.

In one study designed to estimate the mutation rate for the dominant achondroplastic dwarfism gene, of 94,073 children born, 8 were achondroplastic dwarfs born to unaffected parents. We thus estimate the mutation rate of a to A as

$$m = \frac{1}{2}(M/T) = \frac{8}{2} \times 94,073$$
$$m = 4.2 \times 10^{-5}$$

For recessive mutations, estimates of mutation rates are more difficult to come by. The methodology for obtaining mutation rates for recessive alleles is beyond the scope of this book, but a summary of estimated mutation rates for several autosomal and X-linked genes is presented in Table 14–4.

Many new mutations that arise in a population are detrimental to the well-being of the individual and the species. But some are not, and if they are preserved in a population, they may enhance the reproductive potential of the members of the species that carry it. This, after all, is what evolution is all about. Those individuals who leave more offspring behind are more evolutionarily successful.

Natural selection   It is indeed unfortunate that so many people interpret Darwinian natural selection as a series of brutal and violent conflicts in which only the strongest win out. In fact, selection is usually a very subtle process in which various genotypes exhibit differential rates of reproduction over time. In other words, some genotypes reproduce more than others. Violent conflict is not necessarily involved. For example, a female with Turner syndrome (XO) is sterile. She leaves no offspring behind, so selection against the XO genotype is total and set at 1.0. The capacity of a genotype to leave offspring behind is termed **fitness** and is always 1 − selection pressure (or s). The fitness of the XO genotype is 0. Obviously, no violence of any kind is involved.

If genetic variation is the raw material of which species are built, selection is the driving force that shapes the species. Selection is the process by which certain genotypes increase in frequency in a population at the expense of others.

Table 14–4	Estimated mutation rates for some autosomal and X-linked genes

Type of Mutation	Mutations (per Million Genes per Generation)
Autosomal mutations	
Achondroplasia	10
Aniridia	3
Retinoblastoma	8
Dystrophia myotonica	10
Acrocephalosyndactyly (Apert syndrome)	3
Osteogenesis imperfecta	1
Tuberous sclerosis (epiloia)	8
Neurofibromatosis	74
Intestinal polyposis	13
Marfan syndrome	5
Polycystic disease of the kidneys	85
Diaphyseal aclasis (multiple exostoses)	8
von Hippel–Lindau syndrome	0.2
Sex-linked recessive mutations	
Hemophilia	28
Hemophilia A	45
Hemophilia B	3
Duchenne-type muscular dystrophy	67
Incontinentia pigmenti	13
Oculofaciodigital syndrome (OFD)	5

We can see what happens to gene frequencies when one of the genotypes is lethal, as in Tay–Sachs disease. Suppose that in a given generation a recessive lethal allele has a frequency of 0.2 and the normal allele 0.8. The next generation will have the following genotypic ratios:

$$p^2 = 0.64 \ AA$$
$$2pq = 0.32 \ Aa$$
$$q^2 = 0.04 \ aa$$

However, since the fitness of *aa* is 0, *aa* individuals do not produce any progeny. This means that the *reproducing population* consists of *AA* and *Aa* only, and the genotypic frequencies of the reproducing population must be adjusted accordingly:

$$\frac{\text{frequency of AA}}{\text{total reproducing population}} = \frac{p^2}{p^2 + 2pq} = \frac{0.64}{0.64 + 0.32} = 0.66$$

$$\frac{\text{frequency of Aa}}{\text{total reproducing population}} = \frac{2pq}{p^2 + 2pq} = \frac{0.32}{0.64 + 0.32} = 0.34$$

and the *A* and *a* frequencies are 0.83 and 0.17, respectively. This represents a gene-frequency change of 0.03 (0.83 − 0.80 = 0.03; 0.20 − 0.17 = 0.03) for each of the alleles in a single generation. Over several generations, the *A* and *a* frequencies change as shown in Figure 14–10. Selection against a recessive lethal is slow but effective in reducing its frequency, up to a point. You have seen, however, that if the frequency of the recessive lethal is very low, selection is ineffective because most of the alleles are carried in the heterozygotes.

In cases where the recessive allele is detrimental but not lethal, the change in its frequency is even slower. In the example just given, let's say that the fitness of *aa* is 0.5 and assume that half of *aa* are sterile but that otherwise there is random fertilization and equal reproductive ability. The reproducing population in the second generation would be 0.65AA, 0.33Aa, 0.02aa. The *A* and *a* gene frequencies in this case are 0.815 and 0.185, and the gene frequency change is ±0.015 in a single generation. Over several generations, the *A* and *a* alleles change as shown in Figure 14–11.

Natural selection is a creative force in nature; it molds populations by favoring the genotypes that are best adapted to the environment. Since environments are subject to change, so too are predominant genotypes. We can trace the gradual change in genotype frequencies over extended periods. This change is one of our best measures of the actual occurrence of evolution.

Mutation balanced by natural selection   We estimated that the mutation rate for the achondroplastic dwarfism gene is about 1 per 10,000 gametes per

Figure 14–10

Allelic frequency changes when selection against *aa* is 1 (the *aa* genotype is lethal or sterile). After five generations, the *A* frequency has increased from 0.8 to 0.9. The *a* frequency has decreased from 0.2 to 0.1.

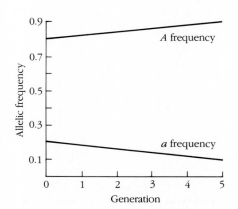

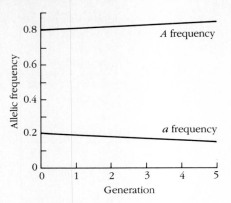

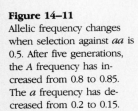

Figure 14–11
Allelic frequency changes when selection against *aa* is 0.5. After five generations, the *A* frequency has increased from 0.8 to 0.85. The *a* frequency has decreased from 0.2 to 0.15.

generation. At that rate, the frequency of the trait in the population should increase from generation to generation, yet it does not. The reason it does not is that dwarfs are less fit than normal-sized people. For various reasons they are not able to produce as many offspring as people who are normal. Thus, selection acts against the dwarf genotype: as new dwarfism genes are pumped into the population, natural selection removes some of them from the mating pool. An equilibrium point is reached when the rate at which new alleles are pumped into a population equals the rate at which they are withdrawn. The frequency of achondroplasia in a population will stabilize after several generations (Figure 14–12).

Genetic drift and migration The Hardy–Weinberg principle of genetic equilibrium applies to populations that are large enough so that random fluctuations in gene frequency do not occur because of sampling errors. But since populations are not always large, fluctuations do occur. This random fluctuation in allelic frequencies is called **genetic drift**. A small religious isolate of the Dunkers, located in south-central Pennsylvania, illustrates the role of genetic drift in an evolving population.

The Dunkers migrated from Germany early in the eighteenth century, and their communities have in large part remained reproductively isolated from one

Figure 14–12
Mutation balanced by natural selection. Without selection, dwarfs accumulate at a constant rate of 1 per 20,000 births per generation. However, if the fitness of dwarfs is 0.2, selection balances mutation and the frequency becomes constant.

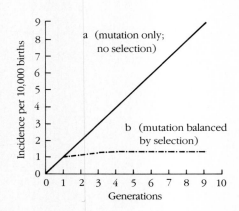

Table 14–5	Comparison of some genotypic frequencies among the Dunkers with the frequencies in the parent and surrounding populations		

| Trait (Blood Groups) | Class Frequency | | |
	U.S.A.	Dunker	Germany
A	0.40	0.60	0.45
B–AB	0.15	0.05	0.15
M	0.30	0.45	0.30
MN	0.50	0.41	0.50
N	0.20	0.14	0.20

another and from the surrounding American population. Marriages between Dunkers and people from other communities have been limited and carefully monitored, and the necessary adjustments have been made in calculating gene frequencies. The drift in allelic frequencies is seen in traits such as blood type, for which there is no obvious or measurable selection pressure—one blood type is as successful reproductively as another (Table 14–5). The frequency of blood group A is 0.6 in Dunkers, but only 0.4 in the surrounding U.S. population and 0.45 in the parent German population. The B and AB groups are almost entirely absent in the Dunkers (0.05) but have a frequency of 0.15 in the U.S. and German populations. Similar fluctuations hold for the M and N alleles. In all cases studied, the Dunkers have frequencies that are unlike either the ancestral or the surrounding population or anything in between. The allelic frequencies have fluctuated widely in this group. No evident selection on the observed traits exists, and this suggests that genetic drift has been at work. In fact, we can actually demonstrate genes caught in the act of drifting in this population of Dunkers. When the population is divided by age group, corresponding roughly to three generations, we see that allelic frequencies fluctuate:

	Age 3–27	Age 27–55	Age 55+
M allele	78	68	55
N allele	22	32	45

Genetic drift has probably been an important force in the evolution of humans. It is reasonable to propose that drift may account for many (not all) of the racial differences we observe among peoples of the world today. Such differences are very difficult to explain in terms of natural selection. Much of human evolution appears to have been operating in small groups split off from larger groups, and such conditions are ideal for the operation of drift.

One other force that can disturb gene frequencies in a population is **migration**, which is simply the movement of organisms (and their genes) into or out of

Box 14–1   A Cultural Effect on the Frequencies of Alleles

Although, as we have discussed in this chapter, certain mechanisms govern the frequencies of alleles in human populations, sometimes humans will introduce their own little quirks into the formula and produce allele frequencies that seem to defy explanation by the usual rules. A fascinating case involves albinism among certain tribes of American Indians.

The frequency of albinism varies widely among different human populations. In Norway it affects about one in 10,000 people; in southern Europe the frequency is more on the order of one in 30,000. One in 20,000 is a usual figure for its occurrence among Caucasians in the United States. But among certain American Indian tribes, notably the Cuna of Panama, the Hopi of Arizona, and the Jemez and Zuni of New Mexico, frequencies of about one albino in every 200 people have been observed repeatedly. What accounts for this incredibly high frequency?

The most detailed studies to date deal with the Hopi, and we shall restrict our further attention to them. Their high frequency of albinism has been recognized for at least a century, but it took research by geneticists from Arizona State University to give the first analytical account of the genetic structure of the population.*

Hundreds of years ago, the tribe sought refuge from its traditional enemies, the Navajo and the Utes, by building homes on the top of high mesas in northeastern Arizona. Indeed, one of their towns, Oraibi, is recognized as the oldest continuously inhabited community in the continental United States, having been established originally about 1150 A.D. During historic times the tribe has never numbered more than a very few thousand, and probably has never been numerous. The population, then, is a small one.

Can one find a means of selection, of reproductive advantage based on some *biologic* superiority, that has favored the high frequency of albinos? All evidence indicates that albinos are not biologically superior. They are known to have higher rates of skin cancer and, often, poor eyesight (myopia and lateral nystagmus). Their extreme sensitivity to sunlight necessitates constant guarding against sunburn in the hot American southwest. Natural selection acting upon some element of biological advantage, then, does not seem to offer any answer.

Could genetic drift be the explanation? The Hopi are divided into three major enclaves of approximately equal size, one on each of three tips of Black Mesa. The three major communities (called First Mesa, Second Mesa, and Third Mesa) are located on a generally east–west line. First Mesa lies more than 11 miles to the east of Second Mesa; Third Mesa lies westerly of Second Mesa at a distance of about 10 miles. Historically, the distances and the rugged terrain have kept the three populations rather separate; they are definitely not the "randomly mating population" needed for a simple treatment of community structure. In the Arizona State study, the 26 albinos studied were found to be associated with only Second Mesa and Third Mesa, but about equally divided among those two. Such a pattern does demonstrate that the populations of the three communities are not randomly intermixed. Would genetic drift account for the high frequency of albinism on two *independent* mesas? It does not seem likely (and the cases of albinism in the Jemez, Zuni, and Cuna tribes further argue against drift as the causal mechanism).

However, the construction of detailed pedigrees revealed considerable inbreeding among the Hopi communities. The Hopis of First Mesa were the least inbred as a population, having had considerable intermixing with a nearby population of Tewa Indians. The Second Mesa population was the most inbred, and did indeed have just slightly more albinos than Third Mesa. Inbreeding, then, seems a rather solid candidate for an explanatory mechanism: remember, it tends to bring recessive genes into homozygous condition. It is recognized also that the tribe likely derives from a rather small group of migrants into the area in the long-distant past; founder effects also must be considered here. We can do little more than recognize their influence, however, since we know virtually nothing of the genetic makeup of the original founders of the tribe.

But there seems to be yet another mechanism operating, one that derives directly from the nature of albinism and the lifestyle of the Hopis. There is little good soil on the tops of the mesas, and the Hopi are an agricultural people. Their farms are located in the broad lowlands to the south and west of the mesas, wherever sufficient water could be found in the arid land. The farms, then, are often many miles from the mesas. The farming was done primarily by the men and boys of the tribe, who would customarily rise very early in the morning, often well before sunrise, and walk and run to the fields. There they would labor throughout the day, and make the return trip to the mesas each evening.

But albinos, it will be remembered, are very sensitive to the sun. They were not expected to work in the fields, and albino males stayed behind on the mesas with the women and children while the menfolk went to work the crops. The albino males had domestic occupations such as weaving, so they did make their contribution to the econ-omy of the tribe. But they spent their long days essentially alone with the women of the tribe. Premarital and extramarital relations are accepted among the Hopi; they are a normal and completely unremarkable part of life. It takes little imagination, then, to understand that the albino males had a marked reproductive advantage over the other males of the tribe.

Possible founder effect, inbreeding, and a reproductive advantage generated purely by the culture of the people—these three mechanisms reduce considerably the mystery of why the Hopi have such a high frequency of albinism. Humans have here added a significant new dimension—a form of cultural selection—to the usual rules governing gene frequency. There are many human genes whose allele frequencies have no immediate explanation based on the usual rules. As we learn more of natural selection and its operation in human societies, we will undoubtedly gain insights to these. But among those insights may well rest a number of uniquely human stories similar to that of the Hopi.

---

*Charles M. Woolf and Frank C. Dukepoo, Hopi Indians, Inbreeding, and Albinism, *Science* 164:30–37, 1969.

---

a population. Migration can and does alter the frequencies of alleles in populations, though we shall not take time to demonstrate these effects.

Evolution is change, specifically change in the genetic composition of populations over time. If there were no mutations, no natural selection, no genetic drift, and no migration, and if all matings were random we would have populations in Hardy–Weinberg equilibrium and no evolution. But all these factors disrupt the equilibrium and constitute the forces for evolutionary change. The principle of Hardy–Weinberg equilibrium, which is the antithesis of evolution in a sense, gives us a means to evaluate the various evolutionary forces. And this is what makes it so fundamental to modern evolutionary theory.

Mechanisms of forming new species

All people occupying the earth today are members of a single species, *Homo sapiens*. We are all classed as the same species because we share many important characteristics and because we are all potentially interbreeding and are isolated reproductively from other species. In other words, we can breed among ourselves, but not with individuals from other species. This limitation is, incidentally, the key criterion for a species: reproductive isolation. The human species exists as a series of groups or subpopulations that are often phenotypically distinct from one another. A group within a species that has characteristic frequencies of certain genes or of certain features of chromosome structure is called a **race**. Racial differences are relative and not absolute; that is, they reflect different *frequencies* of alleles. These alleles may occur in all populations but at different frequencies. This is an especially important point to bear in mind in considering human races.

The human species is subject to the same evolutionary pressures as other species, yet for some reason the existence of races, acknowledged in other species, becomes an emotionally charged issue in human biology. It should not be. The human species is genetically diverse, and it is in this diversity that we find the basis for its evolutionary success. There are very obvious physical differences among human groups. Obviously, Black Africans differ among themselves as well as from Northern Europeans and Asians, who also differ widely among themselves. As we look at more and more human groups, we realize that the differences are of a continuous nature and often involve traits that are not well understood genetically—traits such as hair and skin color, body shape and size, and the size and location of facial features.

Attempts have been made to study human races by looking at genetically well-defined traits, such as the blood groups. We can show that the I^A, I^B, and i alleles exhibit specific frequency differences in various parts of the world (Table 14–5). Other alleles exhibit the same type of variability (Tables 14–6, 14–7, and 14–8). Races can be characterized by gene-frequency differences.

In considering the term *race*, we must bear in mind that it has meaning only in the context of populations and gene frequencies. A race is a population, or an aggregate of populations, with characteristic gene frequencies or features of chromosome structure. It is distinct from other groups or populations within the same species. Racial differences are relative and not in any way absolute, and it is important to remember that members of one race can freely interbreed with members of any other race within the same species. Races *may* become distinct species

---

Table 14–6 Percentage of blood group variation among four different populations

Population	Alleles		B	O	Rh Negative	Duffy Factor
	A_1	A_2				
Whites	5–40	1–37	4–18	45–75	25–46	37–82
Blacks	8–30	1–8	10–20	52–70	4–29	0–6
East Asians (Chinese, Japanese, and related populations)	0–45	0–5	16–25	39–68	0–5	90–100
American Indians	0–20	about 0	0–4	68–100	about 0	22–99

Source: Adapted from Coon, 1965.

Table 14–7	Percentage of PTC tasters in different populations

Group	Percentage of Tasters
Europeans	60–80
Blacks	90–97
East Asians	83–100
Australian aborigines	50–70
Micronesians	70–80

Source: Adapted from Coon, 1965.

if they become reproductively isolated from other groups in the species. This situation occurs through the accumulation of genetic differences.

The process by which species form is a hot topic. One reason is the existence of two very competitive models. The more traditionally accepted, gradualistic, model is diagrammed in Figure 14–13. This model, termed **allopatric** (or **geographic**) **speciation**, portrays a large, genetically and phenotypically variable population inhabiting an ecologically diverse area. The large population is thus composed of races or subspecies, each of which is especially well suited to a particular environment. The partitioning of the population into races, or subspecies, is the primary step in race formation. Though all groups can potentially interbreed, mating tends to be assortative because of geographic or other factors, with mating pairs likely to form from within their own groups. Gene flow is thus restricted and may become more so if additional barriers, such as mountains or rivers or even behaviors, arise that completely cut off or further restrict the flow

Table 14–8	Percentage of population bearing the dry-earwax gene

Population Examined	Percentage Bearing Gene
Northern Chinese	98
Southern Chinese	86
Japanese	92
Melanesians	53
Micronesians	61
Germans	18
American whites	16
American blacks	7

Source: Adapted from Matsunaga, 1962.

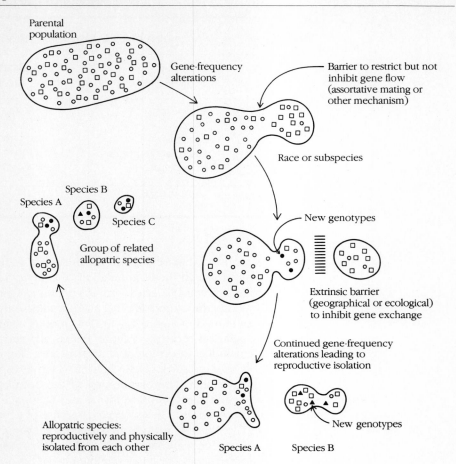

Figure 14–13
Schematic view of allopatric, or geographic, speciation.

of genetic information. In this geographically isolated state, new mutants and new gene combinations are preserved within a mating group and may lead eventually to reproductive isolation and the formation of a new species.

Geographic speciation was conceptualized as gradual, and many question whether this slow tempo adequately explains the entire evolutionary process. One of the prime sources of contention is the fossil record, cited in the past in support of the gradualist position. Often, however, the fossil record does not show a smooth transition from one group to the next. It appears rather as if species remain unchanged, or change in only minor ways, for millions of years, then suddenly disappear to be replaced by something that is substantially different, though clearly related. The scarcity of transitional forms has been traditionally attributed to imperfections in the fossil record, but this view, championed by no less a luminary than Charles Darwin, is now under attack.

The emerging view is that the fossil record is a generally accurate indication of events that took place in the past. The gaps in the fossil record are "real" and seem to show that evolution sometimes proceeds at a very accelerated pace. Transitional fossils will be few—and difficult to find. This new view of evolution, called **punctuated equilibrium**, is compared in Figure 14–14 to the gradualist

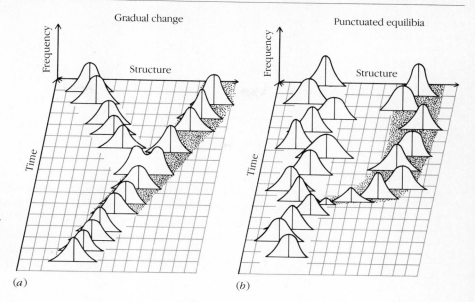

Figure 14–14
Alternative models of evolutionary change. The classic view is of gradual change (*a*); but more abrupt speciation (*b*) may instead be the major process. The graphs are drawn as frequency distributions of morphological structure.

position. It portrays evolutionary change as one of periods during which individual species remain virtually unchanged, punctuated by abrupt genetic events in which a new or descendant species arises from the original stock. Minor changes accumulate before the emergence of a new species, but then a relatively rapid reordering of genetic material occurs and brings a new group into being.

The genetic events that could lead to such major shifts in morphology (the form and structure of organisms) are the subject of much research and discussion. A reasonable possibility is a mutation in a regulator gene that controls the activity of several other genes. Another possibility is a rearrangement of genetic material, such as chromosomal translocations or inversions, that may have major effects on the function of large blocks of genes. A third possibility is genetic drift operating along the margins of the population. Small populations would be pinched off from the main population and would change relatively rapidly as a result.

In species that are more specialized—that is, in those that occupy a narrow ecological niche—we expect to see high rates of speciation coupled with high rates of extinction. That is, a species that is highly adapted to a narrow niche can potentially spread into numerous other narrow niches. But narrow niches also change more rapidly than broader niches, and this means that a species highly specialized for a narrow niche stands a greater chance of extinction when that niche changes. By contrast, in species that are generally adapted to a diverse habitat, the opportunities for developing reproductive isolation are fewer than for species that are more specialized. So there is a slower rate of speciation and extinction among the generalists.

Human evolution may be a consequence of rapid speciation caused by relatively few mutations. Probably our closest living relative is the chimpanzee, and studies of the structural gene differences (genes that produce proteins) between humans and chimps show them to be quite small. Comparing the amino acid

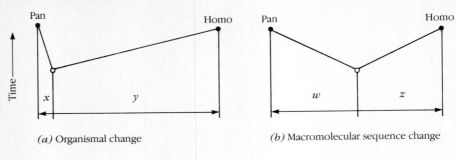

(a) Organismal change

(b) Macromolecular sequence change

Figure 14–15

Biologic evolution in contrast to molecular evolution since the human (Homo) and chimpanzee (Pan) lineages diverged from a common ancestor. (*a*) According to zoological evidence, far more biologic change has occurred in human lineage (*y*) than in chimpanzee lineage (*x*). (*b*) Evidence from both proteins and nucleic acids indicates that chimpanzee genes (*w*) have changed as much as human genes (*z*).

sequences of specific proteins, we find far too few differences to account for the tremendous anatomic, physiologic, behavioral, and ecologic differences between the two species. In fact, studies show that about 99% of human and chimpanzee proteins are identical. Although the two species are usually classified in separate families, their degree of molecular similarity is characteristic of recently diverged species (sibling species) and indicates that they should be placed in a single family.

There is a striking contrast between the biologic and the molecular evolutionary processes that have occurred since the divergence of the human and chimpanzee lineages (Figure 14–15). The two plots show that the biologic, or organismal, change has been tremendous in humans, whereas the chimp has been fairly conservative in its biologic evolution. But molecular evolution, that is, evolution that is manifested as altered amino acid or nucleotide sequences, has been about the same in the two groups. The implication of these findings is that a relatively few changes in regulatory genes may account for the major organismal differences between humans and chimps. Analyses of chromosomal banding patterns suggest that inversions have been a primary source of change in the evolution of the two species.

The emergence of the modern human

To conclude this chapter, we will consider the evolution of humans, a topic that continues to fascinate us. Of necessity, this will be an overview and will emphasize one main point: The hominid line underwent extensive speciation during its evolutionary history.

About 12 million years ago, there were two major families of apelike animals wandering the plains and forests of Africa. The Dryopithecids were woodland apes and the stock from which all the modern apes apparently evolved (Figure 14–16a). The Ramapithecids were also woodland apes, but their tooth structure was more humanlike. We now think that from somewhere in the Ramapithecid family came the first members of the human family, the hominids. A number of Ramapithecids have been suggested as hominid ancestors, including the large, gorillalike animals belonging to the genus *Gigantopithecus* (Figure 14–16b), terrestrial apes that once roamed Asia. However, this group became extinct and in all probability was not the ancestral stock from which the hominids sprang. The most likely hominid ancestor is the genus *Ramapithecus* (Figure 14–16c). This genus diverged rather extensively, and one of the divergent paths led to the emergence of modern humans.

(a)

(c)

(b)

Figure 14–16
(*a*) *Dryopithecus.* (*b*) *Gigantopithecus.* (*c*) *Ramapithecus,* a tentative reconstruction.

It was *Ramapithecus* that stood on the edge of the forest some 8 to 12 million years ago and began to exploit the open grasslands, an environment more dangerous than the forest because of the predators that lived there. The grasslands also presented fresh challenges because of the different food sources. Offering new opportunities as well as new dangers, these grasslands dramatically changed the selection pressures.

From the location of *Ramapithecus* fossils, we think that this animal lived along the fringes of the forest, venturing out to the open grasslands only on occasion. But as the forests dwindled and became more crowded with the apes, *Ramapithecus* spent more time on the grassy plains, sharpening its survival skills in the process. The tooth structure of *Ramapithecus* is unique and suggests that the animal ate foods more commonly associated with the grasslands, such as seeds, fibrous plants, and some meat. During this exploitation of the grasslands, the upright posture (which some investigators think first appeared while the apes were still forest dwellers and before the brain enlarged) continued to evolve, along with manual dexterity for manipulating objects.

About 6 million years ago, the *Ramapithecus* stock diverged. We do not know how this happened or what selection pressures prompted it. We do know that about 3 million years ago there were perhaps four main hominid groups on the scene, all existing at roughly the same time and in the same geographic area of East Africa. The discovery that the hominids had diverged to such an extent has had an explosive impact on the scientific community. It has forced the rejection of the long-held view that the various prehumans and early humans all fit into a simple scheme of steady progression from an apelike stock right through to modern

(a)

(c)

Figure 14–17
(a) *Australopithecus africanus*. (b) *Australopithecus boisei*. (c) *Australopithecus afarensis*. Figure 14–17(c) copyright © drawing by Jay H. Matternes

(b)

humans. This view had been especially popular because it gave the comforting impression that the emergence of modern humans (*Homo sapiens sapiens*) was predetermined. Now we have to contend with the very real possibility that several groups of hominids existed, even coexisted, and that all but one of these groups became extinct—just as commonly happens in other animal groups.

Let's look more closely at the main characters that composed the early hominid family. One group is *Ramapithecus*, a genus we discussed earlier as a possible hominid ancestor. A second group comprised fully erect individuals who stood about four feet tall and whose brains were less than two-thirds the size of the average modern human's. This group has been named *Australopithecus africanus*, or gracile Australopithecine (Figure 14–17a). The taller and stockier *Australopithecus boisei*, or robust Australopithecine (Figure 14–17b), was a third group that shared a common ancestor with *A. africanus*. It is quite possible that *A. boisei* and *A. africanus* coexisted for a considerable time.

A fourth group is the one that evolved into modern humans. Morphologically very similar to *A. africanus*, it probably was behaviorally more sophisticated, since its brain was larger and more complex. We call this group *Homo* and designate *Homo habilis* as its earliest known representative.

The recent discovery of an Australopithecine group called *A. afarensis* (Figure 14–17c) has stirred up a rather heated debate over the existence of a fifth group of hominids. Some consider this group a variant form of *A. africanus*, but others consider it a separate line, perhaps ancestral to *A. boisei*, *A. africanus*, and *Homo* (Figure 14–18).

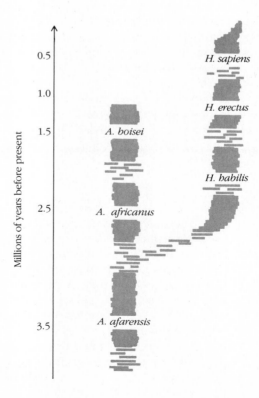

Figure 14–18
Evolutionary relationship of hominids to some other primate families.

Box 14–2   *The Earliest Humans: Pushing the Time Back Farther*

The age-old paradox of which came first, the chicken or the egg, has a sibling paradox in modern studies of human origins: Which came first, the increase in brain size or the upright posture? Anthropologists have for years been frustrated in their efforts to resolve this dilemma in human origins, but recent discoveries of the thigh bone and skull fragment of a 4-million-year-old human may mark the beginning of the end of this problem.

The latest hominid bone discoveries excite researchers and for very good reason. Not only are these bones telling us something about early hominids, but the site of their discovery, the Awash River valley of northern Ethiopia, may tell us more about the environment of the earliest hominids than any other site known.

Until the discovery of the Awash skull fragment and thigh bone, the earliest known upright and walking hominid was "Lucy," immortalized in a book by that name (Johanson and Edey, 1981). Lucy belongs to the hominid group known as *Australopithecus afarensis,* which dates back about 3.7 million years. The new bone discovery puts another 300,000 years onto that age and tells us something very special about these early hominids. Lucy's anatomy clearly suggests that she walked upright but had a very small brain. A careful structural analysis of the skull fragment reveals a brain comparable in size to that of a modern chimpanzee. The structure of the thigh bone clearly indicates upright posture and walking.

These are controversial conclusions and not accepted by all, but the new finds in the arid Awash River valley confirm the idea that Lucy was not an aberrancy. We can now trace the *A. afarensis* group back 4 million years and conjure up images of small hominids who had small brains and walked upright along the margins of grasslands or through the forests that bordered those grasslands.

This discovery reveals that *A. afarensis* was a stable species that persisted for a long time, and that the combination of small brain size and walking was stable. And it indicates that the earlier notion that tool using stimulated upright walking, which in turn stimulated brain enlargement, is probably wrong. Apparently we must rethink our ideas of what prompted the dramatic enlargement of the human brain. The idea of a developmental mutation becomes more appealing with the passage of time.

As this book goes to press, a further expedition is under way to the Awash River valley. Anticipation is high that we may soon be able to lift the shroud of mystery that covers early humans and their way of life. The remarkably complete fossil beds of the Awash promise to tell us much about the world through which the earliest humans walked, lived, loved, and died. Stay tuned!

The only hominid to survive into modern times is *Homo*. All the others became extinct. *Homo* evolved into at least two species before the emergence of modern man: *Homo habilis* and *Homo erectus* (Figures 14–19a and 14–19b). The *Homo sapiens* group consisted of two major subspecies, or races, *Homo sapiens sapiens* and *Homo sapiens neanderthalensis* (Figure 14–19c). These two subspecies coexisted for perhaps 50,000 years, but only *H. sapiens sapiens* exists today. Figure 14–20 summarizes the relationships among some of the hominids we have discussed.

There are many questions we must ask about this evolutionary drama. What forces prompted the diversification of the *Ramapithecus* stock into the hominid line? Why did this occur only in Africa? What made the genus *Homo* so spectacu-

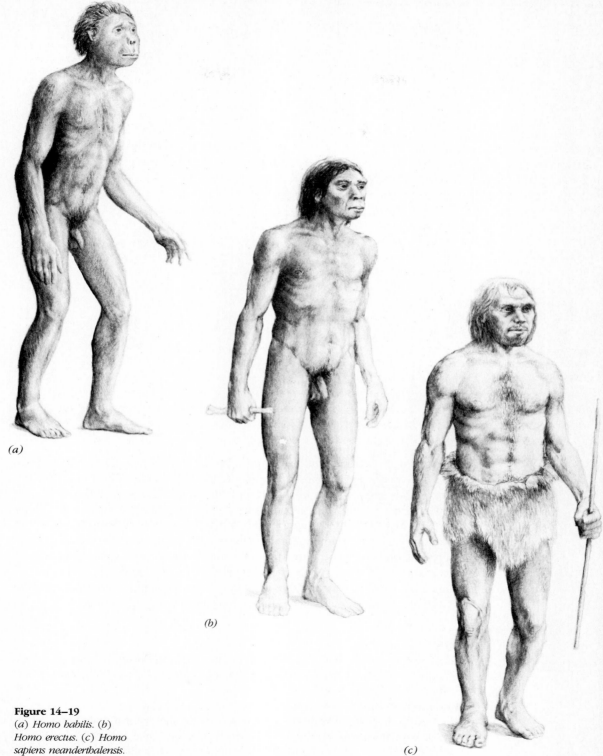

(a)

(b)

(c)

Figure 14–19
(*a*) *Homo habilis.* (*b*)
Homo erectus. (*c*) *Homo
sapiens neanderthalensis.*

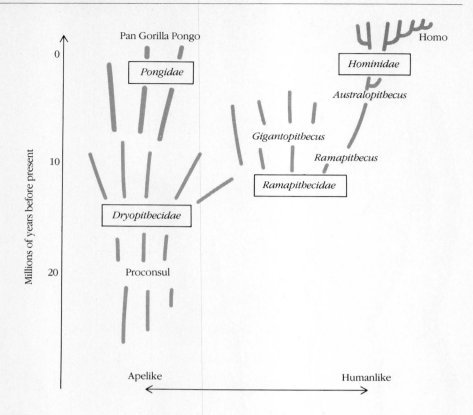

Figure 14–20
An evolutionary scheme of
the hominids.

larly successful? What caused the other hominid groups to become extinct? What
forces led to the extinction of the neanderthal subspecies? Until our research
discovers the answers, our evolutionary history will remain unclear.

The genetic principles discussed in this chapter and throughout the book
have operated over the course of human history. They have been the basis for
generating differences among individuals and among groups, and they led to the
emergence of the modern human. As we come to know more about these genetic
principles, we glimpse more of our evolutionary past. And we can tease our
imagination with an unanswerable question: Barring self-annihilation or an end to
nature's generosity, what will be the evolutionary future of our species?

Summary

1 The human species is highly polymorphic. The **classical model** of this genetic
variation (very few heterozygous loci) has been replaced by the **balance
model** (most loci are heterozygous).

2 **The Hardy–Weinberg principle** can be used to calculate gene frequencies
for recessive alleles, multiple alleles, and sex-linked alleles for a population in
genetic equilibrium—that is, one in which gene and genotypic frequencies

do not change. Genetic equilibrium, or Hardy–Weinberg equilibrium, assumes that mating is random, mutations can be ignored, sampling errors do not occur, all genotypes are equally viable and fertile (no selection), and migrations do not occur.

3 **Inbreeding** is a form of positive assortative mating between individuals who are genetically related. It generally has negative genetic consequences, manifested as increased mortality or increased frequency of rare recessive defects because of increased homozygosity. The effects of inbreeding can be deduced by using the inbreeding coefficient (F).

4 Darwin's theory of evolution through natural selection is based on the idea that individuals with favorable variations are most likely to survive and that the accumulation of variations causes species to change. Thus, selection acts on existing variation in a population. We now recognize **mutation** as the source of this variation.

5 A variant genotype is favorable if it causes one individual to leave more offspring than others, so that the frequency of that genotype increases in each generation.

6 In a population in genetic equilibrium, the rate of unfavorable mutations is balanced by the rate of selection against those mutations, so that neither change nor selection seems to occur.

7 **Gene drift** and **migration** can cause changes in the frequency of genes regardless of whether these genes are favorable or detrimental. Such changes may account for many differences among populations within a species.

8 A **race** is a biological group that has characteristic gene frequencies, or features of chromosomal structure, that distinguish it from individuals belonging to other groups within a species.

9 **Allopatric** (or **geographic**) **speciation** is gradual. It can be contrasted with a mode of evolution called **punctuated equilibrium**, which is characterized by brief periods of rapid change and long periods of relative stability. Both modes may be involved in the evolution of species.

10 Humans may have undergone their most dramatic evolutionary changes as a consequence of changes in only a few genes, probably genes with a gene-regulatory function.

11 Modern humans emerged from a group of hominids spawned by an ancestral stock of hominidlike apes called *Ramapithecus*. All the hominid groups save *Homo* became extinct.

Key Terms and Concepts

Gene pool

Classical model

Balance model

Hardy–Weinberg principle of genetic equilibrium

Inbreeding (consanguineous mating)

Outbreeding

Inbreeding coefficient

Theory of acquired characteristics (theory of use and disuse)

Fitness

Genetic drift

Migration

Race

Allopatric (geographic) speciation

Punctuated equilibrium

Sibling species

Problems

14–1 In estimating the number of color-blind females in a population of 10,000, we arrived at the number 10. What factors may account for the fact that all the color-blind individuals were males, with no observed color-blind females?

14–2 Studies of 108 achondroplastic dwarfs showed that they produced 27 children. These dwarfs had 457 normal siblings who produced 582 children. What is the fitness of the dwarf?

14–3 Working on a column of rock in a Wyoming dig site, a paleontologist uncovers the following sequence of fossils:

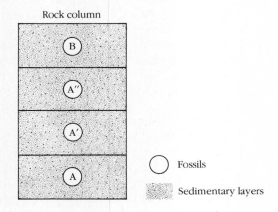

The fossils at the bottom represent Species A. At the next level up (more recent), Species A' is only slightly different. One level further up is A″, which is slightly different from A'. Above the A″ level is a more radically different species that the paleontologist calls B. It is more different from A″ than A″ is from A'. The paleontologist suggests that there was a gradual evolution of A into A' into A″. Between A″ and B was a long period during which sediments were not laid down, so no fossils were made. When conditions once again favored the formation of fossils, evolution had progressed by gradualism to the B stage. Suggest an alternative explanation.

14–4 If hemophilia is due to a sex-linked recessive allele with a frequency of 0.0001, predict the frequencies of the two male genotypes and three female genotypes in the population.

14–5 Criticize the idea that the frequency of an allele in a population is an inverse function of the selection against it.

14–6 Of 82,000 children born to normal parents, 8 were achondroplastic dwarfs. These children would be heterozygous and the result of new mutations. Estimate the mutation rate at this locus.

14–7 Why are all humans considered to be members of the same species?

14–8 Tay–Sachs disease, caused by an autosomal recessive allele, appears about once in every 100,000 births. Assuming Hardy–Weinberg equilibrium, estimate the frequency of this allele in the population and the frequency of carriers (heterozygotes).

14–9 The frequency of *a* in a population is 0.1. In a sample of 40 people, what is the probability that all of them are either *AA* or *Aa*? (Just set up the equation.)

14–10 What are the *A* and *a* gene frequencies in a human population (in equilibrium) in which the *Aa* frequency is 0.5?

14–11 Criticize the contention of a friend who tells you that he can look at a person and tell what race he belongs to.

14–12 The following genotypes were recorded in a sample of people:

AA	*Aa*	*aa*
634	391	85

What are the gene and genotypic frequencies? Is the population in equilibrium?

14–13 About 1 in every 2500 newborn children has cystic fibrosis, a recessive trait that is usually fatal. Estimate the frequencies of the alleles and of the heterozygote in the population.

14–14 A group splits off from the main population and sets up a new population. The group consists of the following genotypes:

MM	*MN*	*NN*
41	38	21

What are the gene frequencies? Is this group representative of an equilibrium population, assuming it is a random sample?

14–15 How will gene frequencies change in three generations if, to begin with, $S_{AA} = 0$, $S_{Aa} = 1.0$, and $p = q = 0.5$?

14–16 Should race be considered a legitimate category of classification?

14–17 In a sample of 219 people, we observe the following genotypes:

AA	*Aa*	*aa*
9	135	75

What are the gene frequencies? Is this population in Hardy–Weinberg equilibrium?

14–18 What proportion of all recessive alleles is found in homozygotes for Tay–Sachs and cystic fibrosis (see Problems 8 and 13)?

14–19 The ability to taste PTC is due to a dominant allele. Two populations were tested for their ability to taste PTC:

	Tasters	Nontasters
Population 1	412	235
Population 2	199	38

Calculate the gene and genotypic frequencies.

14–20 What frequencies of *A* and *a* in a population produce the greatest number of heterozygotes?

14–21 Can a single mutational event give rise to a new species?

14–22 In a population of adult Black Africans, we find the following genotypes for sickle-cell anemia:

AA	*AS*	*SS*
605	390	5

Show that this population is not in Hardy–Weinberg equilibrium, and discuss why it is not.

Answers

See Appendix A.

References

Ayala, F. J. 1976. *Molecular Evolution*. Sinauer, Sunderland, MA.

Ayala, F. J., and J. Valentine. 1979. *Evolving: The Theory and Processes of Organic Evolution*. Benjamin/Cummings, Menlo Park, CA.

Bajema, C. J. 1971. *Natural Selection in Human Populations*. Wiley, New York.

Bodmer, W., and L. L. Cavalli-Sforza. 1976. *Genetics, Evolution, and Man*. Freeman, San Francisco.

Dobzhansky, Th. 1970. *Genetics of the Evolutionary Process*. Columbia University Press, New York.

Dobzhansky, Th., F. J. Ayala, G. L. Stebbins, and J. W. Valentine. 1977. *Evolution*. Freeman, San Francisco.

Gillespie, J. H., and C. H. Langley. 1979. Are evolutionary rates really viable? *J. Mol. Biol.* 13:27–34.

Glass, B. 1954. Genetic changes in human populations, especially those due to gene flow and genetic drift. *Adv. Genet.* 6:95–139.

Gould, S. J., and N. Eldredge. 1977. Punctuated equilibria: The tempo and mode of evolution reconsidered. *Paleobiology* 3:115–151.

Grant, V. 1977. *Organismic Evolution*. Freeman, San Francisco.

Hardy, G. H. 1908. Mendelian proportions in a mixed population. *Science* 28:49–50.

Jenkins, J. B. 1979. *Genetics,* 2nd ed. Houghton Mifflin, Boston.

Johanson, D., and M. Edey. 1981. *Lucy: The Beginnings of Humankind.* Simon and Schuster, New York.

Kimura, M. 1968. Evolutionary rate at the molecular level. *Nature* 217:624–626.

King, M.-C., and A. C. Wilson. 1975. Evolution at two levels: Molecular similarities and biological differences between humans and chimpanzees. *Science* 188:107–116.

Leakey, R. E. 1977. *Origins.* Dutton, New York.

Leakey, R. E. 1981. *The Making of Mankind.* Dutton, New York.

Lerner, I. M., and W. J. Libby. 1976. *Heredity, Evolution, and Society,* 2nd ed. Freeman, San Francisco.

Lewontin, R. C. 1974. *The Genetic Basis of Evolutionary Change.* Columbia University Press, New York.

Lovejoy, C. O. 1981. The origin of man. *Science* 211:341.

Spiess, E. 1977. *Genes in Populations.* Wiley, New York.

Wallace, B. 1968. *Topics in Population Genetics.* Norton, New York.

White, M. J. D. 1977. *Modes of Speciation.* Freeman, San Francisco.

Yule, G. U. 1908. Mendel's laws and their probable relation to intraracial heredity. *New Phytol.* 1:192–207, 222–238.

Yunis, J. J., and O. Prakash. 1982. The origin of man: A chromosomal pictorial legacy. *Science* 215:1525–1530.

Appendix A
Answers

2

2–1 32

2–2 128

2–3 1/16

2–4 each is *AaBbCc*

2–5 27/64

2–6 64 phenotypic classes; 64 genotypic classes

2–7 9 black, 3 red, 3 liver, 1 lemon

2–8 9 white disc, 3 white spheroid, 3 yellow disc, 1 yellow spheroid

2–9 recessive

2–10 recessive; *1 = *Aa*; *2 = *Aa*; *3 = *Aa*

2–11 recessive; mating III8 × III9 is critical; * = *Aa*

2–12 recessive; *Aa*

2–13 dominant; * = *aa*

2–14 recessive

2–15 recessive; * = *Aa*

2–16 dominant; *1 *Aa*; *2 = *aa*

2–17 dominant; * = *aa*

2–18 dominant; * = *aa*

2–19 $\dfrac{n!}{m!\,(n-m)!}\,p^m q^{n-m} = (a):\ \dfrac{5!}{3!\ 2!}\ (\tfrac{3}{4})^3(\tfrac{1}{4})^2 = 270/1024$

$= (b):\ \dfrac{5!}{4!\ 1!}\ (\tfrac{3}{4})^4(\tfrac{1}{4}) = 405/1024$

2–20 1/16

2–21 6/16

2–22 1/32

2–23 ¼ *A-B-*, ¼ *A-bb*, ¼ *aaB-*, ¼ *aabb*

3

3–1 (*a*) and (*b*) are haploid because the chromosomes are not in pairs; (*c*) is diploid. (*a*): metaphase; (*b*): early anaphase; (*c*): late anaphase

3–2 (*a*): anaphase I; (*b*): prophase I; (*c*): metaphase I; (*d*): telophase II; (*e*): anaphase II.

3–3 $2^4 = 16$

3–4 *ab ab AB AB*

3–5 4

3–6 $2^7 = 128$

3–7 $(\frac{1}{2})^{23}$

3–8

Mitotic
prophase

Mitotic
metaphase

Mitotic
metaphase

Meiotic
metaphase I

Meiotic
prophase I

Meiotic
metaphase I

Meiotic
metaphase II

Meiotic
metaphase II

3–9 meiotic metaphase I

3–10 Endoreduplication, a type of endomitosis in which the chromosomes duplicate but the cell fails to divide.

3–11 (*a*) is haploid because each chromosome does not have a homolog; (*b*) is diploid because each chromosome does have a homolog.

3–12 The main purpose of meiosis is to maintain constancy in chromosome number from generation to generation. It does this by the generation of haploid gametes, each containing a representative of each chromosome pair, and by gamete fusion, which reestablishes the diploid state.

3–13 Through independent assortment of chromosomes and crossing over, meiosis produces new gene combinations that maintain variability in a species, which is crucial to the evolutionary success of any species.

3–14 Each generation would have double the number of chromosomes found in the preceding generation.

3–15 In meiosis I, the chromosomes pair and the centromeres do not split; in mitosis, the chromosomes do not pair and the centromeres do split.

3–16 These two divisions are, in their general scheme, the same.

3–17 Yes. Two haploid gametes fuse to produce a diploid zygote, which then undergoes meiosis to produce four haploid cells. Each of these haploid cells divides mitotically to produce the haploid individual. A haploid individual cannot undergo meiosis because its chromosomes do not exist as pairs.

3–18 sperm↘
      ↗zygote   adult<sperm↘
      egg↗   >zygote . . .
                   ↘egg↗

3–19 The nucleus, not the cytoplasm, contains the genetic information; the nuclei in both gametes are genetically equivalent with respect to amount of genetic material; only the cytoplasms differ.

3–20 (a) 46   (e) 23
     (b) 23   (f) 23
     (c) 23   (g) 46
     (d) 46   (h) 46
     (e) 46   (i) 46

3–21 800; 200

3–22 200; 200

4

4–1 1/16

4–2 autosomal dominant; sex-limited to males

4–3 X-linked dominant

4–4 X-linked recessive

4–5 autosomal dominant with incomplete penetrance

4–6 No; if an XX person has H-Y antigen, that person will develop into a male.

4–7 probably autosomal; dominant in males and recessive in females; sex-influenced

4–8 a Y translocation excluding the H-Y gene

4–9 1 XX (female); 2 XY (male); 1 YY (lethal)

4–10 (a) incomplete penetrance
      (b) son was XXY, with the XY from the father

4–11 all females

4–12 X-linked recessive

4–13 (a) $\frac{1}{2}$; (b) 0; (c) 0

4–14 (a) $\frac{1}{2}$; (b) $\frac{1}{2}$; (c) $\frac{1}{2} \times \frac{1}{2} = \frac{1}{4}$

4–15 autosomal; dominant in males, recessive in females

4–16

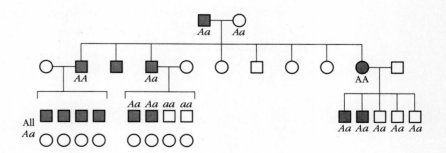

4–17 (a) The mother was Aa; the father was A.
      (b) For a daughter, it is 0; for a son, it is $\frac{1}{2}$.
      (c) Probability of a boy is $\frac{1}{2}$; probability of the disease is $\frac{1}{2}$; so $\frac{1}{2} \times \frac{1}{2} = \frac{1}{4}$.

4–18 (a) Male heterozygotes express the trait, but female heterozygotes do not; only AA females are bald.
      (b) The mother was nonbald, the father was bald.

5

5–1 *A*_____*B*

 *a*_____*b*

5–2 *A*_____*b*
 *a*_____*B*

5–3 *Ab*: 0.4 noncrossover
 aB: 0.4 noncrossover
 AB: 0.1 crossover
 ab: 0.1 crossover

5–4 (a)

        Single   Single
     Crossover Crossover
    *A*_____*B*_____*c*

    *a*_____*b*_____*C*

 (b) *ABC*; *abd*; *aBc*

5–5 Chromosome 18

5–6 G6PD, HGPRT, and PGK gene loci are on the long arm of the X chromosome distal to the break point; the NP locus is on the portion of 14 exclusive of the tip.

5–7 It is on the tip of 2

5–8 The Hpα locus is close to the fragile site.

5–9 See genes listed for 17*q*21 → *q*ter on human gene map (Figure 5–19).

5–10 *A*___**10**___*B*___**6**___*C*

 *B*___**10**___*A*___**16**___*C*

5–11 *A*___**10**___*B*___**6**___*C*

        16

6

6–1 father; meiosis II

6–2 four (X, XY, YY, Y)

6–3 21, 21, XX 21, 21, 21, XX
     21, 21, XY 21, 21, 21, XY
     21, 21, XXY 21, 21, 21, XXY
     21, 21, XYY 21, 21, 21, XYY

6–4 25% would be 21, 21, 21, 21 (tetrasomic); 50% would be 21, 21, 21 (trisomic); 25% would be 21, 21 (disomic).

6–5 Tetrasomy is lethal; the abnormalities are too great to overcome. Therefore, only 75% of the zygotes formed survive to birth.

6–6 Daughter has Turner syndrome and carries a single X from the mother.

6–7 Translocation 7; 13

6–8 Trisomy 21

6–9 XXY

6–10 XYY

6–11

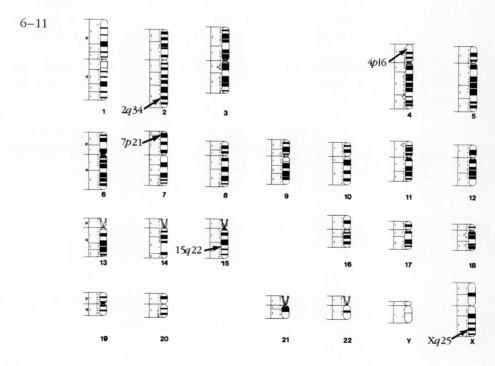

6–12 X chromosome nondisjunction in both meiosis I and meiosis II results in an egg with four X chromosomes. When this egg is fertilized by an X-bearing sperm, an XXXXX zygote forms.

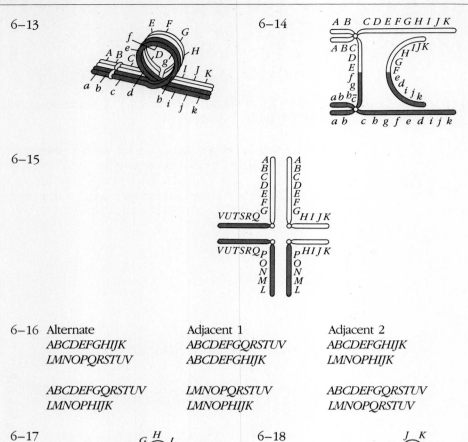

6–13

6–14

6–15

6–16 Alternate

Alternate	Adjacent 1	Adjacent 2
ABCDEFGHIJK	ABCDEFGQRSTUV	ABCDEFGHIJK
LMNOPQRSTUV	ABCDEFGHIJK	LMNOPHIJK
ABCDEFGQRSTUV	LMNOPQRSTUV	ABCDEFGQRSTUV
LMNOPHIJK	LMNOPHIJK	LMNOPQRSTUV

6–17

6–18

6–19 The mother is a balanced translocation heterozygote.

7

These are basically review questions. You can find the answers to most of them by checking the page and figure numbers given.

7–1 233–234

7–2 Figure 7–10; 233

7–3 234

7–4 235; Figure 7–9

7–5 238; Figure 7–13

7–6 237

7–7 242

7–8 One mRNA could be inactivated.

7–9 Some genes code for just RNA.

7–10 243

7–11 242

7–12 241

7–13 243

7–14 244

7–15 244

7–16 245

7–17 247

7–18 246

7–19 247

7–20 249–250

8

8–1 The repressor protein cannot bind to *O*.

8–2 The repressor protein cannot bind to the inducer.

8–3 The repressor protein has two recognition sites: one for the inducer and one for the *O* region.

8–4 constitutive mRNA synthesis

8–5 No. The repressor described in Problem 8–2 would repress both *O* regions.

8–6 a mutation in the *P* region, or a repressor protein defect

8–7 The first observation supports the theory; the second does not.

8–8   Preexisting RNA in the embryo is used up to gastrulation; then new RNA synthesis is blocked by the inhibitor, and development stops.

8–9   No. We need apply the Hayflick limit only to some cells and tissues that control crucial metabolic functions.

8–10   The gene concept has a strong functional component. That is, we have characterized genes by their structure and function. Nuclear transplantation studies add a new dimension to the gene concept by emphasizing differential gene activity in the developing organism.

8–11   The Lyon hypothesis explains this observation.

8–12   not unless the *O* is also affected

9

These are basically review questions. You can find the answers to most of them by checking the page and figure numbers given.

9–1   243, 244

9–2   243–246

9–3   Such a mutation would change more than one amino acid.

9–4   when a mutation increases reproduction

9–5   247, 248

9–6   249

9–7   251

9–8   253

9–9   253

9–10   253

9–11   260

9–12   263

9–13   Mutations can affect different repair enzymes.

9–14   263

9–15 (a) 6 (b) 11

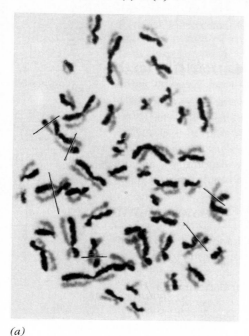

(a)

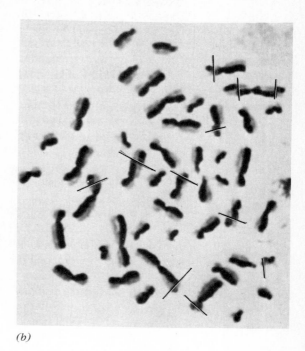

(b)

10

10–1 *B* enters another pathway.

10–2 $ab^+ab^+ \times a^+ba^+b = ab^+a^+b$
where *a* and *b* represent different recessive loci that cause the albino phenotype to appear. In the offspring, the dominant allele (a^+ or b^+) at each locus results in a nonalbino appearance.

10–3 β^t/β^c and β^s/β^c

10–4 See page 326

10–5 See Figure 10–12

10–6 They can become established by conferring a selective advantage to the heterozygote, such as resistance to malaria, as is the case in sickle-cell anemia.

10–7 The α gene has four copies (two copies on each member of chromosome 16), whereas the β gene has only two copies.

10–8 They can go unnoticed if the defective enzyme is involved in a relatively minor metabolic pathway. They are sometimes discovered through special secondary effects, such as drug sensitivity, or routine metabolic studies involving some other clinical problem.

11

These are all review questions, and you can find the answers to them in the chapter. The answer to Problem 11–26 is: best = c; worst = d. For the answer to Problem 11–27, see the Bombay phenotype (page 312).

12

12–1 Epistasis refers to the interaction between nonallelic genes; dominance is the interaction between alleles.

12–2 This is a difficult question with no clear answer. It is important for you to realize that not all problems have clear-cut answers.

12–3 (a) 63.5; 68.1; 65.9; (b) variance = 145.6; SD = 12.1

12–4 (a) This is a trihybrid cross with three independently assorting gene loci. (b) The parental genotypes were $AABBCC \times aabbcc$. The F_1 was $AaBbCc$. We would expect 1/64th of the F_2 to be white, but in a sample of 78, it is so low in frequency that it may not have been produced.

12–5 Wing span: $V_T = V_E + V_G$ Beak length: $V_T = V_E + V_G$
$271.4 = 71.2 + V_G$ $627.8 = 107.3 + V_G$
$V_G = 200.2$ $V_G = 520.5$
$H = \dfrac{200.2}{271.4} = 0.74$ $H = \dfrac{520.5}{627.8} = 0.83$

12–6 10 dominants (1)   4 dominants (210)
   9 dominants (10)   3 dominants (120)
   8 dominants (45)   2 dominants (45)
   7 dominants (120)   1 dominant (10)
   6 dominants (210)   0 dominants (1)
   5 dominants (252)

12–7 MZ = 1; parent = $\frac{1}{2}$; child = $\frac{1}{2}$; first cousin = $\frac{1}{8}$; great-grandparent = $\frac{1}{8}$; uncle = $\frac{1}{4}$.

12–8 Environmental conditions are different between parent and offspring; there is segregation of different modifier genes; and increased heterozygosity in the offspring.

12–9 $AABBcc$; $AABbCc$; $AAbbCC$; $AaBBCc$; $AaBbCC$; $aaBBCC$

12–10 Two different loci product deaf-mutism:

$AAbb \times aaBB$ Deaf-mute
$AaBb$ Normal

12–11 There are genes for disease susceptibility.

12–12 Multifactorial inheritance usually refers to traits that are determined by both genetic and environmental factors, without reference to the genetic factors.

Polygenic inheritance usually refers to traits that are determined by several interacting gene loci, without specific reference to environmental factors. It is common, however, for the two terms to be used synonymously.

12–13 *A/a b/b c/C* × *A/a b/b c/C* (both "brown")
    a/a b/b/C/c ("blue")

12–14 Schizophrenia may be the result of a dominant allele with incomplete penetrance and variable expressivity; or it may be the result of several interacting gene loci. Both models could produce a continuous pattern of variation.

12–15 The definition of phenocopy is given in the glossary; examples of phenocopies are thalidomide-induced phocomelia; the normal condition caused by insulin injection in a diabetic; and the normal condition caused by the antihemophilic factor injected into a hemophiliac.

13

13–1 There is less difference between MZ twins than between DZ twins. You cannot tell much about the environmental influence. The data do not tell us whether the MZ twins were raised together or apart.

13–2 Phenylketonuria; Lesch–Nyhan syndrome; galactosemia; others.

13–3 General population: 0.9
    Parents of schizophrenics: 4.2
    Siblings of schizophrenics: 7.5
    Children of schizophrenics: 9.7
    Second degree relatives: 2.1
    Third degree relatives: 1.7

13–4 *AA* = lethal; *Aa* = zigzag; *aa* = normal. The ratio is 1:2:1.

13–5 Yes; *UuRr* × *UuRr* 1/16 *uurr*

13–6 This term is defined in the glossary and in both this chapter and Chapter 6. Mosaics have been especially useful in studies of *Drosophila* behavior, such as those we discussed in this chapter.

13–7 Two examples are the hyperkinetic and the shaker mutants, both discussed in this chapter.

13–8 Mate the unknown mouse to a homozygous normal mouse; if the offspring are normal, the abnormal mouse is a waltzer; if half of the offspring are abnormal, the abnormal mouse is a twirler.

13–9 This question requires a review of material covered in part of this chapter; it could serve as a topic for a term paper.

13–10 This is a review question.

13–11 This is a review question, and its discussion could be expanded to make a term paper.

14

14–1 An expectation of 10 per 10,000 is quite small, so finding 0 per 10,000 may be the result of sampling. However, this is not the most reasonable explanation. For a female to be color-blind, her mother had to be a carrier and her father color-blind. The probability of such a marriage is about 1.5%, and only half of the offspring of such a mating are expected to be color-blind. Thus, the production of a color-blind female is very unlikely.

14–2 $\dfrac{27/108}{582/457} = 0.197$

14–3 Another way to interpret this fossil series is to suggest that A″ had a wide distribution and underwent a speciation process at some lateral point:
The new species, B, reinvaded the old habitat and forced out A″.

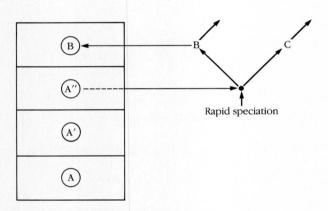

14–4 Since $b = 0.0001$, $H = 0.9999$:

	Genotypes	Frequencies
Males	bY	0.0001
	HY	0.9999
Females	HH	0.9998
	Hb	0.0002
	bb	10^{-8}

14–5 A mutant recessive that is homozygous lethal may be maintained in the heterozygous state if it confers a selective advantage on the heterozygote. Moreover, selection operates against phenotypes, not against individual genes.

14–6 82,000 children = 164,000 alleles. Therefore, the mutation rate is 8/164,000 = 0.0000425.

14–7 All people can interbreed freely.

14–8 $tt = q^2 = 10^{-5}$
$q = 0.003$; therefore $p = 0.997$
$Tt = 2pq = 0.006$

14–9 $AA = 0.81$
$Aa = 0.18$
$aa = 0.01$
The probability that all are either AA or Aa is $(0.99)^{40}$.

14–10 $2pq = 0.5$
$pq = 0.25$
$q = 1-p$
$p(1 - p) = 0.25$
$p^2 - p = 0.25 = 0$
$(p - 0.5)^2 = 0$
$p = q = 0.5$

14–11 Race is a populational phenomenon, involving gene-frequency differences in populations or groups that inhabit different regions of a species distribution area. A single individual does not represent a gene-frequency distribution.

14–12
	AA	Aa	aa	A	a
Frequency:	0.571	0.352	0.077	0.747	0.253

The population is in equilibrium.

14–13 $q^2 = \frac{1}{2500}$, so $q = 0.02$ and $p = 0.98$
$2pq = 0.0392$

14–14 $M = 0.6$, and $N = 0.4$
Expected: $MM = 0.36$, $MN = 0.48$, and $NN - 0.16$
This population is probably in equilibrium.

14–15 They will not change.

14–16 No; it is too arbitrary a unit of classification.

14–17 $A = 0.349$, and $a = 0.651$
	AA	Aa	aa
Expected:	0.12	0.46	0.42
Observed:	0.04	0.62	0.34

This population is not in equilibrium.

14–18 Tay–Sachs = 0.003; cystic fibrosis = 0.02

14–19 Population 1: $t = 0.6$, $T = 0.4$
     $tt = 0.36$, $Tt = 0.48$, $TT = 0.16$
Population 2: $t = 0.4$, $T = 0.6$
     $tt = 0.16$, $Tt = 0.48$, $TT = 0.36$
Note that the frequencies are calculated as expected genotypes and, because of sampling procedures, do not match exactly the observed frequencies.

14–20 $A = a = 0.5$

14–21 A single gene mutation may have profound effects on several other genes if the gene that is mutating is a regulator gene. This type of mutational event could lead to a rapid change in a subpopulation, with reproductive isolation as a possible result. But this outcome would be extremely rare. If the mutational change involved a change in chromosomal number or structure, a very rapid evolution could result.

14–22 $S = 0.2$, and $A = 0.8$
Expected: $AA = 640$, $AS = 320$, $SS = 40$
We find fewer AA than expected because this genotype is more susceptible to malarial infections; we find more AS than expected because this genotype is selectively advantageous in an environment of malaria; the SS genotype is semilethal.

Appendix B
Glossary

Acentric chromosome A chromosome or chromatid lacking a centromere.

Acquired characteristics A theory that argues that traits that an organism acquires by accommodating to the environment are assimilated into the genetic material and transmitted to the next generation.

Acrocentric chromosome A chromosome or chromatid with the centromere near one end.

Allele One of two or more forms of a given gene.

Amino acid The basic building blocks of polypeptides.

Amniocentesis A procedure for sampling the amniotic fluid by direct tap.

Anaphase The stage in mitosis or meiosis when the chromosomes move to opposite poles of the cell.

Androgen A male sex hormone.

Aneuploidy The loss or gain of one or more chromosomes, as compared to the basic chromosome complement.

Antibody An immunoglobulin that is produced in response to exposure to an antigen and reacts very specifically with that antigen.

Anticodon The triplet of tRNA nucleotides that is complementary to, and pairs with, a codon in the messenger RNA.

Antigen Any substance capable of inducing antibodies.

Assortative mating The nonrandom formation of mating pairs; can be positive or negative.

Autoimmunity Antibody production stimulated by a person's own antigens.

Autosome A chromosome other than the sex chromosome.

Backcross The cross of an F$_1$ hybrid with one of the parental lines.

Bacteriophage A bacterial virus.

Bands, chromosome Areas of light or dark staining produced by a variety of techniques.

Barr body The sex chromatin as seen in female somatic cells.

Base analog A DNA or RNA base that resembles the normal bases, but is different, so that it is incorporated into the nucleic acid molecule in place of the normal base.

Base pair In DNA, A must pair with T, and G must pair with C.

Bivalent The figure produced by the pairing of two homologous chromosomes.

Blood group A genetically determined antigen on a red blood cell.

B lymphocytes Small lymphocytes that respond to an antigen by producing circulating or humoral antibodies.

Carcinogen A chemical capable of causing cancer.

Carrier An individual who is heterozygous for a normal allele and for an abnormal allele that is not phenotypically expressed.

Centric fusion Fusion of the long arms of two acrocentric chromosomes at the centromere. See also *Robertsonian translocation*.

Centriole One of the pair of organelles that form the points of focus of the spindle during cell division.

Centromere The region of the chromosome that becomes associated with spindle fibers during cell division; also called the primary constriction or kinetochore.

Chiasma The point at which homologous chromosomes remain attached after pairing has ceased in meiosis (plural: chiasmata).

Chimera An individual composed of cells derived from different zygotes.

Chromatid One of the two daughter strands of a replicated chromosome, joined by the centromere to the other daughter chromatid.

Chromatin The DNA–protein complex of chromosomes.

Chromomere A densely coiled region of chromatin on a chromosome, giving the extended chromosome a beaded appearance.

Chromosome In eukaryotes, a DNA–protein complex that is a linear array of genes.

Chromosome bands See *Bands, chromosome*.

Chromosome lag The failure of a chromosome to remain in its complement during anaphase.

Chromosome map A representation of the linear arrangement of genes on a chromosome, as deduced from genetic and cytological observations.

Chromosome theory of inheritance The established theory that genes and chromosomes are linked and that chromosomes are the carriers of the genetic information.

Cistron A genetic unit of function, usually equated with the term *gene.*

Clone A population of cells originally derived from a single cell by mitosis.

Codominance The expression of both of two different alleles in a heterozygote.

Codon The triplet of nucleotides in DNA or RNA that specifies a particular amino acid in a polypeptide chain.

Coefficient of inbreeding See *Inbreeding coefficient.*

Colchicine A drug that inhibits the formation of the spindle and delays the division of centromeres.

Complement A complex mixture of blood plasma components that are bound in antigen–antibody reactions and that are necessary for cell lysis.

Concordance Similarity in a twin pair, with both members expressing a certain trait.

Congenital Present at birth; may be genetic or environmental.

Consanguinity Relationship through a common ancestor.

Constitutive enzyme An enzyme that is produced at a fixed rate, irrespective of need.

Co-repressor A molecule that combines with a repressor to inhibit the function of an operon.

Correlation coefficient A value from -1 to $+1$ that indicates the degree to which statistical variables vary together.

Crossing over The exchange of genetic material between members of a chromosome pair.

Cytogenetics The study of the relationship between the appearance of chromosomes and the genotype/phenotype of the individual.

Cytokinesis Cytoplasmic division.

Cytoplasmic inheritance The inheritance of traits through the cytoplasm instead of through the nuclear chromosomes.

Delayed-onset trait A trait that does not appear at birth, but does show up later.

Deletion The loss of a chromosome segment.

Deoxyribonucleic acid (DNA) The chemical basis of heredity.

Diakinesis A stage in meiosis in which the chromosomes are maximally condensed.

Dicentric chromosome A chromosome with two centromeres.

Dihybrid cross A cross between individuals differing with respect to two pairs of alleles.

Diploid Autosomes are each present twice; that is, each autosome has a homologous partner.

Diplotene A stage in prophase of meiosis I in which the chromatids become visibly separate.

Discontinuous variation Variation in a population that falls into two or more nonoverlapping classes.

Discordance Dissimilarity in a twin pair, with only one member expressing a particular trait.

Disjunction The separation of chromosomes during cell division.

Dizygotic twins Twins produced by two separate ova, separately fertilized; also known as fraternal twins.

DNA See *Deoxyribonucleic acid.*

Dominant gene An allele that is expressed when only one copy is present.

Dosage compensation The cellular mechanism that compensates for the activity of genes, which, because of their location on the X chromosome, exist in two doses in the female and one dose in the male.

Duplication The occurrence of a chromosomal segment more than once in the genome.

Electrophoresis The differential movement of molecules through a gel under the influence of an electric field; a method for separating molecules.

Endonuclease An enzyme that breaks the bonds between adjacent nucleotides in the interior of a DNA or RNA chain.

Enzyme A protein molecule that catalyzes a specific chemical reaction.

Epistasis The masking by one gene of the expression of another, nonallelic gene.

Equational division The separation of chromosomes into daughter cells with complements similar to the parental cell; describes meiosis II.

Equatorial plane The plane at which the chromosomes align during metaphase of mitosis and meiosis.

Equilibrium In population genetics, the state at which the forces that tend to change gene frequencies are counterbalanced so that there is no net change in gene frequencies from one generation to the next.

Erythrocyte Red blood cell.

Euchromatin The chromosomal region that carries genes and characteristically stains lightly.

Eugenics A philosophy concerned with improving the genetic quality of a population.

Eukaryote An organism characterized by true chromosomes and a membrane-bound nucleus.

Euploid Having the basic haploid complement of chromosomes or a multiple of the basic complement.

Exon A sequence of DNA that is translated into protein.

Exonuclease An enzyme that breaks bonds between adjoining nucleotides only when one of the nucleotides is a terminal one in a DNA or RNA chain.

Expressivity The variation in phenotype associated with a particular genotype.

F See *inbreeding coefficient*.

F_1 The first-generation progeny of a mating.

Feedback inhibition Inhibition of an enzyme by a product of the pathway catalyzed by the enzyme.

Fertilization The fusion of an egg and a sperm to form a zygote.

Fibroblast A predominant cell type in connective tissue.

Fitness A quantitative measure of an organism's ability to survive and transmit its genes to the next generation.

Founder principle A type of genetic drift, in which a small number of individuals from a larger population break off and start a new population. These individuals are not representative of the gene frequencies expressed by the larger parent population.

Fragile site A chromosomal point where breaks may occur spontaneously; it appears to be heritable.

Frameshift mutation A shift in the reading frame during translation, caused by the addition or deletion of bases in a DNA molecule.

Gamete A mature germ cell, either ovum or spermatozoon.

Gene The basic unit of inheritance that occupies a specific locus on the chromosome and has a specific function.

Gene pool The total pool of genes in a population.

Genetic code Sequences of three nucleotides in DNA or RNA that specify amino acids when translated into polypeptides.

Genetic counseling Advising prospective parents of the probability of their having a child with a specific genotype.

Genetic drift Changes in allelic frequencies due to chance sampling of the parent population.

Genetic lethal A genetically determined trait in which the affected individuals do not reproduce.

Genetic marker A single gene trait used to follow the transmission of chromosomes in a mating.

Genetic load The decrease in fitness of a population due to detrimental alleles.

Genetic screening Testing on a population to identify individuals at risk of having a specific genetic disorder or of having a child with such a disorder.

Genome The totality of genes in the haploid set; can also refer to the complete gene complement without regard to the ploidy.

Genotype The genetic constitution of an organism.

Germ line The cell line that produces gametes.

Gonads The reproductive organs in which the germ cells are found and gametes produced (ovaries and testes).

Gynandromorph An individual who is a sexual mosaic, carrying both female tissue and male tissue.

Haploid The chromosome set with only one member of each chromosome pair.

Hardy–Weinberg law A law stating the expected gene frequencies under conditions of random mating.

Hemizygous The condition of having only one set of genes instead of two, as in the case of the genes of the X chromosome in a male.

Heritability A measure of the degree to which the total phenotypic variance is the result of genetic factors and thus can be influenced by selection.

Hemoglobin The oxygen-carrying pigment of red blood cells.

Hemoglobinopathies Abnormal hemoglobins caused by changes in the amino acid composition of the polypeptides.

Heterochromatin Chromatin that remains compacted and stains deeply in interphase.

Heterogametic Producing two types of gametes with respect to sex chromosomes.

Heterogeneity A state of nonuniform composition. If a certain phenotype or similar phenotypes can be produced by different genetic mechanisms, the phenotype is genetically heterogenous.

Heterogenous nuclear RNA (hnRNA) The RNA in the nucleus that still contains introns.

Heterokaryon A cell having two or more genetically different nuclei.

Heterosis The vigor of hybrids, which is greater than

that of the parental lines.

Heterozygous Pertaining to two different alleles at a specific locus in a diploid organism.

Histocompatability Tolerance to transplanted tissue.

Histone A group of basic proteins associated with chromosomes.

Holandric Pertaining to the inheritance of genes linked to the Y chromosome.

Homogametic Producing gametes of a single type; in humans, females are homogametic.

Homologous chromosomes Chromosomes that are identical in their content of gene loci.

Homozygous In a diploid organism, characterized by both alleles being the same at a specific locus.

Human lymphocyte A complex (HLA) A region on chromosome 6 that includes genes determining major transplantation antigens as well as many other genes important in the immune process.

H-Y antigen A gene product, probably Y-linked, that triggers the undifferentiated gonad to develop into testicles.

Hybrid A cross between two distinct types, races, or species.

Hybrid cell A cell derived from two different cultured cell lines that have fused.

Hybridoma A cell hybrid made by the fusion of a cancer cell with a lymphocyte that has been induced to produce antibodies.

Immune reaction A specific reaction between antigen and antibody.

Immunoglobulin The antibody molecule.

Immunologic deficiency disease A disease caused by a deficiency of cells capable of producing antibody.

Inborn error of metabolism A genetically determined biochemical disorder in which a specific enzyme defect produces a metabolic block that has pathological consequences.

Inbreeding A mating between related persons.

Inbreeding coefficient F; the probability that an individual has received both alleles of a pair from an identical ancestral source; or the proportion of loci at which he or she is homozygous.

Incomplete dominance A condition that exists when the phenotype of the heterozygote is intermediate between the two parental extremes.

Independent assortment The distribution that occurs when a pair of alleles located on different chromosome pairs line up and segregate in an independent fashion during meiosis.

Inducer A small molecule that causes a cell to produce larger amounts of the enzyme(s) needed to metabolize that molecule.

Interphase The part of the cell cycle between divisions.

Intersex A member of a bisexual species that has sexual characteristics intermediate between male and female.

Intervening sequence A group of excess DNA bases that interrupts the sequence of amino acid–coding bases at irregular intervals within the gene; the group is eliminated enzymatically from the DNA or mRNA transcript, leaving the amino acid–coding bases in an uninterrupted sequence; synonymous with intron.

Intron See *Intervening sequence.*

Inversion A chromosomal rearrangement in which a central segment produced by two breaks is inverted prior to the repair of the breaks.

In vitro In the test tube; outside of the living body.

In vivo Within the living body.

Isochromosome A chromosome with two identical arms.

Isozyme An enzyme that can be distinguished by some property, usually electrical charge, but that acts enzymatically on the same substance.

Karyotype The chromosome complement.

Kinetochore See *Centromere.*

Leptotene The early stage of prophase in meiosis I before the chromatids are visible as separate structures.

Linkage Association of loci on the same chromosome.

Locus The position of a gene on a chromosome.

Lymphocyte One of the major groups of white blood cells.

Lyon hypothesis The hypothesis that the Barr body is an inactivated X chromosome and that the number of such bodies is equal to the number of X chromosomes less one.

Lysosome A cell organelle containing enzymes that function for the most part in the breakdown of metabolites.

Macrophage A large white blood cell that ingests foreign material in the bloodstream.

Map unit On a linkage map, one map unit is equal to 1% recombination between linked genes.

Mean The sum of all quantities, divided by the number of quantities.

Median The middle value in a group of numbers arranged in order of size.

Meiosis The process by which chromosomes replicate, form homologous pairs, and then segregate into different nuclei to produce the haploid condition.

Meiotic nondisjunction The failure of two members of a chromosome pair to separate during meiosis, so that both members go to one daughter cell, and the other daughter cell gets neither.

Melanin The dark pigment responsible for coloration of skin, hair, and iris of the eye.

Messenger RNA (mRNA) The gene transcript that is translated into polypeptide.

Metacentric chromosome A chromosome with a centromere in or near the middle.

Metafemale A female with three X chromosomes.

Metaphase A stage in cell division when the chromosomes are maximally condensed and arrayed in a plane between the poles of the cell.

Metaphase plate The plane on which metaphase chromosomes are aligned; also known as equatorial plane.

Missense mutation A codon change that results in the substitution of one amino acid for another.

Mitochondria Cytoplasmic organelles associated with oxidative metabolism and energy production.

Mitosis Cell division resulting in the formation of two cells, each with the same chromosome complement as the parent cell.

Mitotic nondisjunction The failure of two members of a chromosome pair to separate at mitosis, so that both go to one daughter cell, and the other daughter cell gets neither.

mRNA See *Messenger RNA*.

Mode The class of numbers that contains the largest number of representatives in a statistical sample.

Modifier gene A gene that modifies the expression of another, nonallelic gene.

Monoclonal Derived from the mitotic division of a single cell.

Monohybrid cross A cross between individuals differing in a single pair of alleles.

Monosomy The existence of only one copy of a particular chromosome instead of two.

Monozygotic twins Twins that arise from a single zygote; also known as identical twins.

Mosaic An individual with two or more genetically different cell lines derived from a single zygote.

Multifactorial Pertaining to a trait influenced by variation at several genetic loci; often synonymous with polygenic.

Multiple alleles One of three or more alternative forms of an allelic series that all map to a specific genetic locus.

Mutagen Any substance that induces mutations.

Mutant A gene in which a mutation has occurred, or the individual expressing that gene.

Mutation The process of changing one allele into another.

Mutation rate The rate at which mutations occur at a given locus, expressed as mutations per gamete per locus per generation.

Natural selection A process of differential fertility, in which some genotypes under the influence of a defined set of environmental parameters are more successful in leaving progeny than others.

Nondisjunction The failure of chromosomes to separate properly in cell division.

Nonsense mutation A mutation that changes a codon specifying an amino acid into one that specifies no amino acid.

Nucleic acid A polymer composed of a sequence of nucleotides; DNA and RNA.

Nucleolus A nuclear body that is associated with specific chromosomal regions and that is involved in the synthesis of ribosomal RNA.

Nucleosome A histone–DNA complex observed in the dissociation of chromsomes.

Nucleotide One of the nucleic acid bases, along with a sugar and a phosphate group, that makes up a unit of DNA or RNA.

Nullisomic Lacking a particular chromosome.

Oocyte The female germ cell during meiosis; primary oocyte refers to the oocyte during meiosis I; secondary oocyte refers to the oocyte after meiosis I is completed.

Oogenesis Formation of the ova.

Oogonium The primordial female germ cell that divides mitotically to eventually produce oocytes.

Operator In the Jacob–Monod operon model, the site at which the repressor molecule binds on DNA, shutting off transcription.

Operon A unit of coordinately controlled genes under the control of an operator and regulatory genes.

Ovum   An unfertilized egg cell.

Pachytene   The stage of prophase in meiosis I during which homologous chromosomes are completely paired and some shortening and coiling are apparent.

Pairing   The side-by-side alignment of homologous chromosomes during the prophase of meiosis I.

Panmixia   See *random mating.*

Pangenesis   A theory proposed by Darwin in which gemmules, modifiable by the environment, filter through the body and are packaged into the gametes; the gemmules are viewed here as the units of heredity.

Paracentric inversion   An inversion that does not include the centromere.

P arm   The short arm of a chromosome (from petite).

Parthenogenesis   Formation of an embryo from an unfertilized egg.

Pedigree   A diagram of the genetic history of an individual or family.

Penetrance   The percentage of individuals of a specific genotype that exhibit an expected phenotype under a specific set of environmental conditions.

Pericentric inversion   An inversion with a break in each of the two arms of a chromosome; the inverted segment includes the centromere.

Pharmacogenetics   The study of genetic variability in response to, and metabolism of, drugs.

Phenocopy   An environmentally induced phenotype that resembles an inherited trait.

Phenotype   The observable properties of an organism, produced by the interaction of the genotype with the environment.

Philadelphia chromosome   The structurally abnormal chromosome 22 that typically occurs in bone marrow cells of patients with chronic myelogenous leukemia. It is a translocation in which part of the long arm of 22 is translocated to the long arm of 9.

Phocomelia   The absence of the proximal portion of a limb or limbs, the hands or feet being attached to the trunk by a single bone.

Pleiotropy   A condition in which a single gene has a wide range of effects.

Point mutation   A mutation affecting a single nucleotide pair.

Polar body   One product of oogenesis; extruded as a small body from the major part of the cytoplasm.

Polygenic   Pertaining to a trait influenced by variation at several gene loci; often synonymous with multifactorial.

Polymerase   An enzyme that catalyzes the formation of a polymer from its constituent building blocks.

Polymorphism   The existence of a trait in two or more forms within a species.

Polypeptide   A chain of covalently linked amino acids; joined by peptide bonds.

Polyploidy   The condition in which the chromosome number of an individual is three or more times the haploid chromosome number.

Position effect   The change in the expression of a gene when its position is changed with respect to neighboring genes.

Proband   The person through whom the pedigree is ascertained; synonymous with propositus or index case.

Prophase   The early phase of cell division from the time the chromosomes first become visible to the beginning of metaphase, when they are maximally condensed.

Prokaryote   An organism that lacks well-defined nuclei and does not undergo meiosis; lacks true chromosomes.

Promoter   In the operon, the region between the operator and the structural genes to which RNA polymerase binds.

Propositus   See *Proband.*

Protein   A polymer of amino acids joined by peptide bonds.

Pure line   A homozygous strain produced by inbreeding.

Q arm   The long arm of a chromosome.

Race   A genetically distinct, and usually geographically distinct, inbreeding division of a species.

Rad   A unit of energy like the roentgen, but based on absorbed energy and applicable to both ionizing and nonionizing radiation; usually equivalent to a roentgen.

Random mating   Selection of a mate without regard to genotype; also called panmixia.

Reading frame   The reading of an RNA sequence as a series of codons of three nucleotides each.

Recessive gene   A gene that is expressed only if the individual is homozygous for it.

Reciprocal cross   A genetic cross that can be symbolized as A male $\times B$ female and B male $\times A$ female, where A and B represent different genotypes.

Reciprocal translocation   Mutual exchange of segments between nonhomologous chromosomes.

Recombinant DNA   Artificially constructed DNA in which a DNA segment from one organism is inserted into the genome of another organism.

Recombination   The formation of a new combination of alleles following meiosis.

Reduction division   The first half of the meiotic process, in which the paired homologs segregate to different nuclei, thus reducing the chromosome number by half.

Regressive variation   Variation in which the offspring's phenotype tends to approach the mean of the population rather than to exceed the parental extremes.

Regulator gene   A gene whose product controls the activity of distant genes.

Rem   The quantity of ionizing radiation that is equivalent in biological damage to one rad.

Repressor   The protein product of a regulator gene that, when bound to the operator, prevents transcription of the operon.

Restriction enzyme   An enzyme that cleaves DNA at specific nucleotide sequences.

Ribonucleic acid (RNA)   A polymer of ribonucleotides.

Ribosome   A cytoplasmic structure composed of RNA and protein, on which polypeptide synthesis from mRNA occurs.

Ring chromosome   A structurally abnormal chromosome in which the end of each arm has been deleted and the broken arms have reunited to form a ring.

RNA   See *Ribonucleic acid.*

Robertsonian translocation   A translocation between acrocentric chromosomes such that both long arms are attached to the same centromere. See also *Centric fusion.*

Roentgen   A unit of ionizing radiation that produces 2×10^9 ion-pairs per cubic centimeter of air or 1.6 ion-pairs per cubic centimeter of water.

Satellite   A chromosomal segment attached to the main part of the arm by a thin chromosomal strand.

Segregation   The separation of the members of a homologous pair of chromosomes into different gametes through the process of meiosis.

Sex chromatin   A chromatin mass in the nucleus of interphase cells of females of most mammalian species, including humans; synonymous with Barr body.

Sex chromosome   The X or Y chromosome.

Sex-influenced   Pertaining to a trait whose expression is modified by the sex of the person in whom it occurs.

Sex-limited   Pertaining to a trait that is expressed in only one sex.

Sex linkage   Linkage of genes located in the X chromosome.

Sibling (sib)   A brother or sister.

Simian crease   A single flexion crease across the palm of the hand.

Sister-chromatid exchange   Exchange of DNA segments between sister chromatids.

Somatic cells   All cells of the body other than the germ cells.

Somatic cell genetics   The study of genetics using cultured somatic cells.

Special creation   The idea that all living things occupying the earth were specially created by God.

Sperm   Spermatozoon; the mature male gamete.

Spermatid   One of the haploid products of meiosis in males before maturation into a spermatozoon.

Spermatocyte   The male germ cell during meiosis. A primary spermatocyte is in meiosis I; a secondary spermatocyte is in meiosis II.

Spermatogonium   A cell in the stem line of male germ cells that divide by mitosis.

Standard deviation   A measure of the variability in a population of N individuals, given by the formula

$$s = \sqrt{\frac{\Sigma(x - \bar{x})^2}{N - 1}}$$

where N is the number of individuals in the sample and $\Sigma(x - \bar{x})^2$ is the sum of the squared deviations from the mean.

Structural gene   The gene that codes for the amino acid sequence of a polypeptide chain.

Structural protein   The product of a structural gene that does not have enzymatic functions.

Suppressor mutation   A mutation that reverses the effects of a mutation at a distant locus.

Synapsis   Pairing of chromosomes at meiosis.

Synaptinemal complex   Observed under the electron microscope, a tripartite structure between paired homologs during the pachytene stage of meiosis I.

Syndrome   A group of symptoms that occur together, characterizing a disease or other morbid state.

Syntenic   Pertaining to loci on the same chromosome. Synteny and linkage are often used synonymously.

Telocentric chromosome   A chromosome that has a terminal centromere.

Telomere   The tip of the chromosome arm.

Telophase   The stage of cell division that begins when the daughter chromosomes reach the poles of the dividing cell and that lasts until the two daughter cells take on the appearance of interphase cells.

Terminating codon   A codon that terminates the elongation of a polypeptide chain.

Testcross   A cross involving a heterozygote crossed back to a homozygous recessive individual.

Tetrad   A bivalent in meiosis consisting of the four chromatids of a homologous chromosome pair.

Thalassemia   A group of hemoglobin diseases characterized by defective hemoglobin synthesis.

Threshold trait   A trait that shows discontinuous variation, but whose inheritance is polygenic.

Transcription   The synthesis of RNA from a DNA template; or, in certain viruses, the synthesis of DNA from an RNA template.

Transfer RNA (tRNA)   A small molecule that binds to mRNA, the ribosome, and an amino acid.

Transformation   The genetic alteration of a cell by the transfer of free DNA in the medium across the membrane and into the cell, where it combines with the host cell's DNA.

Transgressive variation   The production of F_2 phenotypes that exceed the parental extremes.

Transition mutation   A base-pair substitution in which the orientation of the purine and pyrimidine bases on each DNA strand remain the same (i.e., AT to GC).

Translation   The process of converting the information contained in a sequence of RNA bases into a sequence of amino acids.

Translocation   A chromosomal aberration involving the interchange of nonhomologous chromosome segments.

Transversion mutation   A base-pair substitution in which the purine–pyrimidine orientation on each DNA strand is reversed (i.e., AT to TA).

Triploid   Having three sets of the basic haploid chromosome complement.

Trisomy   The diploid condition plus one extra chromosome.

tRNA   See *Transfer RNA.*

Variable expressivity   Different phenotypes produced by the same genotype.

Variance   The mean of the squared deviations from the population mean.

Wild type   The phenotype most frequently observed in nature and the one arbitrarily designated as normal.

X-chromosome inactivation   See *Lyon hypothesis; Barr body.*

X-linked   Pertaining to genes on the X chromosome, or traits determined by these genes.

Y-linked   Pertaining to genes on the Y chromosome, or traits determined by these genes.

Zygote   The diploid cell formed from the fusion of sperm and egg.

Zygotene   The meiotic stage during which homologous chromosomes pair.

Acknowledgments

Chapter Two *Figure 2–12a* courtesy Carl J. Witkop, Division of Human and Oral Genetics, School of Dentistry, University of Minnesota. *Figure 2–13a* courtesy Clyde Keeler, 1964. "The Incidence of Cuña Moon-Child Albinos," *Journal of Heredity,* Vol. 55. By permission of the American Genetic Association. *Figure 2–13b* reprinted from *American Journal of Human Genetics,* 14:391, by C. M. Woolf and R. B. Grant, by permission of the University of Chicago Press. Copyright 1962 by Greene and Stratton for the American Society of Human Genetics.

Chapter Three *Figure 3–16* courtesy T. T. Puck, 1972. *The Mammalian Cell as Microorganism.* Holden-Day: San Francisco, Calif. *Figure 3–19* furnished by Beverly S. Emanuel. *Figure 3–20a* furnished by Beverly S. Emanuel. *Figure 3–20b* furnished by Beverly S. Emanuel. *Figure 3–21* furnished by Beverly S. Emanuel. *Figure 3–22* furnished by Beverly S. Emanuel. *Figure 3–23* modified from J. J. Yunis, 1976. *Science.* 191:1268, Fig. 2–12. Copyright 1976 by the American Association for the Advancement of Science. *Figure 3–25* courtesy Jack Griffith.

Chapter Four *Box 4–1* pedigrees reprinted from *American Journal of Human Genetics* Vol. 9, Fig. 124, by C. Stern, by permission of The University of Chicago Press. Copyright 1957 by the American Society of Human Genetics. *Figure 4–2a* courtesy Mohr, 1932. *Journal of Heredity* 23:345. By permission of the American Genetic Association. *Figure 4–4a* furnished by Judith Bader. *Figure 4–4b* courtesy the *British Medical Journal* and British Medical Association. *Figure 4–6* courtesy Carl J. Witkop, Division of Human and Oral Genetics, School of Dentistry, University of Minnesota. *Figure 4–14b* reprinted with permission: Nyhan, W. L.: Hyperuricemia. 1966. *Alabama Journal of Medical Science* 3:451. *Figure 4–15* reprinted from *American Journal of Human Genetics,* Vol. 16, by Stern, Centerwell, and Sakar, by permission of the University of Chicago Press. Copyright 1964 by Mr. S. D. Sigamoni, Photography Department, Christian Medical College Hospital, Vellore. *Figure 4–20* courtesy Bartalos and Baramki. 1967. *Medical cyto-genetics.* Copyright the Williams and Wilkins Co.: Baltimore, Md. *Figure 4–22* reprinted from *American Journal of Human Genetics* 24:467, by Edwards and Gale, by permission of the University of Chicago Press. Copyright 1972 by the American Society of Human Genetics. *Figure 4–23a* reprinted from *American Journal of Human Genetics* 24:467, by Edwards and Gale, by permission of the University of Chicago Press. Copyright 1972 by the American Society of Human Genetics. *Figure 4–23b* reprinted from *American Journal of Human Genetics* 24:467, by Edwards and Gale, by permission of the University of Chicago Press. Copyright 1972 by the American Society of Human Genetics. *Figure 4–25* courtesy Lynch, et al., 1967. *Archives of Dermatology* 96:629. Copyright 1967 American Medical Association. *Figure 4–27* courtesy Mohr, after Lesser, 1926. *Molecular and General Genetics.* Springer-Verlag: Heidelberg, Germany. Used by permission.

Chapter Five *Box 5–1* Table adapted from L. G. Lundin. 1979. *Clinical Genetics* 16:72–81. Copyright 1979 Munksgaard International Publishers Ltd., Copenhagen, Denmark. *Figure 5–15a* courtesy B. S. Emanuel et al. 1979. *American Journal of Medical Genetics* 4:167. *Figure 5–15b* courtesy B. S. Emanuel et al. 1979. *American Journal of Medical Genetics* 4:167. *Figure 5–15c* courtesy B. S. Emanuel et al. 1979. *American Journal of Medical Genetics* 4:167. *Figure 5–19* courtesy Dr. V. A. McKusick.

Chapter Six *Box 6–1* Art modified from J. J. Yunis and O. Prakash. 1982. "The Origin of Man: A Chromosomal Pictorial Legacy." *Science* 215:1525–1530. *Table 6–6* courtesy E. B. Hook and J. L. Hamerton. The frequency of chromosome abnormalities detected in consecutive newborn studies. *Population Cytogenetics—Studies in Humans.* Copyright 1977 by Academic Press, Inc. *Figure 6–3b* courtesy Kohn et al. 1967. *Pediatric Research* 1:463, by permission of S. Karger AG, Basel, Switzerland. *Figure 6–4* furnished by Beverly S. Emanuel. *Figure 6–6* furnished by Beverly S. Emanuel. *Figure 6–7a* furnished by Beverly S. Emanuel. *Figure 6–10* courtesy D. S. Borgaonkar, Wilmington Medical Center, Delaware. *Figure 6–11* courtesy D. S. Borgaonkar, Wilmington Medical Center, Delaware. *Figure 6–12a* courtesy D. S. Borgaonkar, Wilmington Medical Center, Delaware. *Figure 6–12b* courtesy Zaleski et al., 1966. *Canadian Medical Association Journal* 94:1143–1154, Fig. F. *Figure 6–12c* courtesy Joseph et al. 1964. *Journal of Medical Genetics* 1:95–101. *Figure 6–13a* from *Principles of human genetics,* 3d. ed., by Curt Stern. W. H. Freeman and Co.: San Francisco, Calif. Copyright © 1973. *Figure 6–16a* courtesy J. J. Yunis, M.D. 1980. *The Journal of Pediatrics* 96:1027. Copyright 1980 by C. V. Mosby. *Figure 6–16b* courtesy J. J. Yunis, M.D., Department of Laboratory Medicine and Pathology, University of Minnesota Medical School, Minneapolis, Minn. *Figure 6–16c* courtesy J. J. Yunis, M.D., Department of Laboratory Medicine and Pathology, University of Minnesota Medical School, Minneapolis, Minn. *Figure 6–16d* courtesy J. J. Yunis, M.D. 1980. *The Journal of Pediatrics* 96:1027. Copyright 1980 by C. V. Mosby. *Figure 6–21* courtesy Peter Nowell. *Cancer: a comprehensive treatise,* Vol 1, 2d. ed. Copyright 1982 by Plenum Publishing Corp.: New York. *Figure 6–25* furnished by Beverly S. Emanuel. *Figure 6–26a* furnished by Beverly S. Emanuel. *Figure 6–26b* redrawn from *American Journal of Human Genetics* 31:136, by Sutherland, by permission of The University of Chicago Press. Copyright 1979 by the American Society of Human Genetics. *Figure 6–6Q* furnished by Beverly S. Emanuel. *Figure 6–7Q* furnished by Beverly S. Emanuel. *Figure 6–8Q* furnished by Beverly S. Emanuel. *Figure 6–9Q* furnished by Beverly S. Emanuel.

Chapter Eight *Figure 8–9* after J. B. Gurdon, Transplanted nuclei and cell differentiation. *Scientific American,* December 28, 1968. p. 24. *Figure 8–10* courtesy Arthur Robinson, National Jewish Hospital and Research Center. *Figure 8–12a* courtesy Hastings and Gilford. 1904. *The Practitioner. Figure 8–12b* courtesy Epstein et al. 1966. *Medicine* 45:179.

Chapter Nine *Figure 9–13b* furnished by Beverly S. Emanuel. *Figure 9–18* courtesy David W. Smith, M.D. 1967. *Journal of Pediatrics* 70:463–519. By permission of C. V. Mosby.

Chapter Eleven *Figure 11–16* courtesy Baylor College of Medicine, Houston.

Chapter Twelve *Figure 12–5* courtesy H. B. Taussig, M.D., Johns Hopkins Hospital, and W. Lenz, M.D. *Figure 12–17* courtesy Dr. Holt, Harpenden, Herts, England. *Figure 12–18* courtesy Charles Harbutt/Archives. *Figure 12–19* courtesy Thomas Bouchard. *Science,* November, 1980.

Chapter Thirteen *Figure 13–4* courtesy Ikeda and Kaplan, 1970, *Proceeding of the National Academy of Sciences,* 67:1480. *Figure 13–9* reproduced by permission from the *Differential Aptitude Tests.* Copyright © 1982, 1980, 1975, 1973, 1972 by the Psychological Corporation. All rights reserved. Harcourt Brace Jovanovich: New York.

Chapter Fourteen *Figure 14–1* courtesy Stock, Boston Inc. *Figure 14–14* courtesy the South African Journal of Science, 76:61, 1980. *Figure 14–16a* courtesy Rainbird/Robert Harding Picture Library. London. *Figure 14–16b* courtesy Rainbird/Robert Harding Picture Library. London. *Figure 14–16c* courtesy Rainbird/Robert Harding Picture Library. London. *Figure 14–17a* courtesy Rainbird/Robert Harding Picture Library. London. *Figure 14–17b* courtesy Rainbird/Robert Harding Picture Library. London. *Figure 14–17c* courtesy Jay Matternes. Fairfax, Virginia. *Figure 14–19a* courtesy Rainbird/Robert Harding Picture Library. London. *Figure 14–19b* courtesy Rainbird/Robert Harding Picture Library. London. *Figure 14–19c* courtesy Rainbird/Robert Harding Picture Library. London.

(Note: Every reasonable effort has been made by The Benjamin/Cummings Publishing Company to locate the copyright holders for all of the figures listed above, but in a few instances, this has proved to be impossible.)

Index

(Note: All italicized numbers refer to figures. The letter b *after a number indicates a box; the letter* t *indicates a table.)*